Nelson MindTap + You = Learning amplified

"I love that everything is interconnected, relevant and that there is a clear learning sequence. I have the tools to create a learning experience that meets the needs of all my students and can easily see how they're progressing."

— Sarah, Secondary School Teacher

NELSON
WAmaths

UNITS ① + ②

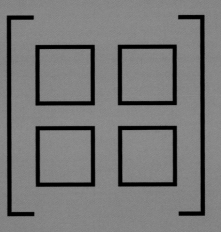

Amanda Pettitt
Dion Alfonsi
Michael Loh
Greg Neal
Dirk Strasser

Contributing authors
Neale Woods
Judy Binns
Sue Thomson

mathematics
applications

11

Nelson WAmaths Mathematics Applications 11
1st Edition
Amanda Pettitt
Dion Alfonsi
Michael Loh
Greg Neal
Dirk Strasser
ISBN 9780170477529

Publisher: Dirk Strasser
Associate product manager: Cindy Huang
Project editor: Tanya Smith
Series text design: Alba Design (Rina Gargano)
Series cover design: Nikita Bansal
Series designer: Nikita Bansal
Permissions researcher: Liz McShane
Content developers: Katrina Stavridis, Roger Walter
Content manager: Alice Kane
Typeset by: Nikki M Group Pty Ltd

Any URLs contained in this publication were checked for currency during the production process. Note, however, that the publisher cannot vouch for the ongoing currency of URLs.

Acknowledgements

TI-Nspire: Images used with permission by Texas Instruments, Inc.

Casio ClassPad: Images used with permission by Shriro Australia Pty. Ltd.

School Curriculum and Standards Authority. Adapted use of 2016–2021 Mathematics Applications and Mathematics Methods examinations, marking keys, and summary examination reports, ATAR 11 and 12 Mathematics Applications and Mathematics Methods syllabus. The School Curriculum and Standards Authority does not endorse this publication or product.

Selected VCE Examination questions are copyright Victorian Curriculum and Assessment Authority (VCAA), reproduced by permission. VCE ® is a registered trademark of the VCAA. The VCAA does not endorse this product and makes no warranties regarding the correctness or accuracy of this study resource. To the extent permitted by law, the VCAA excludes all liability for any loss or damage suffered or incurred as a result of accessing, using or relying on the content. Current VCE Study Designs, past VCE exams and related content can be accessed directly at www.vcaa.edu.au

© 2023 Cengage Learning Australia Pty Limited and Dirk Strasser

Copyright Notice

This Work is copyright. No part of this Work may be reproduced, stored in a retrieval system, or transmitted in any form or by any means without prior written permission of the Publisher. Except as permitted under the *Copyright Act 1968*, for example any fair dealing for the purposes of private study, research, criticism or review, subject to certain limitations. These limitations include: Restricting the copying to a maximum of one chapter or 10% of this book, whichever is greater; providing an appropriate notice and warning with the copies of the Work disseminated; taking all reasonable steps to limit access to these copies to people authorised to receive these copies; ensuring you hold the appropriate Licences issued by the
Copyright Agency Limited ("CAL"), supply a remuneration notice to CAL and pay any required fees. For details of CAL licences and remuneration notices please contact CAL at Level 11, 66 Goulburn Street, Sydney NSW 2000,
Tel: (02) 9394 7600, Fax: (02) 9394 7601
Email: info@copyright.com.au
Website: www.copyright.com.au

For product information and technology assistance,
in Australia call **1300 790 853**;
in New Zealand call **0800 449 725**

For permission to use material from this text or product, please email
aust.permissions@cengage.com

National Library of Australia Cataloguing-in-Publication Data
A catalogue record for this book is available from the National Library of Australia.

Cengage Learning Australia
Level 5, 80 Dorcas Street
Southbank VIC 3006 Australia

For learning solutions, visit **cengage.com.au**

Printed in China by 1010 Printing International Limited.
1 2 3 4 5 6 7 27 26 25 24 23

Contents

To the teacher	v
About the authors	vi
Syllabus grid	vii
About this book	viii

1 Consumer arithmetic — 1

Syllabus coverage		2
Nelson MindTap chapter resources		2
1.1	Wages and salaries	3
1.2	Bonuses and commission	9
1.3	Government allowances and pensions	15
1.4	Budgets	17
1.5	Applying percentages	21
1.6	Interest	29
1.7	Exchange rates	35
1.8	Shares	38
Examination question analysis		41
Chapter summary		44
Cumulative examination: Calculator-free		47
Cumulative examination: Calculator-assumed		48

2 Algebra and matrices — 49

Syllabus coverage		50
Nelson MindTap chapter resources		50
2.1	Algebraic expressions, equations and formulas	51
2.2	Tables and formulas	55
2.3	Introduction to matrices	62
2.4	Matrix addition, subtraction and scalar multiplication	68
2.5	Matrix multiplication and powers	76
2.6	Modelling with matrices	85
Examination question analysis		91
Chapter summary		97
Cumulative examination: Calculator-free		99
Cumulative examination: Calculator-assumed		100

3 Pythagoras' theorem and mensuration — 101

Syllabus coverage		102
Nelson MindTap chapter resources		102
3.1	Measurement	103
3.2	Pythagoras' theorem	105
3.3	Perimeter and area	113
3.4	Volume	126
3.5	Surface area	136
Examination question analysis		139
Chapter summary		144
Cumulative examination: Calculator-free		148
Cumulative examination: Calculator-assumed		149

4 Similar figures and similarity — 152

Syllabus coverage		153
Nelson MindTap chapter resources		153
4.1	Scale factor and similarity	154
4.2	Area and volume of similar figures	160
4.3	Scale drawings	163
Examination question analysis		164
Chapter summary		169
Cumulative examination: Calculator-free		171
Cumulative examination: Calculator-assumed		173

5 Making sense of data — 175

Syllabus coverage		176
Nelson MindTap chapter resources		176
5.1	Introduction to data distributions	177
5.2	Grouped frequency tables and histograms	184
5.3	Dot plots, stem plots and bar charts	194
5.4	The mean and standard deviation	204
5.5	Bell-shaped distributions	211
5.6	Standardised values	219
Examination question analysis		222
Chapter summary		226
Cumulative examination: Calculator-free		231
Cumulative examination: Calculator-assumed		233

6 Comparing data — 237

Syllabus coverage — 238
Nelson MindTap chapter resources — 238
6.1 Five-number summary and outliers — 239
6.2 Box plots — 245
6.3 Back-to-back stem plots and parallel box plots — 250
6.4 Comparing data using measures of centre and spread — 260
Examination question analysis — 262
6.5 Statistical investigation process — 265
Chapter summary — 267
Cumulative examination: Calculator-free — 270
Cumulative examination: Calculator-assumed — 273

7 Applications of trigonometry — 277

Syllabus coverage — 278
Nelson MindTap chapter resources — 278
7.1 The trigonometric ratios — 279
7.2 General applications of right-angled trigonometry — 288
7.3 Angles of elevation and depression — 292
7.4 Bearings and navigation — 295
7.5 Area of a non-right-angled triangle — 302
7.6 The sine and cosine rules for non-right-angled triangles — 309
Examination question analysis — 320
Chapter summary — 329
Cumulative examination: Calculator-free — 332
Cumulative examination: Calculator-assumed — 333

8 Linear equations and graphs — 335

Syllabus coverage — 336
Nelson MindTap chapter resources — 336
8.1 Solving linear equations — 337
8.2 Solving problems using linear equations — 341
8.3 Linear equations in the form $y = ax + b$ — 345
8.4 Interpreting linear equations in the form $y = ax + b$ — 356
Examination question analysis — 359
Chapter summary — 364
Cumulative examination: Calculator-free — 365
Cumulative examination: Calculator-assumed — 366

9 Simultaneous equations and other linear graphs — 369

Syllabus coverage — 370
Nelson MindTap chapter resources — 370
9.1 Simultaneous equations — 371
9.2 Solving simultaneous linear equations algebraically — 375
9.3 Modelling with simultaneous equations — 380
9.4 Other linear graphs — 383
Examination question analysis — 391
Chapter summary — 401
Cumulative examination: Calculator-free — 402
Cumulative examination: Calculator-assumed — 403

Answers — 406
Glossary and index — 434

To the teacher

Now there's a better way to WACE maths mastery.

Nelson WAmaths 11–12 is a new WACE mathematics series that is backed by research into the science of learning. The design and structure of the series have been informed by teacher advice and evidence-based pedagogy, with the focus on preparing WACE students for their exams and maximising their learning achievement.

- Using **backwards learning design**, this series has been built by analysing past WACE exam questions and ensuring that all theory and examples are precisely mapped to the SCSA syllabus.
- To reduce the **cognitive load** for learners, explanations are clear and concise, using the technique of **chunking** text with accompanying diagrams and infographics.
- The student book has been designed for **mastery** of the learning content.
- The exercise structure of **Recap, Mastery, Calculator-free** and **Calculator-assumed** leads students from procedural fluency to **higher-order thinking** using the learning technique of **interleaving**.
- **Calculator-free** and **Calculator-assumed** sections include exam-style questions.
- The cumulative structure of exercise **Recaps** and chapter-based **Cumulative examinations** is built on the learning and memory techniques of **spacing** and **retrieval**.

About the authors

Amanda Pettitt is a Secondary Mathematics teacher at Lesmurdie Senior High School. She teaches ATAR Mathematics in senior school and lower school classes, including extension classes at the lower school level. Prior to teaching Amanda worked as an Analytical Chemist, working locally and in remote parts of Western Australia.

Dion Alfonsi is Head of Mathematics and a Secondary Mathematics Teacher at Shenton College. In the past, he has had the roles of Years 9 & 10 Mathematics Curriculum Leader and Gifted & Talented/Academic Programs Coordinator. Dion has been a Board Member of MAWA, is a frequent presenter at the MAWA Secondary Conference and a teacher of the MAWA Problem Solving Program.

Michael Loh is a Senior Mathematics Teacher at Shenton College, where he also co-ordinates the Gifted and Talented Program. He has taught in the secondary and tertiary sectors for many years and is a regular presenter at various conferences including MAWA, the Google Education Conference and EdTech Summit.

Greg Neal has taught in regional schools for over 40 years and has co-written several senior textbooks for Cengage Nelson. He has been an examination assessor, presents at conferences and has expertise with CAS technology.

Dirk Strasser is an experienced teacher, a former Head of Mathematics, and a lead author and senior publisher of mathematics series for over 30 years. He has published and co-written eight best-selling mathematics series and won several Australian Educational Publishing Awards. He is the Manager of Secondary Mathematics at Nelson Cengage.

Syllabus grid

Topic		Nelson WAmaths Mathematics Applications 11 chapter
Topic 1.1: Consumer arithmetic (20 hours)		
Applications of rates and percentages	1	Consumer arithmetic
Use of spread sheets	1	Consumer arithmetic
Topic 1.2: Algebra and matrices (15 hours)		
Linear and non-linear expressions	2	Algebra and matrices
Matrices and matrix arithmetic	2	Algebra and matrices
Topic 1.3: Shape and measurement (20 hours)		
Pythagoras' theorem	3	Pythagoras' theorem and mensuration
Mensuration	3	Pythagoras' theorem and mensuration
Similar figures and scale factors	4	Similar figures and similarity
Topic 2.1: Univariate data analysis and the statistical investigation process (25 hours)		
The statistical investigation process	6	Comparing data
Making sense of data relating to a single statistical variable	5	Making sense of data
Comparing data for a numerical variable across two or more groups	6	Comparing data
Topic 2.2: Applications of trigonometry (10 hours)		
Applications of trigonometry	7	Applications of trigonometry
Topic 2.3: Linear equations and their graphs (20 hours)		
Linear equations	8	Linear equations and graphs
Straight-line graphs and their applications	8	Linear equations and graphs
Simultaneous linear equations and their applications	9	Simultaneous equations and other linear graphs
Piece-wise linear graphs and step graphs	9	Simultaneous equations and other linear graphs

About this book

In each chapter

Syllabus coverage and extracts are shown at the start of the chapter, along with a listing of **Nelson MindTap chapter resources**.

Important words and phrases are printed in blue and listed in the **Glossary and index** at the back of the book.

Worked examples are explained clearly step-by-step, with the mathematical working shown on the right-hand side.

Important facts and formulas are highlighted in a shaded box.

Exam hacks highlight valuable exam hints and common student errors.

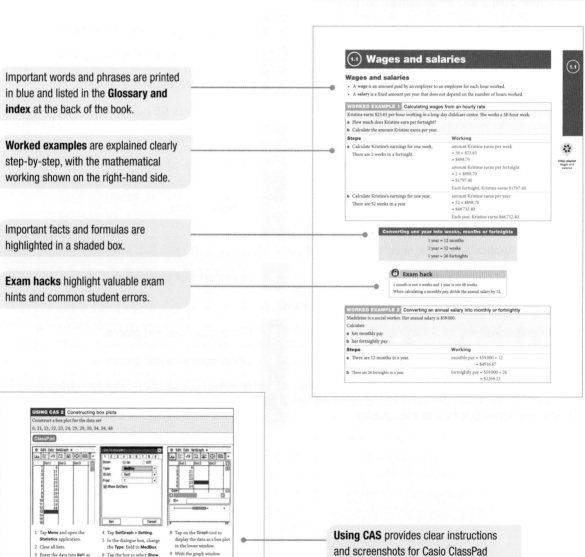

Using CAS provides clear instructions and screenshots for Casio ClassPad and TI-Nspire calculators.

viii Nelson WAmaths Mathematics Applications 11   9780170477529

Graded exercises include **Recap**, **Mastery**, **Calculator-free** and **Calculator-assumed** questions. **Recap** questions revise skills from the previous exercise and function as lesson starters.

Mastery questions provide skill practice linked to worked examples and Using CAS, while **Calculator-free** and **Calculator-assumed** questions apply learned skills to exam-style problems with mark allocation.

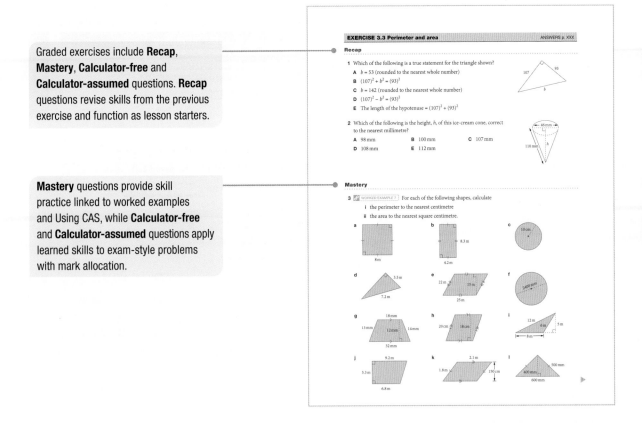

At the end of each chapter

Examination question analysis leads students through an exam-style question that exemplifies the chapter, discussing how to approach the question, providing advice on interpreting the question, common student errors and a full worked solution with a marking key.

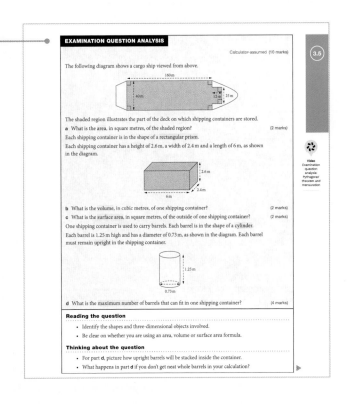

About this book   ix

Chapter summary for easy reference.

Cumulative examination: Calculator-free and **Cumulative examination: Calculator-assumed** are mini-exams based on the format of the WACE examinations, revising work from the chapters in which they appear, as well as previous chapters.

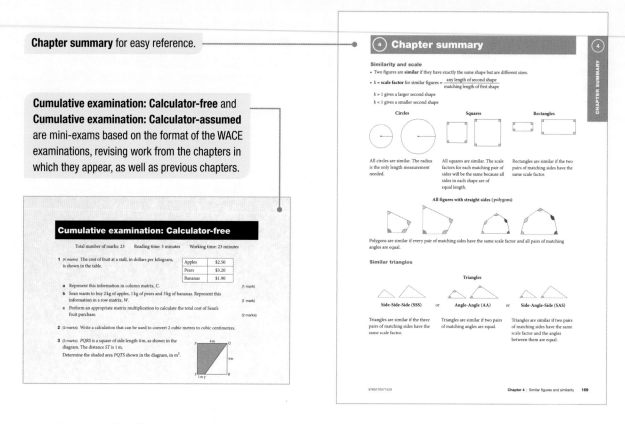

At the end of the book

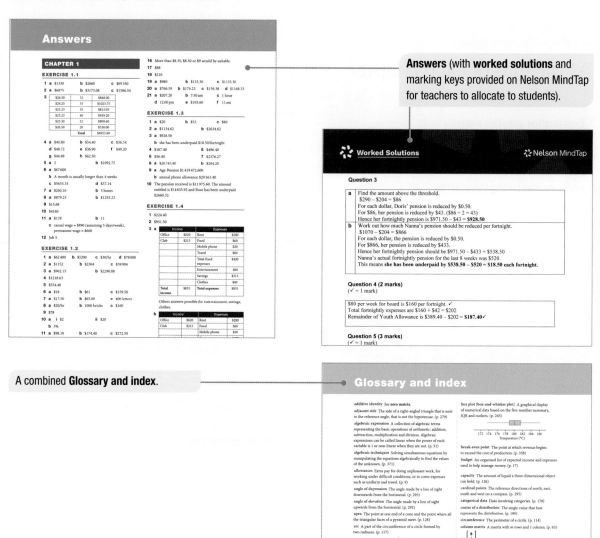

Answers (with **worked solutions** and marking keys provided on Nelson MindTap for teachers to allocate to students).

A combined **Glossary and index**.

Nelson MindTap

Nelson MindTap is an online learning space that provides students with tailored learning experiences. Access tools and content that make learning simpler yet smarter to help you achieve WACE maths mastery.

Nelson MindTap includes an eText with integrated interactives and online assessment.

Margin links in the student book signpost multimedia student resources found on MindTap.

Nelson MindTap for students:

- **Watch** video tutorials featuring expert teacher advice to unpack new concepts and develop your understanding.
- **Revise** using learning checks, worksheets and skillsheets to practise your skills and build your confidence.
- **Navigate** your own path, accessing the content and support as you need it.

Video playlists

Worksheets

Skillsheets

Nelson MindTap for teachers*:

- Tailor content to different learning needs – assign directly to the student, or the whole class.
- Monitor progress using the MindTap assessment tools.
- Integrate content and assessment directly within your school's LMS for ease of access.
- Access topic tests, teaching plans and worked solutions to each exercise set.

*Complimentary access to these resources is only available to teachers who use this book as part of a class set, book hire or booklist. Contact your Cengage Education Consultant for information about access and conditions.

Nelson WAmaths 11–12 series

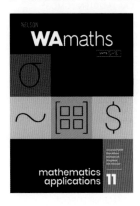 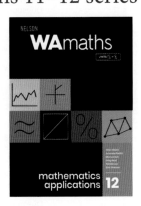 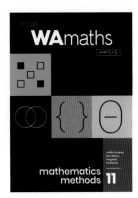 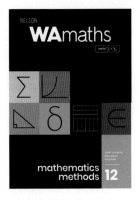

Additional credits

*modified

Chapter 1
Exercise 1.5
Question 11a ©VCAA FM2013 1BRMQ1
Question 11b © VCAA FM2014 1BRMQ1
Question 13 © VCAA FM2014 2BRMQ1ab

Chapter 2
Exercise 2.3
Question 9 © VCAA FM2007 1MQ2*
Question 10 © VCAA FM2009 1MQ3*
Question 11 © VCAA FM2012 1MQ7*
Exercise 2.4
Question 9b © VCAA FM2017 1MQ1*
Question 9b © VCAA FM2009 1MQ1*
Question 10 © VCAA FM2008 1MQ1*
Exercise 2.5
Question 1 © VCAA FM2019 1MQ1
Question 2 © VCAA FM2007 1MQ1
Question 6 © VCAA FM2011 1MQ2*
Question 7a © VCAA FM2014 1MQ1*
Question 7b © VCAA FM2013 1MQ1*
Question 8 © VCAA FM2013 1MQ2*
Exercise 2.6
Question 1 © VCAA FM2006 1MQ2
Question 2 © VCAA FM2012 1MQ2*
Question 10 © VCAA FM2020 1MQ5*
Question 11 © VCAA FM2010 1MQ2
Question 12 © VCAA FM2006 1MQ5*
Question 13 © VCAA FM2009 2MQ1*
Question 14 © VCAA FM2016 2MQ2*
Question 15 © VCAA FM2019 1MQ7*
Cumulative examination: Calculator-free
Question 4 © VCAA FM2008 1MQ2*
Cumulative examination: Calculator-assumed
Question 3 © VCAA FM2012 2MQ1*
Question 4 © VCAA FM2006 2MQ1*

Chapter 3
Exercise 3.2
Question 8 © VCAA FM2018 1GMQ1
Question 9 © VCAA FM2020 1GMQ3
Question 12 © VCAA FM2018 2GMQ3ab
Question 13 © VCAA FM2019 2GMQ3ai
Exercise 3.3
Question 8 © VCAA FM2019 1GMQ1
Question 9 © VCAA FM2015 1GTQ1
Question 10 © VCAA FM2016 1GMQ1
Question 11 © VCAA FM2020 1GMQ2
Question 12 © VCAA FM2014 2GTQ1
Question 13 © VCAA FM2017N 2GMQ1
Exercise 3.4
Question 1 © VCAA FM2021N 1GMQ1
Question 2 © VCAA FM2019N 1GMQ1
Question 6 © VCAA FM2018N 1GMQ5*
Question 7 © VCAA FM2007 1GTQ4
Question 8 © VCAA FM2017 1GMQ4
Question 9 © VCAA FM2015 1GTQ6
Question 10 © VCAA FM2019N 1GMQ6
Question 11 © VCAA FM2006 1GTQ4
Question 12 © VCAA FM2012 1GTQ1
Question 13 © VCAA FM2013 2GTQ2
Examination question analysis
© VCAA FM2019 2GMQ1
Exercise 3.5
Question 2 © VCAA FM2008 1GTQ7
Question 5 © VCAA FM2006 1GTQ6
Question 7 © VCAA FM2013 1GTQ7
Question 8 © VCAA FM2018 1GMQ8
Question 9 © VCAA FM2017N 1GMQ4
Question 10 © VCAA FM2016 2GMQ1
Question 11 © VCAA FM2018N 2GMQ1ab
Question 12 © VCAA FM2008 2GTQ2
Cumulative examination: Calculator-free
Question 1 © VCAA FM2017N 1MQ2
Question 3 © VCAA FM2012 1GTQ2
Cumulative examination: Calculator-assumed
Question 4 © VCAA FM2021N 2GMQ1a-d Q2ab*
Question 6 © VCAA FM2018N 1GMQ7

Chapter 4
Exercise 4.1
Question 4 © VCAA FM2016 1GMQ2*
Question 5 © VCAA FM2019 1GMQ4
Question 6 © VCAA FM2006 1GTQ7*
Question 7 © VCAA FM2003 1GTQ9
Question 8 © VCAA FM2004 1GTQ4*
Question 9 © VCAA FM2002 1GTQ4*
Exercise 4.2
Question 1 © VCAA FM2004 1GTQ4*
Question 4 © VCAA FM2021N 1GMQ2
Question 6 © VCAA FM2010 1GTQ4
Exercise 4.3
Question 1 © VCAA FM2021N 1GMQ2
Cumulative examination: Calculator-free
Question 3 © VCAA FM2009 1GTQ1
Cumulative examination: Calculator-assumed
Question 2 © VCAA FM2017 1GMQ2
Question 3 © VCAA FM2008 1GTQ6

Chapter 5
Exercise 5.1
Question 6 © VCAA FM2020 1CQ7*
Question 7 © VCAA FM2003 1CQ7
Question 9a © VCAA FM2003 1CQ1
Question 9b © VCAA FM2003 1CQ2
Exercise 5.2
Question 8a © VCAA FM2011 1CQ1*
Question 8b © VCAA FM2011 1CQ2
Question 10 © VCAA FM2015 2CQ1bc
Exercise 5.3
Question 8a © VCAA FM2013 1CQ1*
Question 8b © VCAA FM2013 1CQ2
Question 9a © VCAA FM2012 1CQ1*
Question 9b © VCAA FM2012 1CQ2*
Question 10a © VCAA FM2007 1CQ1
Question 10a © VCAA FM2007 1CQ2
Question 11a © VCAA FM2004 1CQ1
Question 11b © VCAA FM2004 1CQ2
Question 12 © VCAA FM2013 2CQ2
Question 13 © VCAA FM2009 2CQ1
Question 14 © VCAA FM2011 2CQ1
Exercise 5.4
Question 9 © VCAA FM2012 1CQ3
Question 10 © VCAA FM2008 1CQ6
Question 11a © VCAA FM2004 1CQ4
Question 11b © VCAA FM2004 1CQ5
Question 12 © VCAA FM2008 1CQ5
Question 13 © VCAA FM2004 1CQ3
Exercise 5.5
Question 1 © VCAA FM2011 1CQ6*
Question 2 © VCAA FM2008 1CQ5
Question 8 © VCAA FM2019 1CQ6
Question 10 © VCAA FM2009 1CQ7
Question 11 © VCAA FM2020 1CQ8
Question 12a © VCAA FM2013 1CQ5
Question 12b © VCAA FM2013 1CQ6
Question 13 © VCAA FM2019 1CQ7*
Question 14 © VCAA FM2007 1CQ4
Question 15 © VCAA FM2012 2CQ1
Examination question analysis
© VCAA FM2017 2CQ1
Exercise 5.6
Question 1 © VCAA FM2017N 1CQ8
Question 2 © VCAA FM2017N 1CQ9
Question 5a © VCAA FM2010 1CQ5
Question 5b © VCAA FM2010 1CQ6
Question 6 © VCAA FM2007 1CQ3
Question 7a © VCAA FM2018 1CQ3
Question 7b © VCAA FM2018 1CQ4
Question 7c © VCAA FM2018 1CQ5
Question 8 © VCAA FM2012 1CQ4
Cumulative examination: Calculator-free
Question 3a © VCAA FM2016S 1CQ4
Question 3b © VCAA FM2016S 1CQ5
Question 3c © VCAA FM2016S 1CQ6
Question 4 © VCAA FM2014 1CQ1
Question 5a © VCAA FM2016 1CQ6
Question 6c © VCAA FM2016 1CQ4*
Question 6d © VCAA FM2016 1CQ5*
Question 7a © VCAA FM2019N 1CQ5*
Question 7b © VCAA FM2019N 1CQ6
Question 7c © VCAA FM2019N 1CQ7
Cumulative examination: Calculator-assumed
Question 1 © VCAA FM2021 2MQ1*
Question 2 © VCAA FM2015 1GTQ2
Question 5 © VCAA FM2016 2CQ1
Question 6 © VCAA FM2020 2CQ2
Question 7 © VCAA FM2007 2CQ1

Chapter 6
Exercise 6.2
Question 6a © VCAA FM2008 1CQ2
Question 6b © VCAA FM2008 1CQ1
Question 6c © VCAA FM2008 1CQ3
Question 7 © VCAA FM2002 1CQ7
Exercise 6.3
Question 1 © VCAA FM2010 1CQ2
Question 8a © VCAA FM2009 1CQ1
Question 8b © VCAA FM2009 1CQ2
Question 8c © VCAA FM2009 1CQ3*
Question 9 © VCAA FM2014 1CQ7
Question 10 © VCAA FM2005 1CQ7
Question 11 © VCAA FM2012 2CQ3*
Examination question analysis
© VCAA FM2019 2CQ3*
Exercise 6.4
Question 1 © VCAA FM2015 2CQ2*
Question 2 © VCAA FM2018N 1CQ6
Cumulative examination: Calculator-free
Question 5 © VCAA FM2002 1CQ5*
Question 6 © VCAA FM2002 1CQ5*
Question 7a © VCAA FM2006 1CQ1
Question 7b © VCAA FM2006 1CQ2
Question 7c © VCAA FM2006 1CQ3
Question 8 © VCAA FM2011 1CQ5
Question 10 © VCAA FM2020 2CQ2
Cumulative examination: Calculator-assumed
Question 2 © VCAA FM2009 1MQ5*
Question 3 © VCAA FM2020 1GMQ8

Question 4 © VCAA FM2004 1GTQ4*
Question 5 © VCAA FM2013 2CQ1
Question 6 © VCAA FM2018N 2CQ1
Question 8 © VCAA FM2011 2CQ1
Question 9 © VCAA FM2021 2CQ1adef
Question 10 © VCAA FM2021 1CQ5

Chapter 7
Exercise 7.1
Question 7 © VCAA FM2007 1GTQ1*
Question 9 © VCAA FM2002 1GTQ1*
Question 10 © VCAA FM2004 1GTQ1*
Exercise 7.2
Question 6 © VCAA FM2002 1GTQ3*
Question 9 © VCAA FM2017N 1GMQ2
Question 12 © VCAA FM2010 1GTQ5*
Exercise 7.3
Question 5 © VCAA FM2009 1GTQ1*
Question 6 © VCAA FM2007 1GTQ2*
Question 9 © VCAA FM2011 1GTQ2*
Exercise 7.4
Question 7 © VCAA FM2010 1GTQ3*
Question 8 © VCAA FM2019N 1GMQ3*
Question 9 © VCAA FM2018N 1GMQ2*
Question 10 © VCAA FM2016 2GMQ2*
Question 11 © VCAA FM2017N 2GMQ2
Exercise 7.5
Question 1 © VCAA FM2019 1GMQ2
Question 2 © VCAA FM2015 1GTQ4
Question 8 © VCAA FM2003 1GTQ8*
Question 9a © VCAA FM2011 1GTQ4*
Question 9b © VCAA FM2004 1GTQ3*
Exercise 7.6
Question 2 © VCAA FM2013 1GTQ2*
Question 11 © VCAA FM2004 1GTQ3*
Question 12 © VCAA FM2011 1GTQ4*
Question 13 © VCAA FM2016 1GMQ6*
Question 14 © VCAA FM2009 2GTQ2ab*
Question 15 © VCAA FM2003 1GTQ8*
Question 16 © VCAA FM2002 1GTQ6*
Question 17 © VCAA FM2016 2GMQ4*
Question 18 © VCAA FM2019N 2GMQ3*
Question 19 © VCAA FM2008 2GTQ3abc*
Question 20 © VCAA FM2005 1GTQ4*
Question 21 © VCAA FM2015 2GTQ2*
Cumulative examination: Calculator-free
Question 2 © VCAA FM2002 1CQ1*
Cumulative examination: Calculator-assumed
Question 2 © VCAA FM2019N 1MQ1*
Question 3 © VCAA FM2021N 1GMQ3
Question 5a © VCAA FM2006 1GTQ1*
Question 5b © VCAA FM2006 1GTQ2*
Question 5c © VCAA FM2003 1GTQ2*
Question 7 © VCAA FM2020 1GMQ10*

Chapter 8
Exercise 8.3
Question 9 © VCAA FM2007 1GRQ1*
Question 10 © VCAA FM2006 1GRQ1*
Exercise 8.4
Question 1 © VCAA FM2018N 1GRQ1*
Question 6 © VCAA FM2018 1GRQ2*
Question 7 © VCAA FM2008 1GRQ2*
Question 8 © VCAA FM2013 1GRQ3*
Question 9 © VCAA FM2020 1GRQ6*
Question 10 © VCAA FM2015 1GRQ8*
Question 11 © VCAA FM2013 1GRQ6*
Question 12 © VCAA FM2015 2GRQ2
Question 13 © VCAA FM2018 2GRQ2
Question 14 © VCAA FM2013 2GRQ3
Cumulative examination: Calculator-assumed
Question 2 © VCAA FM2008 1MQ3*
Question 7a © VCAA FM2017 1CQ1
Question 7b © VCAA FM2017 1CQ2

Chapter 9
Exercise 9.3
Question 1 © VCAA FM2003 1GRQ6*
Question 5 © VCAA FM2020 1GRQ2*
Question 6 © VCAA FM2008 1GRQ6*
Question 7 © VCAA FM2018 1GRQ3ab*
Examination question analysis
© VCAA FM2009 2GRQ1
Exercise 9.4
Question 2 © VCAA FM2005 1GRQ6*
Question 8 © VCAA FM2020 1GRQ3*
Question 9 © VCAA FM2017 1GRQ2*
Question 10 © VCAA FM2020 1GRQ1*
Question 11 © VCAA FM2017 1GRQ4*
Question 12 © VCAA FM2004 1GRQ1*
Question 13 © VCAA FM2004 1GRQ4*
Question 14 © VCAA FM2018N 1GRQ2*
Question 16 © VCAA FM2020 2GRQ1*
Question 17 © VCAA FM2019N 2GRQ2*
Cumulative examination: Calculator-assumed
Question 1 © VCAA FM2017 1MQ2*
Question 2 © VCAA FM2019 1GMQ3
Question 4 © VCAA FM2016S 1CQ2
Question 5a © VCAA FM2007 1CQ6
Question 5b © VCAA FM2007 1CQ5
Question 6 © VCAA FM2015 1GTQ4*

CONSUMER ARITHMETIC

CHAPTER 1

Syllabus coverage
Nelson MindTap chapter resources

1.1 Wages and salaries
Wages and salaries
Spreadsheet calculations
Using CAS 1: Finding wages using a spreadsheet
Overtime

1.2 Bonuses and commission
Bonuses and allowances
Commission and piecework

1.3 Government allowances and pensions

1.4 Budgets

1.5 Applying percentages
Percentage increase or percentage decrease
Mark-up and discount
Finding percentage mark-up or percentage discount
Using CAS 2: Finding the original price
Profit and loss
GST
The unit cost method

1.6 Interest
Simple interest investments and loans
Using CAS 3: Graphing simple interest loans on a spreadsheet
Compound interest
Compounding periods

1.7 Exchange rates

1.8 Shares
Dividend yield
Dividend vs earnings
Price-earnings (P/E) ratio

Examination question analysis
Chapter summary
Cumulative examination: Calculator-free
Cumulative examination: Calculator-assumed

Syllabus coverage

TOPIC 1.1: CONSUMER ARITHMETIC

Applications of rates and percentages

1.1.1 calculate weekly or monthly wage from an annual salary, wages from an hourly rate, including situations involving overtime and other allowances, and earnings based on commission or piecework

1.1.2 calculate payments based on government allowances and pensions

1.1.3 prepare a personal budget for a given income taking into account fixed and discretionary spending

1.1.4 compare prices and values using the unit cost method

1.1.5 apply percentage increase or decrease in contexts, including determining the impact of inflation on costs and wages over time, calculating percentage mark-ups and discounts, calculating GST, calculating profit or loss in absolute and percentage terms, and calculating simple and compound interest

1.1.6 use currency exchange rates to determine the cost in Australian dollars of purchasing a given amount of a foreign currency, or the value of a given amount of foreign currency, when converted to Australian dollars

1.1.7 calculate the dividend paid on a portfolio of shares given the percentage dividend or dividend paid for each share, and compare share values by calculating a price-to-earnings ratio

Use of spread sheets

1.1.8 use a spreadsheet to display examples of the above computations when multiple or repeated computations are required; for example, preparing a wage-sheet displaying the weekly earnings of workers in a fast food store where hours of employment and hourly rates of pay may differ, preparing a budget, or investigating the potential cost of owning and operating a car over a year

Mathematics Applications ATAR Course Year 11 syllabus pp. 8–9 © SCSA

Video playlists (9):
- 1.1 Wages and salaries
- 1.2 Bonuses and commission
- 1.3 Government allowances and pensions
- 1.4 Budgets
- 1.5 Applying percentages
- 1.6 Interest
- 1.7 Exchange rates
- 1.8 Shares

Examination question analysis Consumer arithmetic

Skillsheets (2):
- 1.5 Percentage calculations • Mental percentages

Worksheets (14):
- 1.1 Wages and salaries
- 1.2 Earning money • Pay day
- 1.4 Budgeting scenarios • Budgeting grid • My budget
- 1.5 Profit and loss
- 1.6 Simple interest • Simple interest riddle • Applications of simple interest • What's the interest? • Compound interest table • Compounding periods: spreadsheet
- 1.7 Currency conversion graph

Puzzles (2):
- 1.5 Discounts code puzzle • Best buys puzzle

To access resources above, visit
cengage.com.au/nelsonmindtap

Nelson MindTap

1.1 Wages and salaries

Wages and salaries

- A **wage** is an amount paid by an employer to an employee for each hour worked.
- A **salary** is a fixed amount per year that does not depend on the number of hours worked.

WORKED EXAMPLE 1 — Calculating wages from an hourly rate

Kristine earns $23.65 per hour working in a long-day childcare centre. She works a 38-hour week.

a How much does Kristine earn per fortnight?
b Calculate the amount Kristine earns per year.

Steps	Working
a Calculate Kristine's earnings for one week. There are 2 weeks in a fortnight.	Amount Kristine earns per week = 38 × $23.65 = $898.70 Amount Kristine earns per fortnight = 2 × $898.70 = $1797.40 Each fortnight, Kristine earns $1797.40.
b Calculate Kristine's earnings for one year. There are 52 weeks in a year.	Amount Kristine earns per year = 52 × $898.70 = $46 732.40 Each year, Kristine earns $46 732.40.

Converting one year into weeks, months or fortnights

1 year = 12 months
1 year = 52 weeks
1 year = 26 fortnights

Exam hack

1 month is not 4 weeks and 1 year is not 48 weeks.
When calculating a monthly pay, divide the annual salary by 12.

WORKED EXAMPLE 2 — Converting an annual salary into monthly or fortnightly pay

Madeleine is a social worker. Her annual salary is $59 000.

Calculate
a her monthly pay
b her fortnightly pay.

Steps	Working
a There are 12 months in a year.	Monthly pay = $59 000 ÷ 12 = $4916.67
b There are 26 fortnights in a year.	Fortnightly pay = $59 000 ÷ 26 = $2269.23

Spreadsheet calculations

	A	B	C	D
1	2	5		
2	4	10		
3	6	20		
4				

In a spreadsheet, each cell has a reference given by the column letter and the row number. The number '2' in the spreadsheet above is in cell A1.

To calculate the product of 2 and 5 and put the answer in cell C1:

1. Select cell C1 and type =.
2. Enter **A1*B1** and press **enter**.

	A	B	C	D
1	2	5	10	
2	4	10		
3	6	20		
4				

This calculation can then be copied for the other cells in the column. Using CAS 1 describes how calculations can be performed on ClassPad and TI-Nspire spreadsheets.

USING CAS 1 — Finding wages using a spreadsheet

The following table shows the hours worked during the first week of February and the corresponding pay rates for the employees in a small office.

Employee	Pay rate per hour	Number of hours worked	Pay
Imran	$23.23	20	
Sofia	$24.50	35	
Cathy	$26.75	35	
Mike	$23.23	40	
Anita	$23.23	32	
Ronen	$24.50	20	
		Total wages bill	

The hours each employee works per week and their hourly rate of pay could change. Construct a spreadsheet that calculates each employee's wage and the total office wage bill when the number of hours worked and the rates of pay could change.

ClassPad

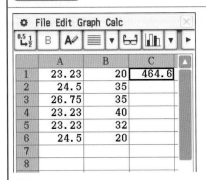

1. Open the **Spreadsheet** application.
2. Enter the the values from the table in columns **A** and **B** as shown above.
3. Place the cursor in cell **C1**.
4. Tap **Edit > Fill > Fill Range**.
5. In the dialogue box, enter the following in the fields, as shown above.

   Formula =A1×B1
   Range   C1:C6

6. Tap **OK**.
7. Place the cursor in cell **C7**.
8. Tap **Calc > List-Calculation > sum**.

   The total wages bill will be displayed.

TI-Nspire

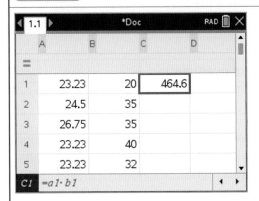

1. Add a **Spreadsheet** page.
2. Enter the values from the table in columns **A** and **B** as shown above.
3. Place the cursor in cell **C1** and enter the formula **=A1*B1**.
4. Press **menu > Data > Fill**.
5. Press the **down arrow** to fill the column up to cell **C6**.
6. Press **enter**.

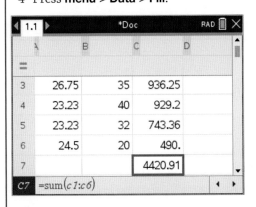

7. Place the cursor in cell **C7**.
8. Enter the formula **=sum(C1:C6)**.
9. Press **enter**. The total wage bill will be displayed.

The total wages bill is $4420.91.

Overtime

Overtime is working beyond usual working hours or days, and is paid at a higher rate, such as 1.5 times the normal pay (**time-and-a-half**) or twice the normal pay (**double time**). Only people who work for a wage are paid for overtime (it doesn't apply to salary earners).

WORKED EXAMPLE 3	Calculating pay at a single rate of overtime

Alyssa's normal junior rate of pay is $16.81 per hour. She is paid time-and-a-half on Saturdays, and double time on Sundays.

a How much does Alyssa earn per hour on Saturdays?

b Calculate the amount Alyssa will earn for working 4 hours on a Sunday.

Steps	Working
a When she works on a Saturday, Alyssa is paid 1.5 times her normal rate.	Pay per hour at time-and-a-half = 1.5 × $16.81 = $25.22
b When Alyssa works on Sunday, she earns 2 × $16.81, or $33.62 per hour.	Pay for 4 hours on Sunday = 4 × $33.62 = $134.48

Overtime pay

Time-and-a-half is 1.5 times the normal pay rate.

Double time is 2 times the normal pay rate.

WORKED EXAMPLE 4	Calculating pay for two rates of overtime

Hasid earns $24 per hour.

a How much will he earn for working a 35-hour week?

b When Hasid works for more than 7 hours per day, he is paid overtime. For the first 3 hours of overtime, he is paid time-and-a-half and any additional overtime hours are paid at double time. How much will Hasid earn for working 12 hours in one day?

Steps	Working
a Multiply hourly rate by 35.	Pay for a 35-hour week = 35 × $24 = $840
b Hasid's 12 hours in one day is divided into 7 hours at normal pay plus 3 hours at time-and-a-half and 2 hours at double time.	Pay = 7 × $24 + 3 × 1.5 × $24 + 2 × 2 × $24 = $372 Hasid's pay for a 12-hour day is $372.

EXERCISE 1.1 Wages and salaries   ANSWERS p. 406

Mastery

1 WORKED EXAMPLE 1 Scott is a qualified ambulance paramedic. He is paid $35 per hour for a 38-hour week.

    a How much does Scott earn per week?

    b How much is Scott paid per fortnight?

    c Calculate Scott's annual pay.

2 **WORKED EXAMPLE 2** Suzanne is a solicitor. Her salary is $82 500 p.a.

   a How much does Suzanne earn per month?

   b Calculate Suzanne's fortnightly pay.

   c How much does Suzanne earn per week?

3 **Using CAS 1** The following table shows the hours worked during the second week in February and the corresponding pay rates for the employees in a small office.

Employee	Pay rate per hour	Number of hours worked	Pay
Imran	$26.50	32	
Sofia	$29.25	35	
Cathy	$23.23	35	
Mike	$23.23	40	
Anita	$25.30	32	
Ronen	$26.50	20	
		Total wages bill	

The hours each employee works per week and their hourly rate of pay could change. Construct a spreadsheet that calculates each employee's wage and the total office wage bill when the number of hours worked and the rates of pay could change.

4 **WORKED EXAMPLE 3** Determine the missing values in the table.

Normal pay per hour	Pay per hour at time-and-a-half	Pay per hour at double time
$27.20	a	b
$24.36	c	d
$24.60	e	f
$31.25	g	h

5 **WORKED EXAMPLE 4** Tuan is a plumber's assistant. He works a 35-hour week at $23.50 per hour. His overtime is paid at time-and-a-half for the first 5 hours of overtime in a week and double time for any hours after that. This week, Tuan worked 42 hours.

   a How many hours did Tuan work at double time?

   b Calculate Tuan's pay for the week.

Calculator-assumed

6 (5 marks) Lance is paid a salary for being an office IT manager. Each week he earns $1300.

   a Calculate Lance's annual salary. (1 mark)

   b Explain why Lance's monthly pay is not $1300 × 4. (1 mark)

   c Divide Lance's annual salary by 12 to calculate his monthly pay. (1 mark)

   d Lance's salary is based on 7 hours work per day, 5 days per week and 52 weeks per year. Calculate the pay rate per hour that is the basis of Lance's salary. (2 marks)

7 (3 marks) Zheng is 18 years old and earns $16.26 per hour at a Chinese take-away store.

    a Last week, Zheng worked 16 hours. How much did he earn? (1 mark)

    b Today, Zheng earned $81.30. How many hours did he work? (2 marks)

8 (3 marks) Ulla receives a yearly stipend of $22 860 from the university to assist her with her postgraduate study and research.

    a How much does Ulla receive per fortnight from the stipend? (1 mark)

    b The stipend isn't enough to cover all of Ulla's living expenses. She also works as a waitress for 4 hours per night, 2 nights per week. She earns $23.50 per hour as a waitress. Calculate Ulla's total fortnightly income. (2 marks)

9 (2 marks) The minimum wage for a trainee pest inspector is $595.70 for a 38-hour week. What is the minimum pay per hour for a trainee pest inspector?

10 (2 marks) Carlos earns $320 per day as a relief teacher. The table shows the number of days he worked during a 5-week period.

Dates	Number of days worked
30 April – 4 May	2
7 May – 11 May	1
14 May – 18 May	5
21 May – 25 May	3
28 May – 1 June	2

How much did Carlos earn over the 5 weeks?

11 (7 marks) Ashok is a casual office worker. He is paid $178 per day irrespective of the number of hours he works. Usually, he works about 12 days per month.

    a How much did Ashok earn for working from 8 am to 1 pm on Monday? Explain your reasoning. (2 marks)

    b During February, Ashok earned $1958. How many days did he work in February? (2 marks)

    c The office offers Ashok a permanent 38-hour a week job at $26 per hour. Do you think he should take the permanent job, or continue to work 12 days per month casually where he gets paid $178 per day? Make your decision based on Ashok's annual pay for each option. (3 marks)

12 (3 marks) Tori is trying to decide which one of three jobs to take.

Job	Conditions	Pay
1	38-hour week, 5 days per week, possibility of overtime	$24/hour
2	75 hours per fortnight, work 9 days per fortnight	$1450 per fortnight
3	Salary based on a 35-hour week	$38 800 p.a.

Ignoring any overtime, which job pays the most per year?

1.2 Bonuses and commission

Bonuses and allowances

Some jobs include **allowances** for doing unpleasant work, for working under difficult conditions, or to cover expenses such as uniform and travel.

Some jobs pay **bonuses** (extra pay) for doing good work, meeting targets or deadlines.

WORKED EXAMPLE 5 | Calculating annual income including allowances

Kaitlyn's basic salary in the navy is $43 434 and she receives an annual $12 128 service allowance, as well as an annual $419 uniform maintenance allowance. When she is at sea, she receives an additional $11 758 annually.

a Calculate Kaitlyn's weekly pay when she is working on land.

b How much does Kaitlyn earn per fortnight when she's at sea?

Steps	Working
a Kaitlyn's annual salary on land = basic salary + service allowance + uniform allowance. Divide by 52 for weekly pay.	Salary = $43 434 + $12 128 + $419 = $55 981 Weekly pay on land = $55 981 ÷ 52 ≈ $1076.56
b Kaitlyn's annual pay at sea = basic salary + service allowance + uniform allowance + sea allowance. Divide by 26 for fortnightly pay.	Salary = $43 434 + $12 128 + $419 + $11 758 = $67 739 Fortnightly pay at sea = $67 739 ÷ 26 ≈ $2605.35

Video playlist
Bonuses and commission

Worksheets
Earning money
Pay day

WORKED EXAMPLE 6 | Calculating income including allowances from an hourly rate

Sonia is paid $25.48 per hour for her work as a security guard. Each week, she receives an additional $61.05 for her guard dog and $6.75 for her torch. She receives $14.15 travel allowance per shift. Sonia works a 4-hour shift, 6 nights per week. How much is she paid per week?

Steps	Working
Sonia's total weekly pay = wages + allowances + dog + torch She is paid travel allowance for each of the six shifts she works.	Wages = 4 × 6 × $25.48 = $611.52 Travel allowance = 6 × $14.15 = $84.90 Sonia's total weekly pay = $611.52 + $84.90 + $61.05 + $6.75 = $764.22

Commission and piecework

Salespeople are often paid by **commission**, which is a percentage of the value of the items they've sold. Sometimes the salespeople are paid a **retainer** plus commission. A retainer is a set payment that does not depend on sales.

Piecework is a type of work where a person is paid per item produced or processed.

WORKED EXAMPLE 7 — Calculating income including commission

Jordan is a used car salesman. He is paid a $2700 monthly retainer plus 5% commission on his monthly sales over $50 000. Calculate his pay for a month when his sales total $80 000.

Steps	Working
1 Find the value of the sales for which he is paid commission.	Sales over $50 000 = $80 000 − $50 000 = $30 000
2 Calculate the commission.	Commission = 5% of $30 000 = $1500
3 Calculate the total pay for the month.	Total earnings = retainer + commission = $2700 + $1500 = $4200

WORKED EXAMPLE 8 — Calculating commission using a sliding scale

Danielle earns commission for selling cosmetics at the following rates:

Commission on monthly sales

First $1000 of sales	5%
On the next $2000	4%
Remainder of sales	3.5%

This month, Danielle's sales totalled $5200. Calculate her commission.

Steps	Working

1 Break the sales of $5200 into the three commission categories.
    First $1000
    Next $2000
    Remaining $2200
Then calculate the commission on each.
	Commission on the first $1000 = 0.05 × $1000
	= $50
	Commission on the next $2000 = 0.04 × $2000
	= $80
	Danielle's remaining sales = $5200 − $1000 − $2000
	= $2200
	Commission on the remaining $2200 = 0.035 × $2200
	= $77

2 Calculate the total commission.	Danielle's total commission = $50 + $80 + $77 = $207

WORKED EXAMPLE 9	Calculating income from piecework

Dimitri works as a tailor altering clothes. He is paid $10 for hemming a skirt and $12 for shortening trousers. Calculate how much Dimitri earns if he hems five skirts and shortens six pairs of trousers.

Steps	Working
1 Multiply the number of skirts by $10 and the number of pairs of trousers by $12.	Pay for hemming skirts = 5 × $10 = $50 Pay for shortening trousers = 6 × $12 = $72
2 Calculate the total earnings.	Dimitri's earnings = $50 + $72 = $122

EXERCISE 1.2 Bonuses and commission

ANSWERS p. 406

Recap

1 Larry earns $1200 each week as a manager of a clothing store.

   a Calculate Larry's annual salary.

   b Calculate Larry's monthly pay.

   c Larry normally works for 38 hours each week. Use his weekly pay of $1200 each week to find his hourly rate of pay, to the nearest cent.

   d Larry's employment contract changes so that he is paid a wage instead of a salary using the rate per hour found in part **c**. If Larry works for 8 hours per day, 5 days per week and 48 weeks per year, find his total wage for the year.

2 Simon is a qualified dental nurse. He is paid $32 per hour for a 36-hour week.

   a How much does Simon earn per week?

   b How much is Simon paid per fortnight?

   c Calculate Simon's annual pay.

Mastery

3 WORKED EXAMPLE 5 Tristan's base salary as an Air Force trainee is $37 485 p.a. In addition, he receives the Australian Defence Force annual allowance of $12 128 and an annual $419 uniform allowance. He also receives $9531 p.a. when he is deployed overseas.

   a Calculate Tristan's weekly pay when he is working in Australia.

   b Determine Tristan's fortnightly pay when he is deployed overseas.

4 WORKED EXAMPLE 6 Zoran works for a pest control company. He is paid $28.75 per hour and receives an extra $28.10 per day for handling poisons. Zoran works for 7.5 hours per day, 5 days per week. Calculate his weekly pay.

5 WORKED EXAMPLE 7 Assam sells window shutters and is paid a retainer of $120 per week to cover his expenses, and a commission of 15% of all sales he makes. Assam's sales for the first week in April totalled $2896. Calculate his pay for that week.

6 **WORKED EXAMPLE 8** Tanika sells cosmetics. She earns commission at the following rates:

Commission on Tanika's monthly sales	
First $500 of sales	5%
On the next $1000	4%
Remainder of sales	3.5%

Calculate Tanika's commission for each of the following monthly sales figures.

a $360

b $1400

c $4200

7 **WORKED EXAMPLE 9** Milan puts letters in envelopes and attaches postage stamps for a marketing company.

He is paid 25 cents per letter. Milan can process 70 letters per hour.

a How much does Milan earn per hour?

b How much will he earn for processing 260 letters?

c How many letters does Milan need to process in order to earn over $100?

Calculator-free

8 (5 marks) Uri works as a labourer cleaning second-hand bricks. He is paid 20 cents for each brick he cleans. He can clean 100 bricks each hour.

a How much can he earn per hour? (1 mark)

b How many bricks does he need to clean to earn $200? (2 marks)

c How much would Uri earn in a day if he works for 8 hours? (2 marks)

9 (2 marks) Jenny's normal pay is $26.00 per hour. How much will she earn when she works 2 hours at time-and-a-half?

10 (4 marks)

a Marco earns 2% commission on all his sales. Find his commission on the following sales:

  i $100 (1 mark)

  ii $1000 (1 mark)

b Find Marco's rate of commission if he earns $6 commission on sales of $200. (2 marks)

Calculator-assumed

11 (3 marks) Sancia's normal pay is $21.80 per hour. She works 3 hours at time-and-a-half and 4 hours at double time?

a How much does she earn at time-and-a-half? (1 mark)

b How much does she earn at double time? (1 mark)

c How much does Sancia earn in total? (1 mark)

12 (7 marks) Mercia has a holiday job supervising children in a resort. She is paid a junior casual rate of $19 per hour Monday to Friday, time-and-a-half on Saturday and double time on Sunday.

    a The table shows the times Mercia worked during one week in January. What are the missing values in the table? (6 marks)

Day	Hours worked	Pay rate per hour	Pay
Weekdays	21	i	iv
Saturday	4	ii	v
Sunday	6	iii	vi

    b Calculate Mercia's pay for the week. (1 mark)

13 (4 marks)

    a Casey earned $108 when he worked for 3 hours at double time. What is Casey's normal pay per hour? (2 marks)

    b How much will Casey earn when he works for 3 hours at time-and-a-half? (2 marks)

14 (5 marks) Emily earns monthly commissions when she sells perfumes according to these rates:

Monthly sales commission

$800	5% of sales
$801 to $1200	$40 plus 4.5% of sales over $800
$1201 and over	$58 plus 4% of sales over $1200

Calculate Emily's commission in a month when her total sales were valued at

    a $360 (1 mark)

    b $998 (2 marks)

    c $5100 (2 marks)

15 (2 marks) Monique has hired Trevor to re-tile her home. Trevor told Monique she required 5 m² of tiles for the kitchen walls, 48 m² of slate for the lounge room floor and 7 m² of slate for the stairs. How much will Trevor charge to lay the tiles and slate if his labour charge is $37.25 per square metre?

16 (3 marks) Holly enjoys cooking scones which she sells at the local Devonshire Tea shop. She buys her ingredients in bulk and it costs her $12 to make 5 dozen scones in 2 hours. She values her labour at $16 per hour.

How much should Holly charge the Devonshire Tea shop for 10 dozen scones?

17 (3 marks) Basam is selling his house for $420 000. The real estate agent's commission is 2% on the first $200 000 and 1.5% on the balance of the sale price. How much will Basam receive from the sale of his house?

18 (2 marks) Elise earned $120 when she worked on Sunday for 2 hours at double time. How much does Elise earn for a normal 7-hour day?

19 (3 marks) Kate is the manager of a fast food chain. She is paid $28 per hour for a 35-hour week. Her bonuses are $8.30 per week laundry allowance, a $30 bonus for every accident-free week at the shop and another $95 bonus per week if the shop makes $100 000 or more in weekly sales. Last week, the shop was accident-free and sales were $110 000. Find

   a the amount Kate earns for working for 35 hours (1 mark)

   b the amount Kate was paid in bonuses last week (1 mark)

   c Kate's wage last week. (1 mark)

20 (8 marks) Zack drives a furniture removal truck. He is paid $23.23 per hour Monday to Friday, time-and-a-quarter on Saturday and double time on Sunday. In addition, he receives a flat fee of $12.59 per day for handling heavy furniture. During the week, Zack delivered heavy furniture for 33 hours Monday to Friday, 6 hours on Saturday and 3 hours on Sunday. Find

   a Zack's pay for Monday to Friday, excluding the bonus for heavy furniture. (2 marks)

   b Zack's pay for Saturday, excluding the bonus for handling heavy furniture. (2 marks)

   c Zack's pay for Sunday, excluding the bonus for handling heavy furniture. (2 marks)

   d Zack's pay for the week, including the bonus for handling heavy furniture. (2 marks)

21 (6 marks) Callum works for the council. He looks after the grass in parks and at sporting venues.

Callum doesn't work any overtime on Monday to Friday. All the hours he works on Saturday are paid at time-and-a-half and his work on Sunday is at double time. For Callum's time and paysheet below, find the values of **a**, **b**, **c**, **d**, **e** and **f**.

Greater Middleton Council (GMC)

PAID BY
Greater Middleton Council (GMC)
215 Main Street,
Middleton WA 6322
ABN: 65428513740

NAME	Callum Jones	**PERIOD STARTING**	08/07/2024
BANK DETAILS	BSB. 084 169 Acc No. 4603 2187	**PERIOD ENDING**	14/07/2024
HOURLY RATE	$25.90	**PAYMENT DATE**	29/07/2024

DAY	START	FINISH	UNPAID BREAKS	PAY
Monday	7 am	3:30 pm	30 minutes	a
Tuesday	b	5:30 pm	1 hour	$220.15
Wednesday	–	–		
Thursday	8 am	4:00 pm	c	$181.30
Friday	7 am	d	1 hour	$103.60
Saturday	7 am	11 am	nil	e
Sunday	8 am	f	nil	$77.70
GROSS PAY				

1.3 Government allowances and pensions

Australian Government payments, pensions and allowances are income amounts that people may receive from a government agency. Commonly these payments are from Services Australia or the Department of Veteran's affairs (DVA).

Taxable government payments, pensions and allowances include:
- Age Pension
- Carer Payment
- Austudy Payment
- JobSeeker Payment
- Youth Allowance
- Defence Force Income Support Allowance
- Veteran Payment
- Invalidity Service Pension
- Disability Support Pension
- Income Support Supplement
- Parenting Payment (Partnered)

The payments stated in this exercise were current at the time of publishing.

Video playlist
Government allowances and pensions

WORKED EXAMPLE 10 | Calculating pension entitlements

To receive a pension, a person needs to pass an income and assets test. Joan is a single, aged pensioner who owns her home. She is allowed to have assets up to $250 000 and retain a full pension of $971.50 per fortnight. For every $1000 over $250 000 in assets, her fortnightly pension reduces by $3. At present, Joan's assets are valued at $198 000 and she is about to inherit $75 000. What effect will the inheritance have on her pension?

Steps	Working
1 Find Joan's assets after her inheritance.	Joan's assets = $198 000 + $75 000 = $273 000
2 Find the value of Joan's assets over $250 000.	Amount = $273 000 − $250 000 = $23 000
3 For every $1000 over $250 000, Joan's pension reduces by $3.	$23 000 ÷ $1000 = 23 Joan's pension = $971.50 − 23 × $3.00 = $902.50 Inheriting $75 000 will decrease Joan's pension to $902.50 per fortnight.

EXERCISE 1.3 Government allowances and pensions

ANSWERS p. 406

[Note: Payments were current at time of publishing.]

Recap

1 Tanika sells health-food products. She earns commission at the following rates:

Commission on Tanika's monthly sales

First $500 of sales	5%
On the next $1000	4%
Remainder of sales	3%

Calculate Tanika's commission for each of the following monthly sales figures.

   a $400   **b** $1200   **c** $2000

2 Kristen's base salary as an Air Force trainee is $45 000 p.a. In addition, she receives the Australian Defence Force annual allowance of $13 500 and an annual $500 uniform allowance. She also receives $9500 p.a. when she is deployed overseas.

   a Calculate Kristen's weekly pay when she is working in Australia.

   b Determine Kristen's fortnightly pay when she is deployed overseas.

Mastery

3   WORKED EXAMPLE 10   The Age Pension Payment for a single person is $971.50 per fortnight but for every dollar of income they receive over $204 per fortnight the pension reduces by 50 cents.

   a Doris is a single aged pensioner and she receives $290 per fortnight from renting out her granny flat. Calculate the value of Doris' fortnightly age pension.

   b Nanna thinks the age pension Centrelink is paying her is incorrect. For the last 8 weeks she has been receiving $520 per fortnight pension and she also receives a fortnightly income of $1070 from her investments. Determine if Nanna has been overpaid or underpaid by Centrelink and the total error in payment for the eight weeks.

Calculator-free

4 (2 marks) Cameron receives $389.40 per fortnight in Youth Allowance. He pays his mother $80 per week for board and his fortnightly public transport costs to travel to TAFE are $42.

How much of his Youth Allowance is left each fortnight after Cameron pays his board and transport expenses?

Calculator-assumed

5 (3 marks) The maximum fortnightly Age Pension payable to eligible people over the age of 65 is $971.50 for singles and $732.30 per person for couples. How much less does a couple receive per fortnight than two single people sharing accommodation?

6 (2 marks) Senior health care cardholders and pensioners are entitled to a government energy payment. Every month they receive a $28.20 Energy Supplement. Calculate the annual value of the Energy Supplement.

7 (2 marks) Jim receives a Disability Support Pension and he lives in a public housing, rent-subsidised unit. Jim must pay 15% of his pension in rent. His pension is $609.30 per fortnight. How much rent does Jim pay per year?

8 (5 marks) Gail receives a Disability Support Pension because she is too sick to work. Her fortnightly pension is $797.90.

  a Calculate Gail's annual pension. (2 marks)

  b Gail's pension includes $6.20 per fortnight for medications and she pays $6.30 per prescription medicine. Gail takes a lot of medication, but after she has paid for 58 prescriptions per year, all further prescriptions are provided free. How much more than her fortnightly medication allowance does Gail have to pay for her medications each year? (3 marks)

9 (4 marks) The Age Pension rates per fortnight are $971.50 for singles with an Energy Supplement of $14.10, and $732.30 each for couples with an Energy Supplement of $10.60.

  a Beryl and Henry are a couple who qualify for a full pension but do not qualify for an Energy Supplement. Determine their total annual pension. (2 marks)

  b Raul is a single pensioner who qualifies for a full pension and an Energy Supplement. Find his annual pension. (2 marks)

10 (5 marks) Brody and Rose are a couple both aged 20 years. Brody's fortnightly taxable income is $721.70 but Rose is unable to work and she receives a Disability Support Pension of $460.60 per fortnight. Recently, Rose received a letter from Centrelink informing her that for the last year she has been overpaid, however, Rose thinks some of these details are incorrect. The details included in the letter are:

- Pension received: $11 975.60
- Entitled: $10 635.92
- Overpaid amount: $1620.85.

The maximum, fortnightly Disability Support Pension for a member of a couple aged under 21 years is $707.60 and this amount reduces by $0.40 for every dollar earned over $360 of the couple's fortnightly taxable income. Are the details included in the letter correct? Justify your answer.

1.4 Budgets

Have you ever wondered what happens to your money? It is a good idea to have a plan so that you don't waste it. A **budget** lists expected income and expenses, and can help you to manage your money.

Income covers all the money earnt.

Expenses covers all the ways money might be spent. There are two types of expenses:

- **Fixed expenses** are costs that are essential and must be paid. Some are the same amount each time, such as rent. Others aren't always the same, such as food and electricity.
- **Discretionary expenses** are amounts that you often spend but which aren't essential, such as entertainment or eating out.

A budget needs to **balance** your income and expenses so that you have enough money for everything you require and some left over to save for special items, such as a car or a holiday.

Video playlist
Budgets

Worksheets
Budgeting scenarios

Budgeting grid

My budget

WORKED EXAMPLE 11 — Creating a budget

Ashleigh works part-time while studying. She receives an allowance from her parents of $200 per week and she earns $245 from her job. She lives in a shared house and pays $120 per week in rent and spends $200 per week on food. She averages $25 per week for her mobile phone and $20 per week on clothes. She divides the remainder equally between entertainment and savings.

a Create a budget for Ashleigh for a week.

b Ashleigh's rent is increased by $25 per week. How would she need to adjust her budget for this increased expense?

Steps	Working
a List income and expenses and make sure total expenses equals total income.	(see table and working below)

Income		Expenses	
Allowance	$200	Rent	$120
Earnings	$245	Food	$200
		Mobile phone	$25
		Clothes	$20
		Entertainment	$40
		Savings	$40
Total income	**$445**	**Total expenses**	**$445**

Total of fixed expenses
= $120 + $200 + $25 + $20
= $365

Remainder available for entertainment and savings
= $445 − $365
= $80

Divided equally
= $80 ÷ 2
= $40

b The increased expense of $25 in rent will mean entertainment and savings will need to reduce by $25 to balance the budget.

The $25 increase in rent means Ashleigh now has $25 less to spend on entertainment and savings. She has $80 − $25 = $55 to divide between entertainment and savings. She could still divide this amount equally between the two ($27.50 each).

EXERCISE 1.4 Budgets

ANSWERS p. 406

Recap

1 Carmel receives $389.40 per fortnight in Youth Allowance. She pays $120 per week for board and her fortnightly public transport costs to travel to TAFE are $45.

How much of her Youth Allowance is left each fortnight after Carmel pays her board and transport expenses?

2 The Age Pension payment for a single person is $971.50 per fortnight; however, for every dollar of income earned over $190 per fortnight the pension reduces by 40 cents.

Jack is a single aged pensioner and he receives $290 per fortnight from working at the local nursery. Calculate the value of Jack's fortnightly Age Pension.

Mastery

3 WORKED EXAMPLE 11   Mitchell works in an office during the week and in the bar at the local club on weekends.

He earns $620 per week from the office job and $215 from the club. He pays $280 per week in rent and spends an average of $60 per week on food. His smartphone costs him $20 per week and travel expenses are $60 per week. The remainder of his income must be divided between entertainment, clothes and savings.

 a Create a budget for Mitchell for one week.

 b Mitchell is considering buying a car. He would no longer have public transport expenses but he would need to allow $100 per week to pay off a loan and $75 for car expenses. Create a new budget for Mitchell.

Calculator-assumed

4 (4 marks) Lily owns a car and has the following expenses each year:
- registration $349
- Compulsory Third Party (CTP) insurance $795
- comprehensive insurance $1110
- maintenance bills of $790.

She spends $53 per week on petrol.

 a How much does Lily spend on her car each year? (2 marks)
 b How much should she set aside in her weekly budget to cover her car expenses? (2 marks)

5 (6 marks) Marko is an apprentice mechanic. His take-home pay is $790 per week. This table shows his weekly expenses.

Example

Item	Amount
Board	$120
Mobile phone	$21
Clothes	$65
Car	$112
Entertainment	$72
Other expenses	$88
Savings	
Total	$790

 a How much does Marko save each week? (2 marks)
 b Calculate his net annual income. (2 marks)
 c How much is Marko able to save each year? (2 marks)

6 (4 marks) This table shows Shania's monthly budget.

Income		Expenses	
Part-time job	$290	Clothes	$140
Babysitting	$130	School needs	$32
		Entertainment	$50
		Mobile phone	$55
		Fares	$23

    a Calculate Shania's total monthly income and expenses. (2 marks)

    b Calculate the amount she can save each month. (1 mark)

    c Calculate the amount she can save each year. (1 mark)

7 (6 marks) Sanjeev has taken a second job to save for a new car. His budget for a week is shown below.

Income		Expenses	
Main job	$750	Rent	$225
Second job		Travel	$56
		Food	$117
		Clothes	$55
		Entertainment	$75
		Bills	$157
		Savings	
Total	$908	**Total**	$908

    a Calculate how much Sanjeev earns from his second job. (1 mark)

    b Calculate how much he can save each week. (1 mark)

    c If the car Sanjeev wants costs $25 000, how long would it take him to save this money? (2 marks)

    d Suggest ways Sanjeev could save more per week so that he can buy the car he wants sooner. Create a new budget for Sanjeev. (2 marks)

1.5 Applying percentages

Percentage increase and percentage decrease

Percentage means 'per 100', which is a number expressed as a fraction of 100.

So, when we say 100% of something, it means it represents the whole of it.

Percentage increase is the increase of a number, amount, or quantity expressed as a percentage of the original value.

Percentage decrease is the decrease of a number, amount, or quantity expressed as a percentage of the original value.

Percentage change

$$\text{percentage increase} = \frac{\text{increase}}{\text{original value}} \times 100\%$$

$$\text{percentage decrease} = \frac{\text{decrease}}{\text{original value}} \times 100\%$$

Consider the situation where a keen bird-watcher counts the number of birds in a forest, over two consecutive months. In December there were 50 birds and in January there were 60. This is an increase of 10 birds which can be expressed as a percentage of the number of birds in December.

The percentage increase in the number of birds = $\frac{10}{50} \times 100 = 20\%$.

Video playlist Applying percentages

Skillsheets Percentage calculations

Mental percentages

WORKED EXAMPLE 12 — Calculating percentage increase or decrease

Find the percentage increase or decrease.

a The price of 2 litres of milk changes from $4.00 to $4.50.

b The number of ripe apples on the tree was 15 last week and 12 this week.

Steps	Working
a The change in the price is an increase. Substitute in the formula: $\text{percentage increase} = \frac{\text{increase}}{\text{original value}} \times 100\%$	price increase = $4.50 − $4.00 = $0.50 $\text{percentage increase} = \frac{0.5}{4.00} \times 100\%$ = 12.5%
b The change in the number is a decrease. Substitute in the formula: $\text{percentage decrease} = \frac{\text{decrease}}{\text{original value}} \times 100\%$	number decrease = 15 − 12 = 3 $\text{percentage decrease} = \frac{3}{15} \times 100\%$ = 20%

Mark-up and discount

When the price of an item is increased by a percentage, it's called a **mark-up**.

When the price of an item is reduced by a percentage, it's called a **discount**.

Mark-up	Discount
For an item marked up by $r\%$: mark-up = original price $\times \dfrac{r}{100}$ new price = original price + mark-up new price = original price $\times \dfrac{100 + r}{100}$	For an item discounted by $r\%$: discount = original price $\times \dfrac{r}{100}$ new price = original price − discount new price = original price $\times \dfrac{100 - r}{100}$

WORKED EXAMPLE 13 — Calculating mark-ups and discounts

a The original price of a pair of jeans is $80. The jeans are marked up by 20%. Find the amount of the mark-up.

b The price of a wetsuit is $200. At the end of season sale, all stock is discounted by 10%. Find the amount of the discount.

Steps	Working
a Find the amount of the mark-up. mark-up = original price $\times \dfrac{r}{100}$	mark-up = $80 \times \dfrac{20}{100}$ = 16 The mark-up is $16.
b Use the formula: discount = original price $\times \dfrac{r}{100}$	discount = $200 \times \dfrac{10}{100}$ = $20 The discount is $20.

Puzzle
Discounts
code puzzle

WORKED EXAMPLE 14 — Calculating the new mark-up price and discount price

a The original price of a pair of headphones is $120. The headphones are marked up by 30%. Find the new price after the mark-up.

b Find the sale price of a shirt marked at $80 after it is discounted by 15%.

Steps	Working
a Use the mark-up price formula: new price = original price $\times \dfrac{100 + r}{100}$	new price = $120 $\times \dfrac{100 + 30}{100}$ = $120 $\times \dfrac{130}{100}$ = $156 The price after mark-up is $156.
b Use the discounted price formula: new price = original price $\times \dfrac{100 - r}{100}$	new price = $80 $\times \dfrac{100 - 15}{100}$ = $80 $\times \dfrac{85}{100}$ = $68 The discounted price is $68.

Finding percentage mark-up or percentage discount

The percentage mark-up uses the same formula as percentage increase.

The percentage discount uses the same formula as percentage decrease.

WORKED EXAMPLE 15 Finding the percentage mark-up or discount

The original price of a pair of jeans is $80.

a Find the percentage mark-up if it has been increased to $100.

b Find the percentage discount if it has been discounted to $30.

Steps	Working
a Use the formula: $$\text{percentage mark-up} = \frac{\text{mark-up}}{\text{original price}} \times 100\%$$	mark-up = 100 − 80 = 20 $$\text{percentage mark-up} = \frac{20}{80} \times 100\%$$ $$= 25\%$$ The percentage mark-up is 25%.
b Use the formula: $$\text{percentage discount} = \frac{\text{discount}}{\text{original price}} \times 100\%$$	discount = 80 − 30 = 50 $$\text{percentage discount} = \frac{50}{80} \times 100\%$$ $$= 62.5\%$$ The percentage discount is 62.5%.

We can also find the original price if we know the new price and r.

USING CAS 2 Finding the original price

Find the original price of a basketball, to the nearest cent, if the new price is $120 after being marked up by 30%.

The formula for the new mark-up price is

$$\text{new price} = \text{original price} \times \frac{100 + r}{100}$$

Substitute new price = 120, $r = 30$ and original price = x.

Solve $120 = x \times \frac{130}{100}$.

ClassPad	TI-Nspire
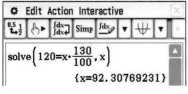 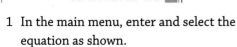	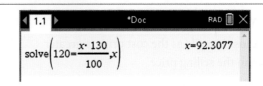
1 In the main menu, enter and select the equation as shown. 2 Tap **Interactive > Advanced > Solve**.	1 On a calculator page, press **Menu > Algebra > Solve**. 2 Enter the equation as shown and press **ctrl + enter**.

The original price of the basketball was $92.31.

Profit and loss

Every company or business works on the basic principles of profit and loss.

Worksheet
Profit and loss

Terminology

Cost price (CP) The price at which a product is purchased.
Selling price (SP) The price at which a product is sold.
Profit (P) If a product is sold at a price greater than the cost price, then the seller makes a profit.
Loss (L) If a product is sold at a price less than the cost price, then the seller makes a loss.

Profit and loss formulas

Profit	Loss
profit = selling price − cost price	loss = cost price − selling price
profit percentage = $P\% = \dfrac{\text{profit}}{\text{cost price}} \times 100$	loss percentage = $L\% = \dfrac{\text{loss}}{\text{cost price}} \times 100$
selling price = cost price $\times \dfrac{100 + P\%}{100}$	selling price = cost price $\times \dfrac{100 - L\%}{100}$

WORKED EXAMPLE 16 — Finding the profit/loss and profit/loss percentage

Find, for each of the following:
  i the profit or loss
  ii the profit or loss percentage.

a A mirror with a cost price of $50 is sold for $75.
b A painting with a cost price of $120 is sold for $80.

Steps	Working
a i Write the given information. A profit is made as the selling price is greater than the cost price.	CP = 50, SP = 75 profit = $75 − $50 = $25
ii Use the formula: profit percentage = $P\% = \dfrac{\text{profit}}{\text{cost price}} \times 100$	profit percentage = $\dfrac{25}{50} \times 100$ = 50 The profit percentage is 50%.
b i Write the given information. A loss is made as the cost price is greater than the selling price.	CP = 120, SP = 80 loss = $120 − $80 = $40
ii Use the formula: loss percentage = $L\% = \dfrac{\text{loss}}{\text{cost price}} \times 100$	loss percentage = $\dfrac{40}{120} \times 100$ = $33\dfrac{1}{3}$ The loss percentage is $33\dfrac{1}{3}\%$.

WORKED EXAMPLE 17 — Finding the selling price given the profit/loss percentage

a A desk with a cost price of $250 is sold at a profit of 40%. Find the selling price.
b A bicycle with a cost price of $800 is sold at a loss of 10%. Find the selling price.

Steps	Working
a Write the given information. Use the formula: $$\text{selling price} = \text{cost price} \times \frac{100 + P\%}{100}$$	CP = 250, P% = 40 $$\text{selling price} = 250 \times \frac{100 + 40}{100}$$ $$= 250 \times \frac{140}{100}$$ $$= 350$$ The selling price is $350.
b Write the given information. Use the formula: $$\text{selling price} = \text{cost price} \times \frac{100 - L\%}{100}$$	CP = 800, L% = 10 $$\text{selling price} = 800 \times \frac{100 - 10}{100}$$ $$= 800 \times \frac{90}{100}$$ $$= 720$$ The selling price is $720.

The **Goods and Services Tax (GST)** is currently a 10% tax on most sales and services in Australia. We can think of this as a mark-up where:

- price without GST = original price
- price with GST = new price
- GST amount = change in price
- $r = 10\% = 0.1$

Working with GST

$$r = 10\% = \frac{10}{100}$$

$$\text{GST} = \text{price without GST} \times \frac{10}{100}$$

$$\text{GST} = \text{price with GST} - \text{price without GST}$$

$$\text{price with GST} = \text{price without GST} \times \frac{110}{100}$$

WORKED EXAMPLE 18 — Working with GST

Round the following answers to the nearest dollar.

a A piece of software has a GST-free price of $190.
  i How much GST is payable?
  ii What is the price with GST?

b A car is advertised at a price of $34 500, which includes GST.
  i What is the price excluding GST?
  ii How much of the advertised price is GST?

Steps	Working
a i Use GST = price without GST × $\frac{10}{100}$. Round to the nearest dollar.	$\text{GST} = 190 \times \frac{10}{100}$ = 19 The GST is $19.
ii Use price with GST = price without GST + GST	Price with GST = 190 + 19 = 209 The price with GST is $209.

b i Use price with GST $= \text{price without GST} \times \dfrac{110}{100}$ and solve using CAS, rounding to the nearest dollar.		price with GST = 34 500 price without GST = x $34\,500 = x \times \dfrac{110}{100}$ $x = 34\,500 \div 1.1$ $= 31\,363.636...$ The price without GST is $31 364.
ii Use GST amount = price with GST − price without GST		GST = 34 500 − 31 364 = 3136 The GST amount is $3136.

The unit cost method

Consider the case where there are two sizes of the same brand of dog food, 7 kg for $29 and 20 kg for $70. How do we know which is the best value? We can calculate the cost for 1 kg, for each size, to know which is the cheapest. This method of comparing the prices of the same product in different sizes, to see which one is the best value, is called the **unit cost method**.

Puzzle
Best buys
puzzle

WORKED EXAMPLE 19 Using the unit cost method to make comparisons

Paul is buying washing powder. His brand comes in three different sizes: 350 g for $10.00, 800 g for $15.40 and 2 kg for $28.70.

a What is the unit cost of each of the sizes? Give your answer correct to four decimal places.
b Which size is the best value?

Steps	Working
a 1 Write the known equality for each of the sizes. If necessary, convert units of measurement so that all the units are the same.	$10.00 for 350 g $15.40 for 800 g $28.70 for 2 kg = 2000 g
2 Divide each price by the number of grams to find the cost of one gram. Give your answer correct to four decimal places.	$10.00 ÷ 350 g = $0.028 57/g $15.40 ÷ 800 g = $0.019 25/g $28.70 ÷ 2000 g = $0.014 35/g
b Compare to find the cheapest and write the answer in words.	The 2 kg for $28.70 is the best value because it is the cheapest per gram.

 Exam hack

When calculating the best value, always makes sure the units of measurement you are comparing are the same.

EXERCISE 1.5 Applying percentages   ANSWERS p. 407

Recap

1 Michael works as a nurse during the week and plays in a band on Saturdays. He earns $750 per week nursing and $100 on Saturdays with the band. He pays $250 per week in rent and spends an average of $150 per week on food. His smartphone costs him $80 per week and travel expenses are $70 per week. The remainder of his income is divided between entertainment and savings. Michael decides to save twice as much as he spends on entertainment.

   a Find the amount of money Michael saves.
   b Create a budget for Michael for one week.

▶ **Mastery**

2 WORKED EXAMPLE 12 Find the percentage increase or decrease.
   a The price of a loaf of bread changes from $3.50 to $4.20.
   b The number of ripe tomatoes on a plant was 20 last week and 15 this week.
   c The price of a t-shirt changes from $40 to $30.
   d The weight of a baby changes from 2.5 kg at the start of the month to 4 kg at the end of the month.
   e The height of a 10-year-old girl is 135 cm and when she is 11 years old her height is 145 cm.

3 WORKED EXAMPLE 13 The original price of a shirt is $60.
   Find the mark-up or discount if the shirt is
   a marked up by 20%
   b discounted by 12%
   c discounted by 9%
   d marked up by 35%.

4 WORKED EXAMPLE 14 The original price of a shirt is $60. Find the new price if the shirt is
   a marked up by 25%
   b discounted by 15%
   c marked up by 6%
   d discounted by 8%.

5 WORKED EXAMPLE 15 The original price of a pair of shorts is $50.
   a Find the percentage mark-up if it is increased to
      i $70   ii $85   iii $60
   b Find the percentage discount if it is discounted to
      i $40   ii $45   iii $25

6 Using CAS 2 Find the original price of a jacket, to the nearest cent, if
   a the new price is $150 and the mark-up is 20%
   b the new price is $45 and the discount is 60%
   c the new price is $99 and the discount is 25%
   d the new price is $200 and the mark-up is 35%.

7 WORKED EXAMPLE 16 Find, for each of the following:
      i the profit or loss
      ii the profit or loss percentage.
   a A lawn mower with a cost price of $350 is sold for $475.
   b A table with a cost price of $200 is sold for $150.
   c A hat costing $120 is sold for $240.
   d A pair of socks with a cost price of $2.50 is sold for $2.00.

8 WORKED EXAMPLE 17
   a A desk with a cost price of $400 is sold at a profit of 50%. Find the selling price.
   b A scooter with a cost price of $1000 is sold at a loss of 20%. Find the selling price.

1.5

9 **WORKED EXAMPLE 18** Round the following answers to the nearest dollar.

   a   A laptop has a GST-free price of $1795.

      i   How much GST is payable?   ii   What is the price with GST?

   b   A haircut has a GST-free price of $89.

      i   How much GST is payable?   ii   What is the price with GST?

   c   A boat is advertised at a price of $159 106, inclusive of GST.

      i   What is the price excluding GST?   ii   How much GST is payable?

   d   A television set is advertised at a price of $2850, inclusive of GST.

      i   What is the price excluding GST?   ii   How much GST is payable?

10 **WORKED EXAMPLE 19** For each pair of sizes

      i   what is the unit cost of each size? Give your answer correct to four decimal places.

      ii   which size is the better value?

   a   150 mL for $2.36 or 1 L for $14.80.

   b   36 cans for $20 or 15 cans for $8.

   c $8.40 for 500 g or $20.99 for 1.2 kg.

Calculator-free

11 (4 marks)

   a   A phone that normally retails for $200 is discounted to $170. Find the percentage discount. (2 marks)

   b   This month, a business charges $1000 to install a water tank. Next month, the charge will increase by 5%. Find the charge next month. (2 marks)

12 (3 marks) A surf shop must mark-up all their stock by 10% at the end of the month. Determine the new price after the mark-up of the following items.

   a   tee shirt $40 (1 mark)

   b   beanie $20 (1 mark)

   c   rash-vest $70 (1 mark)

Calculator-assumed

13 (2 marks) The adult membership fee for a cricket club is $150. Junior members are offered a discount of $30 off the adult membership fee.

   a   Write down the discount for junior members as a percentage of the adult membership fee. (1 mark)

   Adult members of the cricket club pay $15 per match in addition to the membership fee of $150.

   b   If an adult member played 12 matches, what is the total this member would pay to the cricket club? (1 mark)

14 (3 marks) Nina is trying to determine the best value pasta at the market; however each packet has a different price and a different quantity.

   The five different pastas are:

   1   2 kg for $17.80

   2   250 g for $2.40

   3   1.2 kg for $10.85

   4   700 g for $6.30

   5   400 g for $3.65

   a   Find the unit cost for 1 kg for each pasta. (2 marks)

   b   Which pasta is the best buy? (1 mark)

1.6 Interest

Simple interest investments and loans

Interest is the fee for using someone else's money. It applies to both investments and loans.
- For investments, the bank uses our money and pays us the interest.
- For loans, we use the bank's money and we pay the bank the interest.

The amount that is invested or borrowed is called the **principal**.

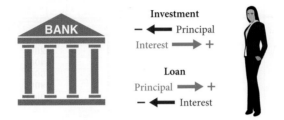

Simple interest is a fixed amount of interest that is paid at regular time periods. When these time periods are years, we use the term **per annum (p.a.)**, which means per year.

> **Simple interest formula**
>
> $I = Prt$
>
> where I is the simple interest
>
> P is the principal
>
> r is the rate per year, and
>
> t is the time in years.

WORKED EXAMPLE 20 — Calculating simple interest

Rohan borrowed $2500 to buy a laptop, with interest charged at 8% per annum, simple interest, over 4 years. Find the

a total interest on the loan.

b total amount owing on the loan after 4 years.

c total amount owing on the loan after 18 months.

Steps	Working
a 1 Summarise the information given and identify the value you need to find.	$P = 2500, r = 8\% = 0.08, t = 4, I = ?$
2 Substitute into $I = Prt$.	$I = Prt$ $= 2500 \times 0.08 \times 4$ $= 800$ The simple interest is $800.
b Total amount owing = principal + interest	Total amount owing = 2500 + 800 = 3300 The amount owing after 4 years is $3300.
c Find t in years by dividing the number of months by 12. Calculate the simple interest and the amount owing on the loan	$P = 2500, r = 8\% = 0.08, t = 18 \div 12 = 1.5$ $I = Prt$ $= 2500 \times 0.08 \times 1.5$ $= 300$ Total amount owing after 18 months = 2500 + 300 = $2800

USING CAS 3 Graphing simple interest loans on a spreadsheet

Albert takes a loan of $4000 with interest charged at 5% per annum, simple interest.

a Create a spreadsheet which calculates the interest each year and the total amount owing on the loan each year for 5 years.

b Draw a graph with the balance of the loan on the vertical axis and the number of years on the horizontal axis.

ClassPad

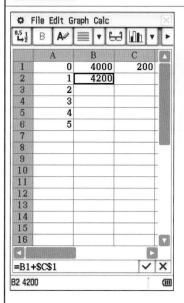

1. Open the **Spreadsheet** application and, enter the years **1** to **6** in column **A**.
2. In cell **B1**, enter the principal **4000**.
3. In cell **C1**, enter =**B1×0.05** (the formula for the interest each year is 4000 × 0.05)
4. In cell **B2**, enter =**B1+$C$1**.
5. Tap **Edit > Fill > Fill Range** and in **range** enter **B1:B6**, then tap **OK**.
6. Select cells **A1:B6** and tap **graph > scatter**.

TI-Nspire

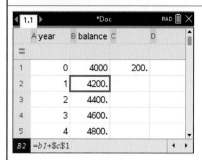
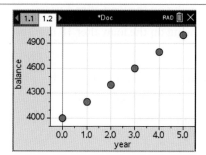

1. Add a **Lists & Spreadsheets** page and name column **A** 'year' and column **B** 'balance'.
2. Enter the years **0** to **5** in column **A**.
3. In cell **B1**, enter the principal **4000**.
4. In cell **C1** enter =**B1×0.05**.
   (the formula for the interest each year is 4000 × 0.05)
5. In cell **B2**, enter =**B1+$C$1**.
6. Press **menu > Data > Fill** and use the touchpad to fill cells **B2** to **B6**, then press **enter**.
7. Add a **Data & Statistics** page.
8. Select year for the horizontal axis and balance for the vertical axis.

The balance shows the amount owing on the loan at the end of that particular year. The graph of the balance vs time under simple interest is linear.

In some cases, we need to find P, r or t when given the simple interest. These problems can be solved using CAS.

- Summarise the information given and let x equal the value we need to find.
- Substitute into $I = Prt$ and solve using CAS.

WORKED EXAMPLE 21 Finding P, r or t for simple interest

a Rumi invested $10 000 at 4% p.a. simple interest and earned a total of $2200 in interest. How long was the money invested?

b An investment of $4000 grows to $5100 over 5 years. Determine the rate of simple interest p.a.

Steps	Working
a 1 Summarise the information given and let x equal what you need to find.	$P = 10\,000$, $r = 4\% = 0.04$, $I = 2200$, $t = x$
2 Substitute into $I = Prt$ and solve for x.	$I = Prt$ $2200 = 10\,000 \times 0.04 \times x$ $\therefore x = 5.5$ The money was invested for 5.5 years.

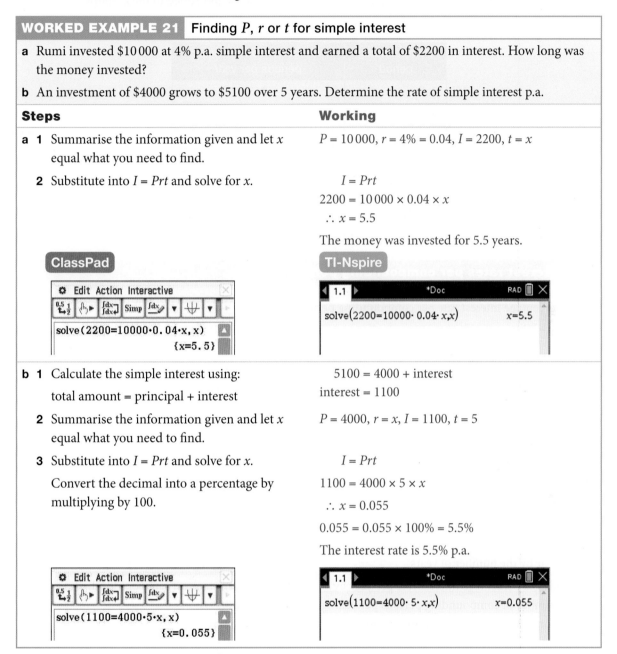

b 1 Calculate the simple interest using: total amount = principal + interest	$5100 = 4000 + \text{interest}$ $\text{interest} = 1100$
2 Summarise the information given and let x equal what you need to find.	$P = 4000$, $r = x$, $I = 1100$, $t = 5$
3 Substitute into $I = Prt$ and solve for x. Convert the decimal into a percentage by multiplying by 100.	$I = Prt$ $1100 = 4000 \times 5 \times x$ $\therefore x = 0.055$ $0.055 = 0.055 \times 100\% = 5.5\%$ The interest rate is 5.5% p.a.

Compound interest

Most investments and loans use **compound interest** where the interest is added to the principal, and the interest for the next time period is calculated using this new balance. The interest is regularly calculated at the end of a certain time period, which is called a **compounding period**.

Worksheets
Compound interest table

Compounding periods: spreadsheet

Compounding periods

Compound interest is always given as a rate per year, but compounding periods can vary.

- Daily compounding means the interest is calculated every day and added to the account.
- Weekly compounding means the interest is calculated every week and added to the account.
- Further compounding periods are shown in the table.

Compounding period	Number of compounding periods per year
Daily	365
Weekly	52
Fortnightly	26
Monthly	12
Quarterly	4
Six-monthly	2
Yearly	1

Interest rates per compounding period

$$\text{percentage interest rate per compounding period} = \frac{\text{percentage interest rate per year}}{\text{number of compounding periods per year}}$$

WORKED EXAMPLE 22 — Working with compounding periods

Sher-Li invests $60 000 at 6% compound interest per annum compounding monthly. Find

a the number of compounding periods per year
b the number of compounding periods over 8 years
c the percentage interest rate per compounding period, written as a decimal.

Steps	Working
a How many compounding periods are there per year?	There are 12 monthly compounding periods per year.
b Multiply the number of compounding periods per year by the number of years.	There are 12 × 8 = 96 monthly compounding periods over eight years.
c Divide the percentage interest rate per year by the number of compounding periods per year and convert this percentage into a decimal by dividing by 100.	The percentage interest rate per month = 6% ÷ 12 = 0.5% 0.5 ÷ 100 = 0.005

Compound interest general rule

The general rule for the accrued value A, after n compounding periods, of a compound interest investment is

$$A = P\left(1 + \frac{r}{n}\right)^{nt}$$

where A = the total value of the investment or loan. (This is also called the **accrued value**.)

P = principal

r = the percentage interest rate per annum as a decimal or fraction

n = the number of compounding periods per year

t = the number of years

Total amount of interest = $A - P$

WORKED EXAMPLE 23 Using the compound interest rule

Ricky invests $38 000 for 2 years in an account where he earns interest of 3% p.a. compounded monthly. Find the

a values of P, r, n and t
b total value of the investment after 2 years
c total interest earned.

Steps	Working
a Write the values of P, r, n and t.	$P = 38\,000$ $r = 3\% = \dfrac{3}{100} = 0.03$ $n = 12$ $t = 2$
b Substitute the values into the compound interest general rule $A = P\left(1 + \dfrac{r}{n}\right)^{nt}$ Round your answer to the nearest cent.	$A = P\left(1 + \dfrac{r}{n}\right)^{nt}$ $= 38000\left(1 + \dfrac{0.03}{12}\right)^{12 \times 2}$ $= 38000\left(1 + \dfrac{0.03}{12}\right)^{24}$ $= \$40\,346.77$
c Substitute into the formula: Total amount of interest $= A - P$	Total amount of interest $= 40\,346.77 - 38\,000$ $= \$2346.77$

EXERCISE 1.6 Interest

ANSWERS p. 407

Recap

1 A shovel with an original price of $40 is marked up by 15%. Find the
   a mark-up
   b price after mark-up.

2 A plumber charges $50 per hour, however, this rate does not include GST. He completes a job which takes 8 hours. Find the
   a GST payable for the 8 hours
   b total bill including GST.

Mastery

3 **WORKED EXAMPLE 20** Mia borrowed $16 500 to buy a car, with interest charged at 5% per annum, simple interest, over 4 years. Find the
   a total interest on the loan
   b total amount owing on the loan after 4 years
   c total amount owing on the loan after 30 months.

4 **Using CAS 3** Lena takes out a loan of $15 000 with interest charged at 6.5% per annum, simple interest.
   a Create a spreadsheet which calculates the interest each year and the total amount owing on the loan each year for 7 years.
   b Draw a graph with the balance of the loan on the vertical axis and the number of years on the horizontal axis.

5 **WORKED EXAMPLE 21**

  a Eilah invested $20 000 at 3.5% p.a. simple interest and earned a total of $6300 in interest. How long was the money invested?

  b An investment of $12 000 grows to $14 000 over 8 years. Determine the rate of simple interest p.a., correct to two decimal places.

6 **WORKED EXAMPLE 22** For each of the following investments, find the

  i number of compounding periods per year

  ii number of compounding periods over nine years

  iii percentage interest rate per compounding period, written as a decimal.

  a Ross invests $62 000 at 3% compound interest per annum compounding weekly.

  b Rachel invests $110 000 at 7% compound interest per annum compounding daily.

  c Monica invests $10 000 at 5% compound interest per annum compounding monthly

  d Joey invests $14 000 at 6% compound interest per annum compounding quarterly.

  e Phoebe invests $22 000 at 8% compound interest per annum compounding fortnightly.

7 **WORKED EXAMPLE 23** For each of the following, find the

  i values of P, r, n and t.

  ii final value of the investment.

  iii total interest earned.

  a Linus invests $75 000 in an account for 5 years, where he earns interest of 6% p.a. compounded monthly.

  b Elke invests $45 000 in an account for 3 years, where she earns interest of 4% p.a. compounded quarterly.

  c Ingrid invests $6000 in an account for 4 years, where she earns interest of 8% p.a. compounded half yearly.

Calculator-free

8 (6 marks) Layla invested $2000 at 5% per annum, simple interest, for 3 years.

  a Express the percentage interest rate as a decimal. (1 mark)

  b Find the amount of interest earned each year. (2 marks)

  c Copy and complete the table to show the balance in Layla's account each year. (3 marks)

Year	Balance
0	2000
1	
2	
3	

▶ **Calculator-assumed**

9 (8 marks) Talia is investing $3000 for four years and wants to compare an investment at 5% p.a. compounding yearly to 5% p.a. simple interest.

  a Copy and complete the following table for $n = 3$ and $n = 4$. (4 marks)

n	Compound Interest ($)	Compound Value of investment ($)	Simple Interest ($)	Simple Value of investment ($)
0	–	3000	–	3000
1	$\frac{5}{100} \times 3000 = 150$	$3000 + 150 = 3150$	$\frac{5}{100} \times 3000 = 150$	$3000 + 150 = 3150$
2	$\frac{5}{100} \times 3150 = 157.50$	$3150 + 157.50 = 3307.50$	$\frac{5}{100} \times 3000 = 150$	$3150 + 150 = 3300$
3				
4				

  b What is the value of the compound interest investment after four years? (2 marks)

  c After four years, how much more is the value of the compound interest investment compared to the simple interest investment? (2 marks)

1.7 Exchange rates

When we want to buy items from overseas, we need overseas money to pay for the goods. Similarly, when overseas companies or individuals want to buy Australian items, they need to pay in Australian dollars.

The amount of overseas currency we can get for 1 Australian dollar (AUD) is called the **exchange rate**.

We know how demand and supply affects prices. The same thing happens with the value of the Australian dollar, whose value changes depending on demand.

This table shows what 1 Australian dollar (AUD) is worth in some countries (2023 figures).

Country exchange rate (currency abbreviation)

USA	0.6738 US dollars (USD)
China	4.8393 renminbi (RMB)
Most of Europe	0.6049 euro (€, EUR)
Hong Kong	5.2631 HK dollars (HKD)
India	55.17 Indian rupee (INR)
Japan	95.461 yen (¥, JPY)
Indonesia	10 124 Indonesian rupiah (IDR)
New Zealand	1.0912 NZ dollars (NZD)
United Kingdom	0.5236 pounds (£, UKP)

Exchange rate calculations

Converting *from Australian dollars* (AUD) to another currency: *Multiply* by the exchange rate

Converting *to Australian dollars* (AUD) from another currency: *Divide* by the exchange rate

WORKED EXAMPLE 24 — Converting currency

Use the exchange rate 1 AUD = 0.74 USD to convert

a $2000 Australian to US dollars

b   US$850 to Australian dollars.

Steps	Working
a Each $1 Australian buys US$0.74. Multiply 2000 by 0.74.	A$2000 = 2000 × 0.74 = US$1480
b To convert back to Australian dollars from US dollars, divide by 0.74.	US$850 = 850 ÷ 0.74 ≈ A$1148.65

WORKED EXAMPLE 25 — Converting currency

The price of a barrel of crude oil is US$97 and the exchange rate is A$1 = US$0.68. Answer the following questions to calculate the price of 1 L of petrol at an Australian petrol station.

a   Convert the price of a barrel of crude oil to Australian dollars.

b   If 1 barrel is equivalent to 159 L, calculate the cost of 1 L of petrol in cents, correct to two decimal places.

c   Add to this cost per litre, the excise (import tax) of 44.2 cents, production, transport and marketing costs of 26.5 cents, and oil company profit of 2.2 cents.

d   Add 10% GST to find the final price of 1 L of petrol, correct to the nearest cent.

Steps	Working
a A$1 = US$0.68. Convert to Australian dollars by dividing by the exchange rate.	Cost of a barrel = US$97 ÷ 0.68 = A$142.65
b Divide the barrel price by 159.	Cost of 1 L of crude oil = $142.65 ÷ 159 = $0.897 1698... ≈ 89.72 c
c Add extra costs.	Cost per litre = 89.72 + 44.2 + 26.5 + 2.2 = 162.62 c
d Add 10% GST.	10% of 162.62 = 16.26 c 162.62 + 16.26 = 178.88 c = $1.7888 ≈ $1.79 The price of 1 L of petrol is approximately $1.79.

EXERCISE 1.7 Exchange rates

ANSWERS p. 408

Recap

1 Derek invests $20 000 into an account for 10 years. The investment earns interest at a rate of 7.5% per annum, simple interest. Find the

   a   interest earned each year

   b   interest earned after 10 years

   c   balance of the account after 10 years.

2 Magdelan invests $8000 in an account for 6 years, where she earns interest of 9% per annum, compounded quarterly. Find the

   a values of P, r, n and t
   b final value of the investment
   c total interest earned.

Mastery

3 **WORKED EXAMPLE 24** Use the exchange rate A$1 = US$0.74 to copy and complete each statement.

   a A$900 = US$_____
   b A$1325 = US$_____
   c A$_____ = US$2350
   d A$_____ = US$650

4 Use the exchange rate A$1 = £0.54 (UK pounds) to copy and complete each statement.

   a A$200 = £_____
   b A$5400 = £_____
   c A$_____ = £468
   d A$_____ = £5560

5 **WORKED EXAMPLE 25** The price of a barrel of crude oil is US$80 and the exchange rate is A$1 = US$0.63. Answer the following questions to calculate the price of 1 L of petrol at an Australian petrol station.

   a Convert the price of a barrel of crude oil to Australian dollars.
   b If 1 barrel = 159 L, calculate the cost of 1 L of petrol in cents, correct to two decimal places.
   c Add to this cost per litre the excise (import tax) of 44.2 cents, production, transport and marketing costs of 26.5 cents, and oil company profit of 2.2 cents.
   d Add 10% GST to find the final price of 1 L of petrol, correct to the nearest cent.

Calculator-assumed

6 (3 marks) Claire buys a book online from the US using a debit card for US$24.

   a Find the price in Australian dollars when the exchange rate is A$1 = US$0.75. (1 mark)
   b Claire's bank charges a foreign currency fee, equivalent to 3% of the Australian dollar value, for any online overseas purchases. Calculate the total amount in Australian dollars that Claire will pay for the book. (2 marks)

7 (4 marks) Jonty regularly imports materials from America for his leadlight business. The supplies costs US$420. The price Jonty pays for his supplies in Australian dollars changes as the exchange rate changes.

Copy and complete this table to show the cost of Jonty's supplies at different exchange rates. The first line has been completed as an example.

Exchange rate	Calculation	Cost of supplies (to nearest $)
A$1 = US$0.55	420 ÷ 0.55 = 763.64	$764
A$1 = US$0.65		
A$1 = US$0.75		
A$1 = US$0.85		
A$1 = US$1.05		

8 (4 marks) The price of petrol is affected by the supply of crude oil and the current exchange rate. The majority of Australia's crude oil comes through the Singapore market and is priced in American dollars. This table shows the costs involved.

The cost of petrol

Crude oil	50%
Taxes and GST	38%
Production, transport and marketing costs	10%
Oil company profit	2%

Alex paid $1.60 per litre for petrol. How much of the $1.60 went to pay for

a crude oil? (1 mark)

b taxes and GST? (1 mark)

c production, transport and marketing costs? (1 mark)

d oil company profit? (1 mark)

9 (12 marks) If 1 barrel is equivalent to 159 L, calculate the cost of 1 L of petrol in Australian dollars, correct to two decimal places, for each of the following situations.

	Price of a barrel of crude oil	Exchange rate	
a	US$75	A$1 = US$0.73	(3 marks)
b	US$65	A$1 = US$0.85	(3 marks)
c	US$60	A$1 = US$0.80	(3 marks)
d	US$80	A$1 = US$0.70	(3 marks)

1.8 Shares

Video playlist
Shares

A person with money to invest can buy a part, or **share**, of a company, and then receive a share of the company's profits, called a **dividend**. The **share market** (or stock market) is the place where shares in companies are bought and sold, mostly online these days.

The financial report on a daily news program describes how different shares performed, whether their prices rose or fell. The stock markets around the world usually have an influence on each other. For example, the New York Stock Exchange is a world leader because of its large size. Gains and falls on this exchange are usually followed by similar changes in other share markets.

This graph shows the changes in the Australian ASX 200 (blue graph, statistics based on the average price of Australian shares) and the US S&P 500 (red graph, statistics based on the average price of US shares) over 3 weeks. The Australian market followed the US market but not to the same extent.

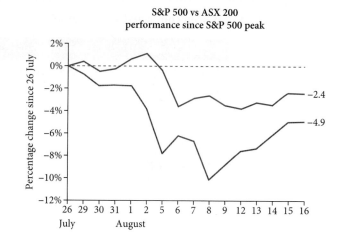

The 'missing' days are weekend days, when the share market is closed.

WORKED EXAMPLE 26 — Calculating dividends from the dividend per share

Ali owns 2650 shares in Primary Aviation.

a The company paid a dividend of $1.23 per share. How much did Ali receive?

b Ali bought the shares for $8.24 then sold them for $11.50 each. What percentage profit did Ali make, correct to one decimal place? Ignore any fees Ali may have been charged on the sale.

Steps	Working
a Multiply the dividend per share by the number of shares.	Amount Ali received = 2650 × $1.23 = $3259.50
b Calculate Ali's profit by subtracting the amount Ali paid for the shares from the amount he receives from selling them. Use the formula: $\dfrac{\text{profit}}{\text{original price}} \times 100\%$	Amount Ali paid for each share = $8.24 Amount Ali received from the sale of each share = $11.50 Ali's profit per share = $11.50 − $8.24 = $3.26 percentage profit = $\dfrac{3.26}{8.24} \times 100\%$ = 39.5631…% ≈ 39.6%

Dividend yield

The **dividend yield** of a share is the dividend as a percentage of its market price. It is like an interest rate for shares and is used by investors to compare the performance of a share.

Dividend yield

$$\text{dividend yield} = \dfrac{\text{dividend}}{\text{share price}} \times 100\%$$

WORKED EXAMPLE 27 — Calculating dividend yield

The market value of Owen's mining shares is $38.75 and they paid a dividend of $1.96. Calculate its dividend yield, correct to one decimal place.

Steps	Working
dividend yield = $\dfrac{\text{dividend}}{\text{share price}} \times 100\%$	dividend yield = $\dfrac{1.96}{38.75} \times 100\%$ ≈ 5.1%

WORKED EXAMPLE 28 — Calculating dividends from the dividend yield

Calculate the total dividend to someone who owns 3000 shares at $24.30 per share if the dividend yield is 2.5%.

Steps	Working
1 Multiply the number of shares by the price per share.	Total value of the shares = 3000 × 24.30 = $72 900
2 Use dividend = share price × dividend yield	dividend = 72 900 × 2.5% = 1822.50 The total dividend is $1822.50.

Dividend vs earnings

A dividend is a share of a company's profits that is paid to shareholders, whereas earnings are the profits a company generates. For shareholders, dividends provide a way of receiving a regular income from their investment. The earnings or profits a company makes can be used to reinvest in the company, pay dividends or repurchase shares.

Price earnings (P/E) ratio

Price earnings ratio

$$\text{P/E ratio} = \frac{\text{market price per share price}}{\text{annual earnings per share}}$$

The **price earning (P/E) ratio** gives investors an indication of how expensive the company's shares are in relation to their profits. The higher the P/E ratio, the more expensive a stock is compared to its earnings. P/E ratios provide a way for investors to compare shares.

Consider a stock with a market value of $24 and earnings of $4. The P/E ratio is 24 ÷ 4 = 6. Put another way, investors are paying $6 for every $1 of earnings.

WORKED EXAMPLE 29 — Calculating the price earnings ratio

Calculate the P/E ratio, correct to two decimal places, for a company that had a stock price of $16.40, if its earnings per share from the previous 12 months were $2.20.

Steps	Working
Use $\text{P/E ratio} = \dfrac{\text{market price per share price}}{\text{annual earnings per share}}$	$\text{P/E ratio} = \dfrac{16.40}{2.20}$ ≈ 7.45

WORKED EXAMPLE 30 — Calculating the P/E ratio from the dividend yield

Cassandra paid $44 000 for 1000 shares with a dividend yield of 4%. The company pays 30% of its earnings each year in dividends. Find

a the market price of the shares
b the dividend per share
c the earnings per share
d the P/E ratio, correct to one decimal place.

Steps	Working
a Divide the total value of the shares by the number of shares.	Market price per share = $\dfrac{44000}{1000}$ = $44
b Use the percentage yield to calculate the dividend per share. dividend = share price × dividend yield	Dividend per share = $44 × 4% = $1.76
c 30% of the earnings is paid as dividends. Create an equation and solve using CAS.	Let x = earnings per share $x \times 30\% = 1.76$ $0.3x = 1.76$ $x = 5.87$ The earnings per share is $5.87.
d Use $\text{P/E ratio} = \dfrac{\text{market price per share price}}{\text{annual earnings per share}}$	$\text{P/E ratio} = \dfrac{44}{5.87}$ ≈ 7.5

EXAMINATION QUESTION ANALYSIS

Calculator-assumed **(11 marks)**

Simon has $5000 that he wants to invest for a period of time without touching it.

He is currently deciding between two options and wishes to compare them.

Option A: Invest the $5000 in an account earning simple interest at the rate of 6.2% per annum for 5 years.

Option B: Invest the $5000 in an account earning compound interest at the rate of 6% per annum, with interest paid monthly for 5 years.

a Option A:
  i Find the interest earned on the simple interest investment after 5 years. (2 marks)
  ii Find the closing balance of Simon's account after 5 years. (1 mark)

b Option B:
  i Copy and complete the table for the 3rd month of the compound interest investment. (3 marks)

Month	Amount at the beginning of month	Interest for the month	Amount at the end of month
1	5000.00	25.00	5025.00
2	5025.0	25.13	5050.13
3	A	B	C

  ii Find the closing balance of Simon's account after 5 years. (2 marks)
  iii Find the interest earned on the compound interest investment after 5 years. (2 marks)

c Which option is the most profitable for Simon? (1 mark)

Video
Examination question analysis: Consumer arithmetic

Reading the question

- This question requires an understanding of simple and compound interest. These problems can be done using a formula or a table.
- Highlight the compounding period for the interest rate. This can be daily, weekly, fortnightly, monthly or yearly.
- Highlight the type of answer required in each part. This may be an account balance or an amount of interest.
- Look at the number of marks allocated to each part of the question as this will indicate the number of steps needed to show in your working.

Thinking about the question

- This question requires an understanding of simple and compound interest.
- Each type of interest can be found using a formula; however, you will need to know how each variable in the formula is defined.
- You will need to be able to calculate the interest and the closing balance of an account.

Worked solution (✓ = 1 mark)

a **i** Simple interest: $P = 5000$, $r = 6.2\% = 0.062$, $t = 5$

$I = Prt$
$= 5000 \times 0.062 \times 5$ ✓
$= 1550.00$

Interest is $1550.00 ✓

ii Closing balance = 5000 + 1550 = 6550

The closing balance after 5 years is **$6550.00**. ✓

b **i** Month 3:

Interest rate per month = 6% ÷ 12 = 0.06 ÷ 12 = 0.005

A = opening balance = **$5050.13** ✓

B = interest = 5050.13 × 0.005 = **$25.25** ✓

C = closing balance = 5050.13 + 25.25 = **$5075.38** ✓

ii Compound interest: $P = 5000$, $r = 6\% = 0.06$, $t = 5$, $n = 12$ ✓

$$A = P\left(1 + \frac{r}{n}\right)^{nt}$$

$$= 5000\left(1 + \frac{0.06}{12}\right)^{12 \times 5}$$

$$= 5000(1.005)^{60}$$

$$= 6744.25$$

The closing balance after 5 years with compound interest is **$6744.25**. ✓

iii Interest = 6744.25 − 5000 ✓ = 1744.25 ✓

The compound interest is $1744.25.

c Option B, 6% p.a. compounded monthly, is the most profitable for **Simon**. ✓

EXERCISE 1.8 Shares

ANSWERS p. 408

Recap

1 Use the exchange rate A$1 = £0.55 (UK pounds) to copy and complete each statement.

  a A$200 = £ _____
  b A$6000 = £ _____
  c A$ _____ = £550
  d A$ _____ = £11 000

2 Find the exchange rate for one Australian dollar to US dollars if A$950 = US$665.

Mastery

3 WORKED EXAMPLE 26   Rachel has 4200 shares in NVM Limited.

  a The company paid a dividend of $0.65 per share. How much did Rachel receive?
  b Rachel paid $5.30 each for the shares. How much did the package cost her?
  c Rachel sold the shares for $6.95 each. How much profit did she make?
  d Calculate Rachel's percentage profit, correct to one decimal place.

4 **WORKED EXAMPLE 27** Calculate the dividend yield for each dividend, correct to two decimal places.

	Share value	Dividend
a	$12.60	$1.05
b	$103.40	$3.55
c	$89.72	$2.10
d	$50.30	$0.96

5 **WORKED EXAMPLE 28** Calculate the total dividend to someone who owns 1500 shares at $32.20 per share if the dividend yield is 6.5%.

6 **WORKED EXAMPLE 29** Calculate the P/E ratio, correct to two decimal places, for a company that had a stock price of $25.50 when its earnings per share from the previous 12 months were $5.50.

7 **WORKED EXAMPLE 30** Olympus paid $35 000 for 5000 shares with a dividend yield of 4.5%. The company pays 20% of its earnings each year in dividends. Find the
   a market price of the shares
   b dividend per share
   c earnings per share
   d P/E ratio, correct to one decimal place.

Calculator-free

8 (5 marks) Last week, Cengage Bank shares were worth $10.00 each. Mia purchased 50 shares last week and today they are worth $11.00 each. Calculate
   a Mia's profit on each share (1 mark)
   b the total purchase price of the shares (1 mark)
   c Mia's total profit on the 50 shares (1 mark)
   d Mia's percentage profit on the Cengage shares. (2 marks)

Calculator-assumed

9 (10 marks) Eva had $48 000 to invest. Term deposit rates were very low at only 0.75% p.a. so she decided to invest only half in a term deposit and the remainder in shares.
   a The shares she selected cost $3.20 each. How many shares did she buy? (2 marks)
   b At the end of the year, the shares paid a 55 c dividend. Calculate the total value of Eva's dividend payment. (2 marks)
   c Calculate the dividend yield, correct to one decimal place. (2 marks)
   d Calculate the interest Eva earned for the year on the term deposit. (2 marks)
   e Which was the better investment and by how much? (2 marks)

10 (7 marks) Cupid paid $40 000 for 2000 shares with a dividend yield of 12.0%. The company pays 40% of its earnings each year in dividends. Find the
   a market price of the shares (1 mark)
   b dividend per share (2 marks)
   c earnings per share (2 marks)
   d P/E ratio, correct to one decimal place. (2 marks)

Chapter summary

Wages and salaries
- A **wage** is an amount paid by an employer to an employee for each hour worked.
- A **salary** is a fixed amount per year that does not depend on the number of hours worked.

Converting a year into weeks, months or fortnights:

1 year = 12 months

1 year = 52 weeks

1 year = 26 fortnights

Overtime pay
Time-and-a-half is 1.5 times normal pay.

Double time is 2 times normal pay.

Commission and piecework
- **Commission** is a percentage of the value of the items sold. Sometimes, the salespeople are paid a retainer plus commission.
- A **retainer** is a set payment that does not depend on sales.
- **Piecework** is a type of work where a person is paid per item produced or processed.

Budgets
- A **budget** lists expected income and expenses, and can help to manage money.
- Income covers all the money earnt.
- Expenses cover all the ways money might be spent. There are two types of expenses: fixed and discretionary.

Percentage change

$$\text{percentage increase} = \frac{\text{increase}}{\text{original value}} \times 100\%$$

$$\text{percentage decrease} = \frac{\text{decrease}}{\text{original value}} \times 100\%$$

Mark-up and discount
When the price of an item is increased by a percentage, it's called a **mark-up**.

When the price of an item is reduced by a percentage, it's called a **discount**.

Mark-up	Discount
For an item marked up by $r\%$:	For an item discounted by $r\%$:
mark-up = original price $\times \frac{r}{100}$	discount = original price $\times \frac{r}{100}$
new price = original price + mark-up	new price = original price − discount
new price = original price $\times \frac{100 + r}{100}$	new price = original price $\times \frac{100 - r}{100}$

Profit and loss formulas

Profit	Loss
profit = selling price − cost price	loss = cost price − selling price
profit percentage = P% = $\dfrac{\text{profit}}{\text{cost price}} \times 100$	loss percentage = L% = $\dfrac{\text{loss}}{\text{cost price}} \times 100$
selling price = cost price × $\dfrac{100 + P\%}{100}$	selling price = cost price × $\dfrac{100 - L\%}{100}$

GST

The **Goods and Services Tax** is currently a 10% tax on most sales and services in Australia.

- price without GST = original price
- price with GST = new price
- GST amount = change in price
- $r = 10\% = 0.1$

GST = price without GST × $\dfrac{10}{100}$

GST = price with GST − price without GST

price with GST = price without GST × $\dfrac{110}{100}$

Simple interest formula

$I = Prt$

where I is the simple interest
  P is the principal
  r is the rate per year, and
  t is the time in years.

Compound interest general rule

The general rule for the **accrued value** A, after n compounding periods, of a compound interest investment is

$$A = P\left(1 + \dfrac{r}{n}\right)^{nt}$$

where A is the total value of the investment or loan (this is also called the accrued value).
  P is principal
  r is the percentage interest rate per annum
  n is the number of compounding periods per year.
  t is the number of years

Total amount of interest = $A - P$

Exchange rate

The amount of overseas currency we can get for 1 Australian dollar (AUD) is called the **exchange rate**.

Exchange rate calculations
Converting *from Australian dollars* (AUD) to another currency: *Multiply* by the exchange rate
Converting *to Australian dollars* (AUD) from another currency: *Divide* by the exchange rate

Shares

The **dividend yield** of a share is the dividend as a percentage of its market price. It is like an interest rate for shares and is used by investors to compare the performance of a share.

Dividend yield

$$\text{dividend yield} = \frac{\text{dividend}}{\text{share price}} \times 100\%$$

Price earnings (P/E) ratio

$$\text{P/E ratio} = \frac{\text{market price per share price}}{\text{annual earnings per share}}$$

Cumulative examination: Calculator-free

Total number of marks: 9   Reading time: 1 minute   Working time: 9 minutes

1 (3 marks) The percentage mark-up on a scarf with a cost price of $20 is 10%.
Find the

   a amount of the mark-up (2 marks)

   b selling price of the scarf. (1 mark)

2 (2 marks) Renata sells cars and is paid a commission of 2% on the total value of cars sold each month. At the end of February, Renata has sold cars with a total value of $100 000.

Find the amount of commission Renata earns.

3 (4 marks) This table shows Julie's weekly balanced budget.

Income		Expenses	
Allowance	$100	Rent	$80
Earnings	$160	Food	$30
		Mobile phone	$10
		Clothes	$20
		Entertainment	$40
		Savings	
Total income		**Total expenses**	

Find

   a Julie's total weekly income (1 mark)

   b Julie's total weekly expenses (1 mark)

   c the amount of money Julie saves in the week. (2 marks)

Cumulative examination: Calculator-assumed

Total number of marks: 23   Reading time: 3 minutes   Working time: 23 minutes

1 (6 marks) Brit invests $35 000 in an account, for 5 years, where she earns interest of 8% p.a. compounded quarterly. Find the

   a values of P, r, n and t (3 marks)

   b final value of the investment (2 marks)

   c total interest earned. (1 mark)

2 (7 marks) Dion earns commission for selling surfboards at the following rates:

Commission on monthly sales	
First $5000 of sales	5%
On the next $4000	4%
Remainder of sales	3.5%

His sales for summer are

   December: $8500

   January: $12 400

   February: $4600

   a What is Dion's commission for each month? (6 marks)

   b Find Dion's total commission for the summer. (1 mark)

3 (10 marks) Angeline had $58 000 to invest. Term deposit rates were very low at only 0.75% p.a. so she decided to invest only half in a term deposit and the remainder in shares.

   a The shares she selected cost $5.80 each. How many shares did she buy? (2 marks)

   b At the end of the year, the shares paid a 75 c dividend. Calculate the total value of Angeline's dividend payment. (2 marks)

   c Calculate the dividend yield, correct to one decimal place. (2 marks)

   d Calculate the interest Angeline earned for the year on the term deposit. (2 marks)

   e Find the total return on Angeline's investment. (2 marks)

CHAPTER 2

ALGEBRA AND MATRICES

Syllabus coverage
Nelson MindTap chapter resources

2.1 Algebraic expressions, equations and formulas
Linear and non-linear expressions
Using CAS 1: Evaluating expressions
Linear equations
Using single- and multi-variable formulas
Using CAS 2: Evaluating formulas

2.2 Tables and formulas
Tables of values
Using CAS 3: Using spreadsheets
Practical problems involving formulas

2.3 Introduction to matrices
Features of a matrix
Types of matrices
Storing and displaying information in $m \times n$ matrices

2.4 Matrix addition, subtraction and scalar multiplication
Addition and subtraction
Scalar multiplication
Using CAS 4: Addition, subtraction and scalar multiplication of matrices
Equal matrices

2.5 Matrix multiplication and powers
Matrix multiplication
The order of multiplication
Powers of matrices
Using CAS 5: Multiplying matrices

2.6 Modelling with matrices
Costing and pricing problems
Communication networks
Using CAS 6: Solving problems involving multi-step communication matrices

Examination question analysis
Chapter summary
Cumulative examination: Calculator-free
Cumulative examination: Calculator-assumed

Syllabus coverage

TOPIC 1.2: ALGEBRA AND MATRICES

Linear and non-linear expressions

1.2.1 substitute numerical values into algebraic expressions, and evaluate (with the aid of technology where complicated numerical manipulation is required)

1.2.2 determine the value of the subject of a formula, given the values of the other pronumerals in the formula (transposition not required)

1.2.3 use a spreadsheet or an equivalent technology to construct a table of values from a formula, including tables for formulas with two variable quantities; for example, a table displaying the body mass index (BMI) of people of different weights and heights

Matrices and matrix arithmetic

1.2.4 use matrices for storing and displaying information that can be presented in rows and columns; for example, databases, links in social or road networks

1.2.5 recognise different types of matrices (row, column, square, zero, identity) and determine their size

1.2.6 perform matrix addition, subtraction, multiplication by a scalar, and matrix multiplication, including determining the power of a matrix using technology with matrix arithmetic capabilities when appropriate

1.2.7 use matrices, including matrix products and powers of matrices, to model and solve problems; for example, costing or pricing problems, squaring a matrix to determine the number of ways pairs of people in a communication network can communicate with each other via a third person

Mathematics Applications ATAR Course Year 11 syllabus p. 9 © SCSA

Video playlists (7):

2.1 Algebraic expressions, equations and formulas
2.2 Tables and formulas
2.3 Introduction to matrices
2.4 Matrix addition, subtraction and scalar multiplication
2.5 Matrix multiplication and powers
2.6 Modelling with matrices
Examination question analysis Algebra and matrices

Worksheets (2):

2.4 Addition and subtraction of matrices
2.5 Multiplying matrices

Nelson MindTap

To access resources above, visit
cengage.com.au/nelsonmindtap

2.1 Algebraic expressions, equations and formulas

Linear and non-linear expressions

An **algebraic expression** is a collection of terms representing the basic operations of arithmetic: addition, subtraction, multiplication and division. Expressions can involve a single algebraic **variable**; that is, one unknown value, or multiple variables. For example, $3x + 5$ is an expression involving one variable, whereas $3x + 5y$ has two unknown values. Both of these expressions are called **linear expressions** because the power of each of the variables is 1. Expressions are called non-linear when the power of the variables is not 1; for example, $4x^2 - x$ and $\frac{8}{x} + \sqrt{y}$ are **non-linear expressions**.

Given that variables are used to represent numerical values, we can evaluate an algebraic expression by substitution.

WORKED EXAMPLE 1 — Evaluating an algebraic expression

Evaluate each of the following expressions when $x = -1$ and $y = 4$.

a $3x + 5$
b $3x + 5y$
c $4x^2 - x$
d $\frac{8}{x} + \sqrt{y}$

Steps	Working
1 Substitute the value of x and y into the expressions. 2 Evaluate the expression using the correct order of operations.	a $3x + 5$ $= 3(-1) + 5$ $= -3 + 5$ $= 2$
	b $3x + 5y$ $= 3(-1) + 5(4)$ $= -3 + 20$ $= 17$
	c $4x^2 - x$ $= 4(-1)^2 - (-1)$ $= 4(1) + 1$ $= 5$
	d $\frac{8}{x} + \sqrt{y}$ $= \frac{8}{-1} + \sqrt{4}$ $= -8 + 2$ $= -6$

> **Exam hack**
>
> Always show the substitution line of working out before evaluating the expression.

CAS can also be used to evaluate single-variable and multi-variable expressions.

USING CAS 1 — Evaluating expressions

Evaluate $\dfrac{3x + 6y}{z^2}$ when $x = 1$, $y = 2$ and $z = 5$.

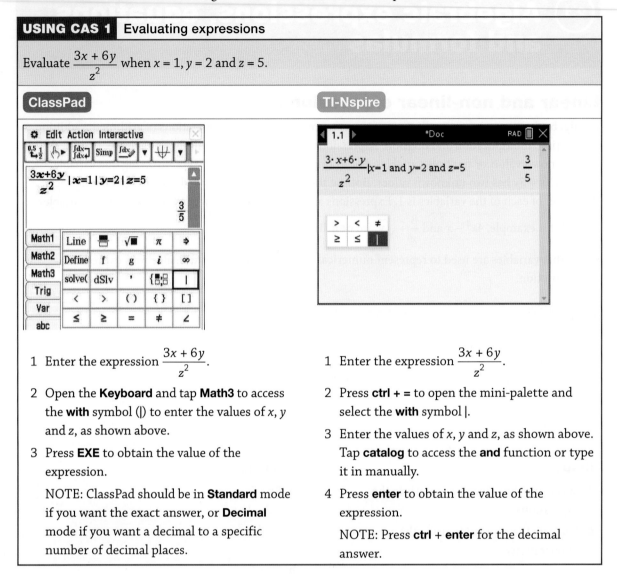

ClassPad

1. Enter the expression $\dfrac{3x + 6y}{z^2}$.
2. Open the **Keyboard** and tap **Math3** to access the **with** symbol (|) to enter the values of x, y and z, as shown above.
3. Press **EXE** to obtain the value of the expression.

   NOTE: ClassPad should be in **Standard** mode if you want the exact answer, or **Decimal** mode if you want a decimal to a specific number of decimal places.

TI-Nspire

1. Enter the expression $\dfrac{3x + 6y}{z^2}$.
2. Press **ctrl + =** to open the mini-palette and select the **with** symbol |.
3. Enter the values of x, y and z, as shown above. Tap **catalog** to access the **and** function or type it in manually.
4. Press **enter** to obtain the value of the expression.

   NOTE: Press **ctrl + enter** for the decimal answer.

Linear equations

When an algebraic expression is equal to a numerical value or another algebraic expression, it is called an **equation**. Equations can be linear or non-linear, and they can also involve single or multiple variables. You may be required to use your prior knowledge of solving simple single-variable **linear equations** throughout various topics in the Year 11 Mathematics Applications course; however, you will revisit this topic formally in Chapter 8. Worked example 2 on the following page shows a simple example that may arise when we explore the idea of equal matrices later in this chapter.

WORKED EXAMPLE 2 — Solving a simple linear equation

Solve the following linear equations for the given variable.

a $4x - 8 = 12$
b $5a + 1 = 2a + 7$

Steps	Working
a 1 Apply the inverse operations to isolate the value of x.	$4x - 8 = 12$
2 Add 8 to both sides of the equation.	$4x = 20$
3 Divide both sides of the equation by 4.	$x = 5$
b 1 Collect the like terms on the same side of the equation using inverse operations.	$5a + 1 = 2a + 7$ $5a - 2a = 7 - 1$
2 Divide both sides of the equation by 3.	$3a = 6$ $a = 2$

Using single- and multi-variable formulas

When an equation has a physical representation or purpose, it is often referred to as a **formula**. We have seen many formulas (sometimes referred to as formulae in plural) in the context of measurement, such as perimeter, area, surface area and volume formulas. The purpose of the formula is often to calculate the value of a specific variable, which is called the **subject** of the formula. For example, the subject of the formula $V = \pi r^2 h$ is V, which represents the volume of a cylinder with radius r and perpendicular height h.

WORKED EXAMPLE 3 — Evaluating a formula

Consider the formula
$$T = 2\pi \sqrt{\frac{L}{g}}$$

a Identify the subject of the formula.
b Evaluate the formula to two decimal places when $L = 10$ and $g = 9.8$.

Steps	Working
a Identify the variable that the formula calculates.	The subject of the formula is T.
b 1 Substitute in the known values of L and g.	$T = 2\pi \sqrt{\frac{L}{g}}$ $= 2\pi \sqrt{\frac{10}{9.8}}$
2 Evaluate and answer correct to two decimal places.	$= 6.35$

Expressions, equations and formulas

- An **expression** does not have an equal sign.
  For example, $8x + 3$.
- An **equation** does have an equal sign, representing that the left-hand side has the same value as the right-hand side.
  For example, $8x + 3 = 2x - 3$.
- **Formulas** are equations that generally have a purpose.
  For example, $c^2 = a^2 + b^2$.

CAS can be used to store and evaluate formulas given different values of unknown variables.

USING CAS 2 Evaluating formulas

Consider the formula

$$A = \frac{1}{2}(a+b)h$$

Evaluate the formula for $a = 10$, $b = 15$, $h = 20$.

ClassPad	TI-Nspire
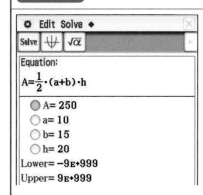	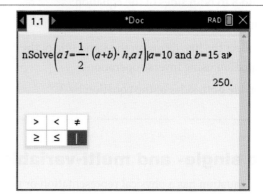
1 Tap **Menu > NumSolve**.	1 Press **menu > Algebra > Numerical Solve**.
2 Tap **Edit > Clear All** to delete any existing formulas.	2 Enter the equation $a1 = \frac{1}{2}(a+b) \times h$.
3 In the **Equation** line, enter $A = \frac{1}{2}(a+b)h$, as shown above.	NOTE: all letters in TI-Nspire are in lower case so use **a1** or something besides **a** for the area.
4 Enter the values of **a**, **b** and **h**, as shown above.	3 Press **ctrl + =** to open the mini-palette and select the **with** symbol |.
5 Tap to select **A**, then tap **Solve**.	4 Enter the values of **a**, **b** and **h**. Tap **catalog** to access the **and** function or type it in manually.
6 When the Result screen appears, tap **OK**.	5 Press **enter** to obtain the area.
$A = 250$	

EXERCISE 2.1 Algebraic expressions, equations and formulas   ANSWERS p. 408

Mastery

1 WORKED EXAMPLE 1 Evaluate each of the following expressions when $a = 3$ and $b = -2$.

  a $-2a + 4$ **b** $9a + 3b$ **c** $3a - \dfrac{b^2}{2}$ **d** $\dfrac{36}{ab}$

2 Using CAS 1 Evaluate $\dfrac{4y - 7x}{z^2}$ when $x = -1$, $y = 5$ and $z = -3$.

3 WORKED EXAMPLE 2 Solve the following linear equations for the given variable.

  a $5z + 11 = -29$ **b** $3b - 1 = 7 - b$

4 **WORKED EXAMPLE 3** Consider the formula

$$D = \sqrt{x^2 + y^2 + z^2}$$

   a Identify the subject of the formula.
   b Evaluate the formula to two decimal places when $x = 3$, $y = 2$ and $z = 5$.

5 **Using CAS 2** Consider the formula

$$A = 2\pi r^2 + 2\pi rh$$

Evaluate the formula to two decimal places for $r = 8$ and $h = 12$.

Calculator-free

6 (4 marks) Evaluate the following algebraic expressions using the given values.

   a $6x - y + \dfrac{4}{z}$ for $x = 100$, $y = 40$, $z = 2$ (2 marks)

   b $2a^2(b - c)$ for $a = -2$, $b = 10$, $c = 8$ (2 marks)

7 (4 marks) Compare the value of the two expressions below when $x = 5$.

$$\dfrac{\sqrt{4 + x}}{x^2} \qquad \sqrt{\dfrac{4 + x}{x^2}}$$

8 (4 marks) Consider the formula $x = \left(\dfrac{u + v}{2}\right)t$.

   a Determine the value of x when $u = 2$, $v = 13$ and $t = 8$. (2 marks)
   b The formula can be rearranged to make t the subject, such that $t = \left(\dfrac{2}{u + v}\right)x$.

   Determine the value of t when $u = 1$, $v = -7$, and $x = -24$. (2 marks)

2.2 Tables and formulas

Video playlist
Tables and formulas

Tables of values

When a rule or formula needs to be evaluated for multiple different values of a variable, it is often useful to display this information in a table.

WORKED EXAMPLE 4 Constructing a table of values for a single-variable formula

Consider the formula $V = 4\pi r^2$. Complete the table of values for the given values of r, leaving your answers in terms of π.

r	1	2	3	4
V				

Steps	**Working**
Substitute and evaluate the formula for the different values of r, leaving your answers exact.	When $r = 1$, $V = 4\pi$. When $r = 2$, $V = 4\pi(2)^2 = 16\pi$. When $r = 3$, $V = 4\pi(3)^2 = 36\pi$. When $r = 4$, $V = 4\pi(4)^2 = 64\pi$. | r	1	2	3	4
---	---	---	---	---
V	4π	16π	36π	64π	

Chapter 2 | Algebra and matrices   55

When an equation or formula involves two variables, then a two-way table can be used, whereby the values of one variable are represented by the rows and the values of the second variable are represented by the columns. Such a table would be structured like the one shown below for the formula $C = \sqrt{A^2 + B^2}$.

C (Subject of the formula)		B (Second variable)			
		4	5	6	7
A (First variable)	1				
	2				
	3	5			

Each entry in the table is then the value of C when the formula is evaluated. For example, when $A = 3$ and $B = 4$, $C = \sqrt{3^2 + 4^2} = \sqrt{25} = 5$.

WORKED EXAMPLE 5 — Constructing a table of values for a two-variable formula

Consider the formula $C = \dfrac{A}{2} + B - 1$. Copy and complete the table of values for the given values of A and B. Some of the values have been completed for you.

C		B			
		−1	1	3	5
A	2				5
	6	1			
	10			7	

Steps — Working

Apply the formula $\dfrac{A}{2} + B - 1$ to complete the missing entries in the table.

C		B			
		−1	1	3	5
A	2	−1	1	3	5
	6	1	3	5	7
	10	3	5	7	9

WORKED EXAMPLE 6 — Finding unknown values in tables

The value of z is determined by the formula $z = 2x + 3y$. A table of values of z is shown on the right.

a Determine the values of A and B in the table shown.

b Describe the pattern in the values of z when $x = y$.

z		y			
		0	1	2	3
x	0	0	3	6	A
	1	2	5	8	11
	2	4	B	10	13

Steps — Working

a 1 Locate A in the table and identify the corresponding values of x and y. Determine z for these values.

$z = A$ when $x = 0$, $y = 3$
$A = 2(0) + 3(3)$
$A = 9$

2 Locate B in the table and identify the corresponding values of x and y. Determine z for these values.

$z = B$ when $x = 2$, $y = 1$
$B = 2(2) + 3(1)$
$B = 7$

b 1 Locate the values of z when x and y are the same; that is, in the diagonal entries of the table.

The values of z when $x = y$ are 0, 5, 10 …

2 Observe and describe the pattern.

Each time x and y increase by 1, the value of z increases by 5.

Spreadsheets can also be used to display the results of a formula, however, the difference between a spreadsheet and a two-way table is that the rows are usually labelled as numbers {1, 2, 3, 4 …} and the columns are usually labelled as capital letters {A, B, C, D …}. Each entry in the spreadsheet, called a **cell**, is then referred to by its corresponding letter and number. For example, consider the spreadsheet below.

	A	B	C	D
1	A1	B1	C1	D1
2	A2	B2	C2	D2
3	A3	B3	C3	D3
4	A4	B4	C4	D4

These cell references can then be used to create formulas in other cells. For example, if we wanted cell B2 to show an average of cells A1, B1 and A2, we could write the formula

$$= \frac{A1 + B1 + A2}{3}$$

inside cell B2.

CAS has the capability to generate spreadsheets of values using different formulas.

USING CAS 3 **Using spreadsheets**

The first four entries of column A of a spreadsheet contain the numbers 100, 200, 300 and 400.

	A	B
1	100	
2	200	
3	300	
4	400	

Suppose that column B is to contain the values of column A halved and then increased by 50.

a Write a spreadsheet formula for the entry in cell B1 in terms of A1.

b Complete column B in the spreadsheet.

Steps	Working
a Use the operations to express B1 in terms of A1.	$B1 = \frac{A1}{2} + 50$

ClassPad

b

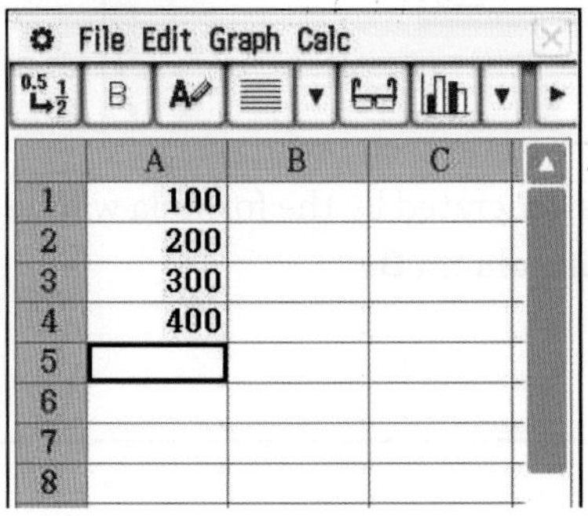

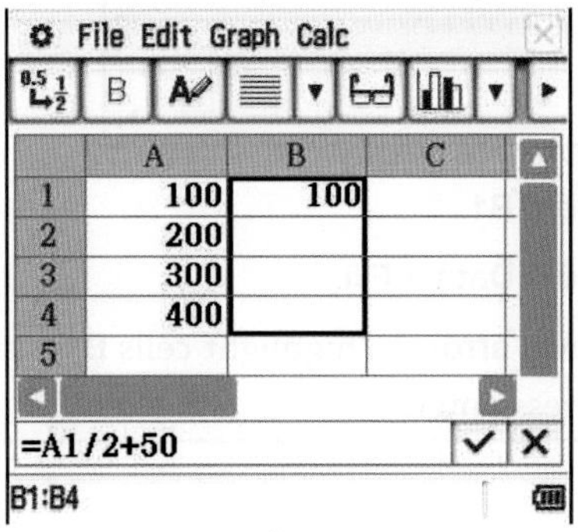

1 Tap **Menu** > **Spreadsheet**.
2 In column **A**, enter the values **100, 200, 300** and **400**, as shown above.
3 In cell **B1**, enter **=A1/2+50** and press **EXE**.
4 Highlight cells **B1** to **B4**.
5 Tap **Edit** > **Fill** > **Fill Range**.

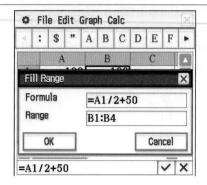 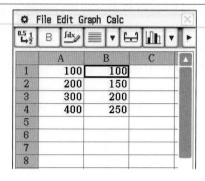

6 In the dialogue box, check that the Formula and the Range are correct, then tap **OK**.

7 The values generated by the formula will be displayed in column **B**.

TI-Nspire

b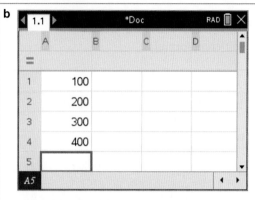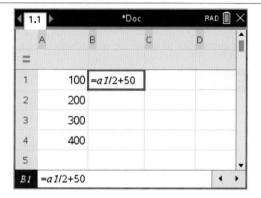

1 Add a **Lists & Spreadsheet** page.

2 In column **A**, enter the values **100, 200, 300** and **400**, as shown above.

3 In cell **B1**, enter **=a1/2+50**, then press **enter**.

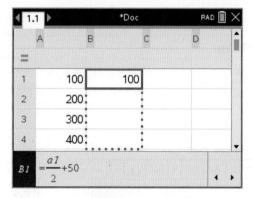 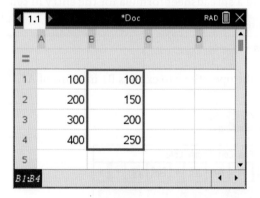

4 Highlight cell **B1**.

5 Press **menu > Data > Fill**.

6 Use the **down arrow** to highlight cells **B1** to **B4**, then press **enter**.

7 The values generated by the formula will be displayed in column **B**.

Practical problems involving formulas

Displaying the values of a formula in a table can be useful in solving practical problems in context, whereby the formula may or may not be provided to you.

> 🔒 **Exam hack**
>
> If a formula is not provided, but a partially complete table of values is given, use the values in the table to verify whether your formula is correct.

WORKED EXAMPLE 7 — Solving practical problems using tables

A café makes a new special coffee blend from two different types of coffee beans. The cost of the new blend depends on how much of the two types of beans are used. Blend A costs 5 cents per gram and Blend B costs 7 cents per gram. Let C be the cost (in \$) of a packet of the special coffee blend, which is determined by the formula

$$C = \frac{5A + 7B}{100}$$

The cost of different sized packets of the special blend are shown in the table below.

C (\$)		B (grams)			
		600	800	1000	1500
A (grams)	400	62	76	90	X
	800	82	96	110	145
	2000	142	Y	170	205

a Explain the significance of the 100 on the denominator of the formula.
b Calculate the values of X and Y in the table above.
c Use the table to determine the cheapest possible packet of the special coffee blend that weighs at least 2 kg.

Steps	Working
a Check the units in the question and use the units to explain the value 100.	The cost of Blends A and B are in cents per gram. The cost of the special blend packet is in dollars. The 100 on the denominator of the formula converts the cost in cents to dollars.
b 1 Substitute the values of A and B into the formula for C.	$C = X$ when $A = 400$ and $B = 1500$. $$X = \frac{5(400) + 7(1500)}{100}$$ $= 125$
2 Evaluate the answer.	$C = Y$ when $A = 2000$ and $B = 800$. $$Y = \frac{5(2000) + 7(800)}{100}$$ $= 156$ Therefore, $X = 125$ and $Y = 156$.
c 1 Identify all values in the table for which $A + B \geq 2000$.	The values of C for which $A + B \geq 2000$ are shown shaded in green.

C (\$)		B (grams)			
		600	800	1000	1500
A (grams)	400	62	76	90	125
	800	82	96	110	145
	2000	142	156	170	205

2 Find the lowest value of C.	The cheapest value of C is 142, which is when the blend is made from 2000 grams of Blend A and 600 grams of Blend B.
3 Interpret the answer in the context of the question.	

EXERCISE 2.2 Tables and formulas

ANSWERS p. 408

Recap

1 The value of the expression $\dfrac{(5-x)(y+1)}{2}$ when $x = 1$ and $y = 3$ is

   A 16   **B** 8   **C** 4   **D** −4   **E** 2

2 The subject of the formula $C = \dfrac{5}{9}(F - 32)$ is

   A 5   **B** 9   **C** F **D** C **E** −32

Mastery

3 WORKED EXAMPLE 4   Consider the formula $C = \dfrac{5}{9}(F - 32)$. Copy and complete the table of values below for the given values of F, answering to two decimal places where appropriate.

F	32	60	104	120
C				

4 WORKED EXAMPLE 5   Consider the formula $C = \dfrac{A + B}{3} - 1$. Copy and complete the table of values below for the given values of A and B. Some of the values have been completed for you.

	C	\multicolumn{4}{c}{B}			
			2	5	8	14
A		4		2		
		7			4	
		13	4			

5 WORKED EXAMPLE 6   The value of z is determined by the formula $z = 2x - 4y$. A table of values of z is shown below.

	z	\multicolumn{4}{c}{y}			
			0	1	2	3
		0	0	−4	−8	−12
x		1	A	−2	−6	−10
		2	4	0	B	−8

  a Determine the values of A and B in the table shown.

  b Describe the pattern in the values of z when $x = y$.

6 Using CAS 3   The first four entries of column A of a spreadsheet contain the numbers 16, 30, 48 and 50.

Suppose that column B is to contain the values that are 20 less than 150% of column A.

  a Write a spreadsheet formula for the entry in cell B1 in terms of A1.

  b Complete column B in the spreadsheet on CAS.

	A	B
1	16	
2	30	
3	48	
4	50	

7 **WORKED EXAMPLE 7** A chocolatier makes its own nutty-chocolate mix using chocolate that costs $2.50 per 100 grams and nuts that cost $5 per 100 grams. Let C be the amount of chocolate (in grams) in the mix and N be the amount of nuts (in grams) in the mix. Let P be the selling price (in $) of a bag of the nutty-chocolate mix, determined by the formula

$$P = \frac{2.5C + 5N}{100}$$

The selling price of different sized bags of the nutty-chocolate mix are shown in the table below.

P ($)		\multicolumn{4}{c}{N (grams)}			
		150	200	250	300
C (grams)	100	10	12.5	15	17.5
	200	X	15	17.5	20
	400	17.5	20	Y	25

a Explain the significance of the 100 on the denominator of the formula.

b Calculate the values of X and Y in the table above.

c Use the table to determine the cheapest possible bag of the nutty-chocolate mix that weighs at least 500 g.

Calculator-free

8 (8 marks) Stephen has two sources of monthly income, Source A and Source B. The income that Stephen receives is not fixed and so changes month to month. As part of his personal savings scheme, at the end of each month he decides to put 25% of his income from Source A and 60% of his income from Source B into his savings account. Let S be the amount of money (in $) that Stephen transfers into his savings account at the end of a given month, and A and B be the amount of monthly income (in $) from each respective source.

a State a formula for S in terms of A and B to represent this situation. (2 marks)

b Knowing the possible amounts of income he typically receives from Sources A and B, Stephen decides to create a table to show the relationship between S, A and B.

S		\multicolumn{4}{c}{B}			
		1000	1200	1500	2000
A	1800	1050	1170	1350	1650
	2000	1100	X	1400	1700
	2400	1200	1320	1500	1800
	Y	1300	1420	1600	1900

i Determine the value of X in the table, and interpret the result in the context of the situation. (3 marks)

ii Determine the value of Y in the table. (3 marks)

> **Calculator-assumed**

9 (9 marks) An individual's body mass index (BMI) is a measure of weight relative to height and is sometimes used to identify health risks. The BMI value is calculated using the formula

$$\text{BMI} = \frac{W}{H^2}$$

where W is the individual's weight in kilograms and H is the individual's height in metres.

a Brody weighs 65.2 kg and is 178 cm tall. Calculate his BMI, correct to one decimal place. (3 marks)

b A table of values containing some BMIs for different weights and heights is shown below.

BMI		H (m)		
		1.51	1.82	1.95
W (kg)	54	23.7	16.3	
	75	32.9		19.7
	81		24.5	21.3

  i Copy and complete the table of values. (2 marks)

  A healthy BMI is often considered as between 18.5 and 24.9.

  ii What proportion of the nine BMIs in the table above would be considered healthy? (2 marks)

  iii If a person has a height of 180 cm, determine their largest possible weight to still fall within the healthy BMI range. Answer correct to the nearest kilogram. (2 marks)

2.3 Introduction to matrices

Video playlist
Introduction to matrices

Features of a matrix

Two-way tables are an effective way to store and display numerical information for two variables, however, they are not the only way! The entries in the rows and columns of a table can also be represented in a **matrix**, which is a rectangular array with m rows and n columns containing numerical entries, called **elements**. Matrices are displayed using square brackets and are often named using a capital letter. For example, matrix Q below has 4 rows, 2 columns and, hence, 8 elements.

$$Q = \begin{bmatrix} 5 & 12 \\ 9 & 0 \\ 17 & 22 \\ 1 & 19 \end{bmatrix} \begin{matrix} \leftarrow \text{Row 1} \\ \leftarrow \text{Row 2} \\ \leftarrow \text{Row 3} \\ \leftarrow \text{Row 4} \end{matrix}$$

The number of rows, m, and the number of columns, n, of a matrix make up its **size** (also called **dimensions** or **order**). For example, the dimensions of matrix Q is 4×2 (pronounced as *four by two*). You may also see this written as $\dim(Q) = 4 \times 2$. We can also use the notation q_{ij} to refer to the entry in matrix Q in the ith row and the jth column. So, for matrix Q, we would have $Q = \begin{bmatrix} q_{11} & q_{12} \\ q_{21} & q_{22} \\ q_{31} & q_{32} \\ q_{41} & q_{42} \end{bmatrix}$. For example, $q_{31} = 17$ as it is the entry in the 3rd row and the 1st column.

> **Order and entries of a matrix**
>
> If a matrix A has m rows and n columns, then it has order $m \times n$. This can be written as $\dim(A) = m \times n$. The entry in the ith row and jth column of a matrix A is represented as a_{ij}.

> **WORKED EXAMPLE 8** Determining the size and entries of matrices

Consider the matrices A and B shown below.

$$A = \begin{bmatrix} 2 & 1 \\ -2 & 3 \\ 0 & 1 \end{bmatrix} \qquad B = \begin{bmatrix} 1 & 2 & 3 \\ 3 & 6 & 5 \end{bmatrix}$$

a State the order of matrix A.
b State the value of a_{22}.
c State the order of matrix B.
d State the value of b_{23}.

Steps	Working
a 1 Count the number of rows and columns.	$\dim(A) = 3 \times 2$
2 State the order in the form $m \times n$.	
b Identify the entry in the 2nd row and 2nd column.	$a_{22} = 3$
c 1 Count the number of rows and columns.	$\dim(B) = 2 \times 3$
2 State the order in the form $m \times n$.	
d Identify the entry in the 2nd row and 3rd column.	$a_{23} = 5$

Types of matrices

We can name specific types of matrices depending on the order and the values of the entries.

Type	Description	Example and order
Row matrix	A matrix with one row and n columns.	$A = \begin{bmatrix} 0 & 1 & 4 \end{bmatrix}$ $\dim(A) = 1 \times 3$
Unit row matrix	A row matrix where all entries are 1s.	$B = \begin{bmatrix} 1 & 1 & 1 & 1 \end{bmatrix}$ $\dim(B) = 1 \times 4$
Column matrix	A matrix with m rows and one column.	$C = \begin{bmatrix} -2 \\ 5 \\ 1 \end{bmatrix}$ $\dim(C) = 3 \times 1$
Unit column matrix	A column matrix where all entries are 1s.	$D = \begin{bmatrix} 1 \\ 1 \end{bmatrix}$ $\dim(D) = 2 \times 1$
Square matrix	A matrix that has the same number of rows and columns. That is, an $n \times n$ matrix.	$E = \begin{bmatrix} 3 & 1 \\ 4 & 2 \end{bmatrix}$ $\dim(E) = 2 \times 2$
Zero matrix	A matrix where all entries are 0s. This is also called the **additive identity** matrix.	$F = \begin{bmatrix} 0 & 0 & 0 \\ 0 & 0 & 0 \end{bmatrix}$ $\dim(F) = 2 \times 3$
Identity matrix	A square matrix that has all entries in its **leading diagonal** as 1s, and the rest 0s. The leading diagonal of a square matrix starts from the top left entry and runs to the bottom right entry. This is also called the **multiplicative identity** matrix and is typically written as I_n where n is the number of rows (or columns).	$I_2 = \begin{bmatrix} 1 & 0 \\ 0 & 1 \end{bmatrix}$ $I_4 = \begin{bmatrix} 1 & 0 & 0 & 0 \\ 0 & 1 & 0 & 0 \\ 0 & 0 & 1 & 0 \\ 0 & 0 & 0 & 1 \end{bmatrix}$

WORKED EXAMPLE 9 — Identifying types of matrices

For each of the matrices shown below, state its order and classify the type of matrix.

a $\begin{bmatrix} 0 & 0 \\ 0 & 0 \\ 0 & 0 \\ 0 & 0 \\ 0 & 0 \\ 0 & 0 \end{bmatrix}$
b $\begin{bmatrix} 14 \\ 50 \\ -91 \\ 35 \end{bmatrix}$
c $\begin{bmatrix} 0 & 0 & 0 \\ 0 & 0 & 0 \\ 0 & 0 & 0 \end{bmatrix}$
d $\begin{bmatrix} 1 & 0 & 0 & 0 & 0 & 0 \\ 0 & 1 & 0 & 0 & 0 & 0 \\ 0 & 0 & 1 & 0 & 0 & 0 \\ 0 & 0 & 0 & 1 & 0 & 0 \\ 0 & 0 & 0 & 0 & 1 & 0 \\ 0 & 0 & 0 & 0 & 0 & 1 \end{bmatrix}$
e $\begin{bmatrix} 1 & 1 & 1 & 1 & 1 \end{bmatrix}$

Steps	Working
For each of the following: 1 State the order in the form $m \times n$. 2 Observe the shape and entries of the matrix to classify the matrix.	a 6×2, zero matrix b 4×1, column matrix c 3×3, square matrix, zero matrix d 6×6, square matrix, identity matrix e 1×5, unit row matrix

Storing and displaying information in $m \times n$ matrices

A benefit of storing information in a matrix is that it does not necessarily need detailed labels like a table does. Once the significance of the rows and columns have been defined, all that matters are the numerical entries and what they represent.

WORKED EXAMPLE 10 — Representing information in a matrix

The table below shows the quantity (in grams) of the main ingredients used in the baking of different types of muffins.

		Type of muffin				
		Apple	Banana	Berry	Chocolate	Savoury
Ingredient	Butter	90	160	175	125	130
	Flour	130	175	150	250	150
	Sugar	80	75	100	120	50

Construct a matrix to show

a all of the information above. Let this matrix be M and state its order.
b the amount of flour in each of the types of muffins. Let this matrix be F and state its order.
c the quantity of each type of ingredient needed for chocolate muffins. Let this matrix be C and state its order.
d the amount of butter needed for savoury muffins. Let this matrix be B and state its order.

Steps	Working
a 1 Construct a matrix with three rows and five columns. 2 Label the matrix M. 3 Input all the corresponding entries. 4 State the order of the matrix.	$M = \begin{bmatrix} 90 & 160 & 175 & 125 & 130 \\ 130 & 175 & 150 & 250 & 150 \\ 80 & 75 & 100 & 120 & 50 \end{bmatrix}$ $\dim(M) = 3 \times 5$

b 1 Construct a matrix with one row and five columns. 2 Label the matrix F. 3 Input all the corresponding entries. 4 State the order of the matrix.	$F = \begin{bmatrix} 130 & 175 & 150 & 250 & 150 \end{bmatrix}$ dim$(F) = 1 \times 5$
c 1 Construct a matrix with three rows and one column. 2 Label the matrix C. 3 Input all the corresponding entries. 4 State the order of the matrix.	$C = \begin{bmatrix} 125 \\ 250 \\ 120 \end{bmatrix}$ dim$(C) = 3 \times 1$
d 1 Construct a matrix with one row and one column. 2 Label the matrix B. 3 Input the corresponding entry. 4 State the order of the matrix.	$B = [130]$ dim$(B) = 1 \times 1$

EXERCISE 2.3 Introduction to matrices   ANSWERS p. 409

Recap

1 The following spreadsheet formula is used to compute the value of cell D1.

$$= \frac{A1}{2} + \frac{B1}{3} + \frac{C1}{4}$$

If A1 = 12, B1 = 15 and C1 = 12, then the value of D1 is

A 0   **B** 1   **C** 12   **D** 14   **E** 15

2 If 50% of Substance A (in grams) and 30% of Substance B (in grams) were mixed together to create C kilograms of Substance C, then assuming no additional mass is lost or gained, the formula for C in terms of A and B is

A $C = 0.5B + 0.2A$ **B** $C = \dfrac{50A + 30B}{100}$ **C** $C = \dfrac{0.5A + 0.3B}{1000}$

D $C = 0.5A + 0.3B$ **E** $A = 0.5B + 0.3C$

Mastery

3 **WORKED EXAMPLE 8** Consider the matrices A and B shown below.

$$A = \begin{bmatrix} 1 & 2 & 5 \\ -1 & 1 & 12 \\ 19 & 10 & 2 \end{bmatrix} \qquad B = \begin{bmatrix} 2 \\ -3 \end{bmatrix}$$

   a State the order of matrix A.   **b** State the value of a_{23}.
   c State the order of matrix B.   **d** State the value of b_{21}.

4 **WORKED EXAMPLE 9** For each of the matrices shown below, state its order and classify the type of matrix.

a $\begin{bmatrix} 1 \\ 0 \end{bmatrix}$

b $\begin{bmatrix} 1 & 0 & 0 \\ 0 & 1 & 0 \\ 0 & 0 & 1 \end{bmatrix}$

c $\begin{bmatrix} 0 & 0 & 0 & 0 \end{bmatrix}$

d $\begin{bmatrix} 1 & 0 & 0 \\ 0 & 1 & 0 \\ 1 & 0 & 1 \end{bmatrix}$

e $\begin{bmatrix} 1 \\ 1 \\ 1 \end{bmatrix}$

f $\begin{bmatrix} 1 & 0 & 0 & 0 \\ 0 & 1 & 0 & 0 \\ 0 & 0 & 1 & 0 \\ 0 & 0 & 0 & 0 \end{bmatrix}$

5 **WORKED EXAMPLE 10** The table below shows how many of each type of tickets were sold to a school production of *The Jungle Book* and the way in which it was sold.

	Sold by school office	Sold online	Sold at the door
Student ticket	172	67	30
Adult ticket	3	139	10
Concession ticket	0	65	9
Teacher ticket	11	15	17

Construct a matrix to show

a all of the information above. Let this matrix be T and state its order.

b the amount of adult tickets sold in different ways. Let this matrix be A and state its order.

c the amount of each type of ticket sold online. Let this matrix be O and state its order.

d the amount of concession tickets sold at the door. Let this matrix be C and state its order.

Calculator-free

6 (5 marks) Identify

  a the number of entries in a 6 × 10 matrix (1 mark)

  b the number of rows in a 5 × 7 matrix (1 mark)

  c the number of columns in a 4 × 1 matrix (1 mark)

  d the sum of the entries in a 3 × 3 identity matrix (1 mark)

  e the possible dimensions of a square matrix with less than 15 entries. (1 mark)

7 (5 marks) Consider the matrices C and D below.

$$C = \begin{bmatrix} 2 & 1 \\ 2 & 1 \\ 1 & 2 \end{bmatrix} \quad D = \begin{bmatrix} 1 & 0 & d_{13} \end{bmatrix}$$

  a State the order of matrix C. (1 mark)

  b Determine the value of $c_{11} + c_{21} + c_{32}$. (2 marks)

  c State the value of d_{13} if it is equal to the sum of the entries in matrix C. (2 marks)

8 (2 marks)

  a Construct I_2. (1 mark)

  b Construct the zero matrix with the same number of rows as I_2 but half as many columns. (1 mark)

9 (2 marks) The number of tourists visiting three towns, Augusta, Dunsborough and Geraldton, was recorded for three consecutive years. The data is summarised in the table.

	2020	2021	2022
Augusta	975	1002	1390
Dunsborough	2105	1081	1228
Geraldton	610	1095	1380

Construct a matrix that could be used to show the number of tourists visiting each of the three towns in 2021, stating its order.

10 (2 marks) The number of people attending the morning, afternoon and evening sessions at a local cinema is given in the table below. The admission charges (in $) for each session are also shown in the table.

	Session		
	Morning	Afternoon	Evening
Number of people attending	25	56	124
Admission charge ($)	15	19	22

Construct a matrix that could be used to show the admission charges for each of the three sessions.

11 (2 marks) A store has three outlets, A, B and C. These outlets sell dresses, jackets and skirts made by a local brand. The table lists the number of dresses, jackets and skirts that are currently held at each outlet, made by the local brand.

	Size 10	Size 12	Size 14	Size 16
Outlet A	2 dresses	3 jackets	1 skirt	4 jackets
Outlet B	1 skirt	1 jacket	3 jackets	1 dress
Outlet C	2 skirts	2 dresses	2 dresses	1 jacket

Construct a 4×3 matrix C to display the stock across the three outlets, whereby the rows represent the sizing options and the columns represent the type of item.

12 (4 marks) The prices (in $) to fly one-way with StarJet between Perth (P), Broome (B) and Esperance (E) is shown in the table below.

		To		
		Perth	Broome	Esperance
From	Perth	0	531	324
	Broome	421	0	866
	Esperance	448	898	0

a Construct a matrix that shows the prices to fly from Perth to either Broome or Esperance. (2 marks)

b Construct a matrix that shows the prices to fly between Perth and Esperance. (2 marks)

2.4 Matrix addition, subtraction and scalar multiplication

Addition and subtraction

Suppose the matrix M shown represents the quantity of small, regular and large milkshakes that were sold at a school canteen on Monday.

$$M = \begin{bmatrix} 8 & 15 & 20 \end{bmatrix}$$

Now, suppose the matrix T shown represents the quantity of small, regular and large milkshakes that were sold at the same school canteen on Tuesday.

$$T = \begin{bmatrix} 4 & 11 & 12 \end{bmatrix}$$

What could we do to represent the combined number of small, regular and large milkshakes that were sold at the school canteen over Monday and Tuesday?

Given that both matrices have an order of 1 × 3, we could construct a new matrix A defined by $M + T$ that sums the number of small, regular and large milkshakes on each day, respectively.

$$A = \begin{bmatrix} 8 & 15 & 20 \end{bmatrix} + \begin{bmatrix} 4 & 11 & 12 \end{bmatrix} = \begin{bmatrix} 12 & 26 & 32 \end{bmatrix}$$

However, this matrix addition can only be done because the order of both matrices M and T are the same. A similar process could be carried out with matrix subtraction if we wanted to know how many more of each size of milkshake were sold on Monday compared to Tuesday.

$$S = \begin{bmatrix} 8 & 15 & 20 \end{bmatrix} - \begin{bmatrix} 4 & 11 & 12 \end{bmatrix} = \begin{bmatrix} 4 & 4 & 8 \end{bmatrix}$$

Matrix addition and subtraction

Only matrices that have the same order can be added or subtracted. To do this, we add or subtract each pair of corresponding elements. The answer we get has the same order as the original two matrices.

$$\begin{bmatrix} 2 & -7 \\ 11 & 3 \end{bmatrix} + \begin{bmatrix} 1 & 3 \\ 5 & 12 \end{bmatrix} = \begin{bmatrix} 2+1 & -7+3 \\ 11+5 & 3+12 \end{bmatrix} = \begin{bmatrix} 3 & -4 \\ 16 & 15 \end{bmatrix}$$

	A	+	B			=	C
Order:	2 × 2	+	2 × 2			=	2 × 2

$$\begin{bmatrix} 2 & -7 \\ 11 & 3 \end{bmatrix} - \begin{bmatrix} 1 & 3 \\ 5 & 12 \end{bmatrix} = \begin{bmatrix} 2-1 & -7-3 \\ 11-5 & 3-12 \end{bmatrix} = \begin{bmatrix} 1 & -10 \\ 6 & -9 \end{bmatrix}$$

	A	−	B			=	C
Order:	2 × 2	−	2 × 2			=	2 × 2

When a matrix addition or subtraction is not possible, because the orders of the matrices are not the same, then we can say that the resulting matrix *cannot be defined*.

Recall that the zero matrix is also referred to as the additive identity matrix. This is because when an $m \times n$ zero matrix is added to any other $m \times n$ matrix, say A, the resulting matrix is just A. Much like the number 0 is the additive identity for regular numbers, because $0 + x = x$. The term additive identity means that when added to another, it does not change the result.

WORKED EXAMPLE 11 — Adding and subtracting matrices

Consider the matrices A, B, C and D shown below.

$$A = \begin{bmatrix} 2 & 1 \\ -2 & 3 \\ 0 & 1 \end{bmatrix} \quad B = \begin{bmatrix} 1 & 2 & 3 \\ 3 & 6 & 5 \end{bmatrix} \quad C = \begin{bmatrix} 4 & -1 & 2 \\ 0 & 5 & 1 \end{bmatrix} \quad D = \begin{bmatrix} -1 \\ 0 \\ 1 \end{bmatrix}$$

For each of the following matrix additions or subtractions, state whether the resulting matrix is defined. If so, determine the resulting matrix. If not, explain why.

a $B + C$ **b** $A + D$ **c** $B - C$
d $C - B$ **e** $C + B$ **f** $C + C$

Steps	Working
1 Write the order of each matrix.	$\dim(A) = 3 \times 2$
2 Check to see if the orders of the two matrices being added or subtracted are the same.	$\dim(B) = 2 \times 3$
	$\dim(C) = 2 \times 3$
3 If they are the same, state defined, and add or subtract each pair of entries. If not, use the dimensions to explain that the matrix is not defined.	$\dim(D) = 3 \times 1$
	a Defined. $$B + C = \begin{bmatrix} 1 & 2 & 3 \\ 3 & 6 & 5 \end{bmatrix} + \begin{bmatrix} 4 & -1 & 2 \\ 0 & 5 & 1 \end{bmatrix} = \begin{bmatrix} 5 & 1 & 5 \\ 3 & 11 & 6 \end{bmatrix}$$
	b Not defined, as the orders are not the same.
	c Defined. $$B - C = \begin{bmatrix} 1 & 2 & 3 \\ 3 & 6 & 5 \end{bmatrix} - \begin{bmatrix} 4 & -1 & 2 \\ 0 & 5 & 1 \end{bmatrix} = \begin{bmatrix} -3 & 3 & 1 \\ 3 & 1 & 4 \end{bmatrix}$$
	d Defined. $$C - B = \begin{bmatrix} 4 & -1 & 2 \\ 0 & 5 & 1 \end{bmatrix} - \begin{bmatrix} 1 & 2 & 3 \\ 3 & 6 & 5 \end{bmatrix} = \begin{bmatrix} 3 & -3 & -1 \\ -3 & -1 & -4 \end{bmatrix}$$
	e Defined. $$C + B = \begin{bmatrix} 4 & -1 & 2 \\ 0 & 5 & 1 \end{bmatrix} + \begin{bmatrix} 1 & 2 & 3 \\ 3 & 6 & 5 \end{bmatrix} = \begin{bmatrix} 5 & 1 & 5 \\ 3 & 11 & 6 \end{bmatrix}$$
	f Defined. $$C + C = \begin{bmatrix} 4 & -1 & 2 \\ 0 & 5 & 1 \end{bmatrix} + \begin{bmatrix} 4 & -1 & 2 \\ 0 & 5 & 1 \end{bmatrix} = \begin{bmatrix} 8 & -2 & 4 \\ 0 & 10 & 2 \end{bmatrix}$$

In the above example, you may have notice that the resulting matrix $B + C$ was the same as the matrix $C + B$. This is because matrix addition is **commutative**; that is, the order in which two matrices are added together is not important.

You may then notice that $B - C \neq C - B$, meaning that matrix subtraction is NOT commutative, but there is a relationship between the entries of $B - C$ and the entries of $C - B$. All of the values in $B - C$ are -1 times the corresponding entries in $C - B$. You might also see that all the entries in $C + C$ are 2 times the corresponding entries in C. These observations represent a **scalar multiplication** of a matrix.

Scalar multiplication

A **scalar** is any real number, not a matrix. When a matrix is multiplied by a scalar value, each element in that matrix is multiplied by that scalar, and so the resulting matrix has the same order as the initial matrix. Note that it is not necessary to write a multiplication symbol between the scalar and the matrix. For example, $C + C = 2 \times C = 2C$.

Scalar multiplication

For a matrix A and a scalar k, the entry a_{ij} in A becomes $k \times a_{ij}$ in the matrix kA.

$$4 \times \begin{bmatrix} 5 & 2 \\ 7 & -3 \\ 12 & 0 \end{bmatrix} = \begin{bmatrix} 4 \times 5 & 4 \times 2 \\ 4 \times 7 & 4 \times (-3) \\ 4 \times 12 & 4 \times 0 \end{bmatrix} = \begin{bmatrix} 20 & 8 \\ 28 & -12 \\ 48 & 0 \end{bmatrix}$$

Order: $4A$ $=$ C
         3×2 3×2

WORKED EXAMPLE 12 — Multiplying matrices by a scalar value

Consider the matrices A, B, C and D shown below.

$$A = \begin{bmatrix} 6 & 2 \\ -2 & 10 \\ 0 & 4 \end{bmatrix} \quad B = \begin{bmatrix} -1 & 2 & -2 \\ 1 & 0 & 2 \end{bmatrix} \quad C = \begin{bmatrix} 4 & -1 & 2 \\ 0 & 5 & 1 \end{bmatrix} \quad D = \begin{bmatrix} 0 \\ 0 \\ 0 \end{bmatrix}$$

Determine

a $2C$ **b** $100D$ **c** $-3B$ **d** $\dfrac{1}{2}A$

Steps	Working
For each of the examples of scalar multiplication: **1** Write the scalar multiple at the front of the matrix. **2** Apply the scalar multiplication to each entry in the matrix.	**a** $2C = 2\begin{bmatrix} 4 & -1 & 2 \\ 0 & 5 & 1 \end{bmatrix}$ $= \begin{bmatrix} 8 & -2 & 4 \\ 0 & 10 & 2 \end{bmatrix}$
Exam hack A scalar multiple of a zero matrix is still a zero matrix.	**b** $100D = \begin{bmatrix} 0 \\ 0 \\ 0 \end{bmatrix}$
	c $-3B = -3\begin{bmatrix} -1 & 2 & -2 \\ 1 & 0 & 2 \end{bmatrix}$ $= \begin{bmatrix} 3 & -6 & 6 \\ -3 & 0 & -6 \end{bmatrix}$
	d $\dfrac{1}{2}A = \dfrac{1}{2}\begin{bmatrix} 6 & 2 \\ -2 & 10 \\ 0 & 4 \end{bmatrix}$ $= \begin{bmatrix} 3 & 1 \\ -1 & 5 \\ 0 & 2 \end{bmatrix}$

When matrix expressions are formed using a mixture of operations such as addition, subtraction and scalar multiplication, it can be referred to as a **linear combination**. In linear combinations, the standard order of operations apply.

WORKED EXAMPLE 13 — Evaluating linear combinations of matrices

If $A = \begin{bmatrix} 0 \\ -2 \\ 10 \end{bmatrix}$, $B = \begin{bmatrix} 12 \\ -4 \\ 8 \end{bmatrix}$, $C = \begin{bmatrix} 2 & -5 & 0 \\ -2 & 3 & 4 \end{bmatrix}$ and $D = \begin{bmatrix} 3 & 0 & 7 \\ 1 & 3 & 4 \\ -1 & 1 & 0 \end{bmatrix}$, determine the following matrices, where defined.

a $A - 2B$

b $C + 5D$

c $3B + 2A$

Steps	Working
For each of the linear combinations:	a $\dim(A) = 3 \times 1$
1 Identify the orders of the matrices.	$\dim(B) = 3 \times 1$
2 If the orders are the same, write out the linear combination with the scalar multiples in front of the matrix.	$A - 2B = \begin{bmatrix} 0 \\ -2 \\ 10 \end{bmatrix} - 2\begin{bmatrix} 12 \\ -4 \\ 8 \end{bmatrix}$
3 Apply the scalar multiples.	$= \begin{bmatrix} 0 \\ -2 \\ 10 \end{bmatrix} - \begin{bmatrix} 24 \\ -8 \\ 16 \end{bmatrix}$
4 Carry out the addition or subtraction.	
5 If the orders are not the same, state that the matrix is not defined.	$= \begin{bmatrix} -24 \\ 6 \\ -6 \end{bmatrix}$
	b $\dim(C) = 2 \times 3$
	$\dim(D) = 3 \times 3$
	Not defined.
	c $3B + 2A = 3\begin{bmatrix} 12 \\ -4 \\ 8 \end{bmatrix} + 2\begin{bmatrix} 0 \\ -2 \\ 10 \end{bmatrix}$
	$= \begin{bmatrix} 36 \\ -12 \\ 24 \end{bmatrix} + \begin{bmatrix} 0 \\ -4 \\ 20 \end{bmatrix}$
	$= \begin{bmatrix} 36 \\ -16 \\ 44 \end{bmatrix}$

Matrices can also be defined using CAS and as a result, calculations can be carried out as long as the resulting matrix is defined.

USING CAS 4 — Addition, subtraction and scalar multiplication of matrices

Given that $Q = \begin{bmatrix} 3 & -5 & 10 \\ 11 & 6 & 17 \\ -2 & 15 & 0 \end{bmatrix}$ and $R = \begin{bmatrix} 12 & 2 & 1 \\ -3 & 4 & 7 \\ 22 & 14 & -7 \end{bmatrix}$, evaluate $4Q - 2R$ and $3Q + 7R$.

ClassPad

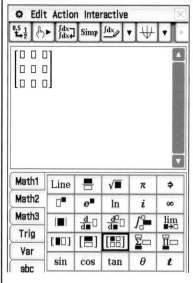

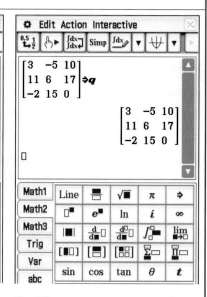

1. Open the **Main** application and clear all values.
2. Open the **Keyboard** and tap **Math2**.
3. Tap on the **2×2** matrix template twice to create a **3×3** matrix.
4. Enter the values for matrix Q.
5. Tap the **store** arrow to store the matrix as the variable q.
6. Press **EXE**.
7. The matrix is now stored as the variable q.

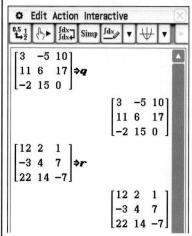

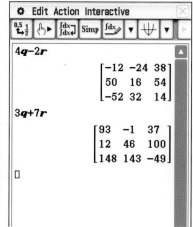

8. Create a new **3×3** matrix and enter the values for matrix R.
9. Tap the **store** arrow to store the matrix as the variable r.
10. Use the matrices stored in q and r to complete the matrix operations.

TI-Nspire

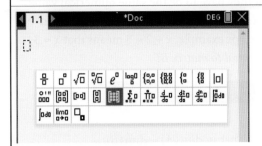

1. From a **Calculator** page, press the **template** key.
2. Select the **3×3** matrix template.

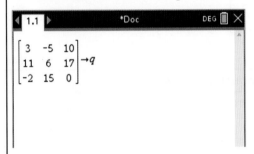

5. Enter the values for matrix Q.
6. Press **ctrl > var** to store the matrix as the letter **q**.
7. Press **enter**.

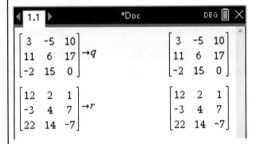

9. Create a new **3×3** matrix and enter the values for matrix R.
10. Press **ctrl > var** to store the matrix as the letter **r**.

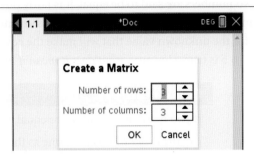

3. In the **Create a Matrix** screen, keep the default values of **3** rows and **3** columns.
4. Select **OK**.

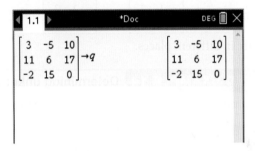

8. The matrix is now stored as the letter **q**.

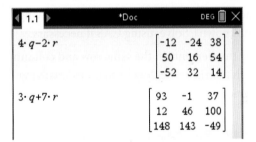

11. Use the matrices stored in **q** and **r** to complete the matrix operations.

Equal matrices

When two numbers are said to be equal, it is fairly obvious that they have to be the same numerical value. For example, 1 = 1 is true because they are the same number. In algebraic equations, we are often asked to solve for the unknown value that will make the equation true. For example, $x + 2 = 5$ means that $x = 3$ because 5 = 5 is true. However, *what does it mean if matrix A is equal to matrix B?*

> **Equal matrices**
>
> If two matrices, A and B, are equal, then:
> - the order of A is the same as the order of B. That is, $\dim(A) = \dim(B)$.
> - every entry a_{ij} is the same as every entry b_{ij}. That is, all the elements of A are in the same place and have the same value as the elements of B.

For example, if $A = \begin{bmatrix} 1 & 2 \\ -1 & 3 \end{bmatrix}$, $B = \begin{bmatrix} 2 & 3 \\ -1 & 1 \end{bmatrix}$ and $C = \begin{bmatrix} 1 & 2 \\ -1 & 3 \end{bmatrix}$, then $A = C$, but $A \neq B$ and $B \neq C$ as the entries are in different places.

WORKED EXAMPLE 14 Determining unknowns in equal matrices

Find the value of x, y and z in each of the following matrix equations.

a $\begin{bmatrix} 6 & 12 \\ x & 7 \end{bmatrix} - \begin{bmatrix} 6 & -5 \\ 1 & 2z \end{bmatrix} = \begin{bmatrix} 0 & y \\ 4 & -3 \end{bmatrix}$

b $3\begin{bmatrix} x & 2 & -1 \end{bmatrix} + 2\begin{bmatrix} 4 & 5 & y \end{bmatrix} = \begin{bmatrix} 14 & z & 11 \end{bmatrix}$

Steps	Working		
a Using the elements in the same row and column of each matrix, write down equations involving the unknowns. Solve, using CAS if necessary.	$x - 1 = 4$ $x = 5$	$12 - (-5) = y$ $y = 17$	$7 - 2z = -3$ $2z = 10$ $z = 5$
b Using the elements in the same row and column of each matrix, write down equations involving the unknowns. Solve, using CAS if necessary.	$3x + 8 = 14$ $3x = 6$ $x = 2$	$-3 + 2y = 11$ $2y = 14$ $y = 7$	$6 + 10 = z$ $z = 16$

EXERCISE 2.4 Matrix addition, subtraction and scalar multiplication   ANSWERS p. 409

Recap

1 Consider the following matrices:

$$J = \begin{bmatrix} 3 & 5 \\ 6 & 2 \end{bmatrix} \quad K = \begin{bmatrix} -2 \\ 5 \\ 1.1 \end{bmatrix} \quad L = \begin{bmatrix} 2 & 6 & 3 \\ 4 & 1 & -2 \\ 5 & 2 & 8 \end{bmatrix} \quad M = \begin{bmatrix} 12 \end{bmatrix}$$

Which statement is **true**?

A All of these matrices are square matrices.
B K is a 1×3 matrix.
C L has 10 elements.
D J is an identity matrix.
E M is a 1×1 matrix.

2 Which of the following is **not** true about $M = \begin{bmatrix} 7 & 0 & 0 \\ 0 & 0 & 0 \\ 0 & 0 & 9 \end{bmatrix}$?

A M is a 3×3 matrix.
B M is a column matrix.
C M is a square matrix.
D M has 9 elements.
E M has seven zero elements.

Mastery

3 WORKED EXAMPLE 11 Consider the matrices A, B, C and D shown below.

$$A = \begin{bmatrix} -1 & 7 \\ 2 & 0 \end{bmatrix} \quad B = \begin{bmatrix} 3 & -12 \end{bmatrix} \quad C = \begin{bmatrix} -1 & 4 \end{bmatrix} \quad D = \begin{bmatrix} 0 & 5 \\ -2 & 1 \end{bmatrix}$$

For each of the following matrix additions or subtractions, state whether the resulting matrix is defined. If so, determine the resulting matrix. If not, explain why.

a $A + B$ **b** $B - C$ **c** $A + D$ **d** $D - B$

4 WORKED EXAMPLE 12 Consider the matrices A, B, C and D shown below.

$$A = \begin{bmatrix} -3 \\ 2 \\ 1 \end{bmatrix} \quad B = \begin{bmatrix} 10 \\ 20 \end{bmatrix} \quad C = \begin{bmatrix} 4 & 8 & 0 \\ -6 & 10 & 20 \\ -2 & 50 & 2 \end{bmatrix} \quad D = \begin{bmatrix} 1 & 1 & 1 \end{bmatrix}$$

Determine

a $\frac{1}{2}C$ **b** $10D$ **c** $-\frac{1}{5}B$ **d** $2A$

5 WORKED EXAMPLE 13 If $A = \begin{bmatrix} 2 & -1 \end{bmatrix}$, $B = \begin{bmatrix} -4 & 2 \end{bmatrix}$, $C = \begin{bmatrix} 1 & -2 \\ 3 & 3 \\ 2 & 4 \end{bmatrix}$ and $I = \begin{bmatrix} 1 & 0 \\ 0 & 1 \end{bmatrix}$, determine the following matrices, where defined.

a $3A + 2B$ **b** $C + 5I$ **c** $\frac{1}{2}B - 2A$

6 Using CAS 4 Given that $M = \begin{bmatrix} -2 & 3 \\ 4 & 5 \\ 1 & 6 \end{bmatrix}$ and $N = \begin{bmatrix} 1 & 0 \\ 1 & 2 \\ 0 & 1 \end{bmatrix}$, evaluate the following using CAS.

a $3M + 3N$ **b** $5M + 4N$ **c** $6M - 4N$ **d** $12N - 3M$

7 WORKED EXAMPLE 14 Find the value of x, y and z in each of the following.

a $\begin{bmatrix} 5 \\ y \\ 1 \\ 3 \end{bmatrix} + \begin{bmatrix} x \\ 7 \\ 8 \\ 4 \end{bmatrix} = \begin{bmatrix} 9 \\ 3 \\ 9 \\ z \end{bmatrix}$ **b** $\begin{bmatrix} 7 & x \\ 9 & 4 \end{bmatrix} - \begin{bmatrix} 1 & 4 \\ y & 3 \end{bmatrix} = \begin{bmatrix} 6 & 5 \\ 10 & z \end{bmatrix}$

c $\begin{bmatrix} x & 3 \\ 7 & -2 \end{bmatrix} + 2\begin{bmatrix} 5 & y \\ 11 & 8 \end{bmatrix} = \begin{bmatrix} 25 & 37 \\ 57 & z \end{bmatrix}$ **d** $3\begin{bmatrix} 10 & 5 \\ 4 & x \\ 8 & 7 \end{bmatrix} + 4\begin{bmatrix} 1 & 9 \\ 0 & 6 \\ y & 2 \end{bmatrix} = 3\begin{bmatrix} 34 & 51 \\ z & 45 \\ 19 & 29 \end{bmatrix}$

Calculator-free

8 (8 marks) For each of the following matrix expressions, justify whether they are defined. If defined, evaluate the matrix. If not, explain why.

a $\begin{bmatrix} 6 & -3 \end{bmatrix} - \begin{bmatrix} -7 & 9 \end{bmatrix}$ (2 marks)

b $3\begin{bmatrix} 10 \\ 3 \end{bmatrix} + 2\begin{bmatrix} -1 & 8 \end{bmatrix}$ (2 marks)

c $2\begin{bmatrix} 6 & 0 \\ 12 & -7 \\ 3 & 5 \end{bmatrix} - 6\begin{bmatrix} 9 \\ 0 \\ 4 \end{bmatrix}$ (2 marks)

d $\begin{bmatrix} 0 & -4 \\ 2 & 5 \end{bmatrix} + \begin{bmatrix} 5 & 4 \\ -2 & 2 \end{bmatrix} + \begin{bmatrix} 15 \\ 12 \end{bmatrix}$ (2 marks)

9 (7 marks) Determine the following resulting matrices.

a $\quad 12\begin{bmatrix} 1 & 3 \\ 0 & 2 \end{bmatrix}$ (1 mark)

b $\quad \begin{bmatrix} 4 & 2 \\ 1 & 5 \end{bmatrix} + 3\begin{bmatrix} 4 & 2 \\ 1 & 5 \end{bmatrix}$ (2 marks)

c $\quad 2\begin{bmatrix} 3 & 2 \\ 0 & 4 \end{bmatrix} + 3\begin{bmatrix} -1 & 0 \\ 1 & 6 \end{bmatrix}$ (2 marks)

d $\quad 3\begin{bmatrix} 2 & 1 \\ 0 & 3 \end{bmatrix} + 2\begin{bmatrix} -1 & 0 \\ 2 & -7 \end{bmatrix}$ (2 marks)

10 (5 marks) Solve for the value of d in the following matrix equations.

a $\quad \begin{bmatrix} 1 & 0 \\ 0 & 1 \end{bmatrix} + \begin{bmatrix} 3 & 7 \\ 8 & d \end{bmatrix} = \begin{bmatrix} 4 & 7 \\ 8 & 11 \end{bmatrix}$ (2 marks)

b $\quad 4\begin{bmatrix} 2 & 1 & -1 \\ 1 & d & 0 \end{bmatrix} + 2\begin{bmatrix} 3 & 2 & 1 \\ -1 & d-1 & 5 \end{bmatrix} = \begin{bmatrix} 14 & 8 & -2 \\ 2 & 40 & 10 \end{bmatrix}$ (3 marks)

11 (7 marks) Consider the matrix equation $2 \times \begin{bmatrix} 3 & 0 \\ 4 & -1 \end{bmatrix} + W = \begin{bmatrix} 6 & 2 \\ 7 & 0 \end{bmatrix}$.

a State the order of matrix W. Justify your answer. (2 marks)

b Find W. (2 marks)

c If $W - A = 4I_2$, determine the matrix A, where I_2 is the 2 × 2 identity matrix. (3 marks)

2.5 Matrix multiplication and powers

Video playlist
Matrix multiplication and powers

Worksheet
Multiplying matrices

Matrix multiplication

Matrix multiplication is not the same as standard numerical multiplication; it is a newly defined operation called **row-by-column multiplication**. Recall the example whereby on a Monday, 8 small milkshakes, 15 regular milkshakes and 20 large milkshakes were sold at a school canteen, represented by the following row matrix.

$$M = \begin{bmatrix} 8 & 15 & 20 \end{bmatrix}$$

Now, suppose we want to know the total amount of money made by the school canteen just from the milkshakes sold. If a small milkshake costs \$3, a regular milkshake costs \$4 and a large milkshake costs \$5, then we need the calculation to be 8 × 3 + 15 × 4 + 20 × 5.

To do so, we define a 3 × 1 column matrix C representing the cost of each type of milkshake, $C = \begin{bmatrix} 3 \\ 4 \\ 5 \end{bmatrix}$, and we multiply M by C.

$$\begin{bmatrix} 8 & 15 & 20 \end{bmatrix} \times \begin{bmatrix} 3 \\ 4 \\ 5 \end{bmatrix} = [24 + 60 + 100] = [184]$$

This resulting 1 × 1 matrix represents that the school canteen made \$184 dollars on Monday from selling milkshakes.

This is a new type of multiplication that you will have to learn! Just remember to multiply *row-by-column*; that is, each individual entry in the row gets multiplied by the corresponding entry in the column, and then the individual results are added together. However, the matrix product is only defined when a specific condition on the dimensions is satisfied.

> **Matrix multiplication**
>
> Let A have the dimensions of $m \times n$ and B have the dimensions of $p \times q$. The matrix product AB is only defined if the number of columns in matrix A is the same as the number of rows in matrix B. That is, $n = p$. If this condition is met, then the resulting matrix AB has the dimensions of $m \times q$. If this condition is not met, then the matrix product is not defined.
>
> For $A = \begin{bmatrix} a_{11} & a_{12} \\ a_{21} & a_{22} \\ a_{31} & a_{32} \end{bmatrix}$ and $B = \begin{bmatrix} b_{11} & b_{12} & b_{13} \\ b_{21} & b_{22} & b_{23} \end{bmatrix}$, then $\dim(A) = 3 \times 2$ and $\dim(B) = 2 \times 3$.
>
> Then $AB = \begin{bmatrix} a_{11}b_{11} + a_{12}b_{21} & a_{11}b_{12} + a_{12}b_{22} & a_{11}b_{13} + a_{12}b_{23} \\ a_{21}b_{11} + a_{22}b_{21} & a_{21}b_{12} + a_{22}b_{22} & a_{21}b_{13} + a_{22}b_{23} \\ a_{31}b_{11} + a_{32}b_{21} & a_{31}b_{12} + a_{32}b_{22} & a_{31}b_{13} + a_{32}b_{23} \end{bmatrix}$ which has $\dim(AB) = 3 \times 3$.
>
> $$AB = C$$
>
> Order: $(m \times n)(n \times q) = (m \times q)$
>
> Product is defined

A numerical example showing how a 2×3 matrix is multiplied by a 3×2 matrix to form a 2×2 matrix.

$(1 \times 7) + (2 \times 8) + (3 \times 9) = 50$

$\begin{bmatrix} 1 & 2 & 3 \\ 4 & 5 & 6 \end{bmatrix} \times \begin{bmatrix} 7 & 10 \\ 8 & 11 \\ 9 & 12 \end{bmatrix} = \begin{bmatrix} 50 & \end{bmatrix}$

$\begin{array}{ccccc} A & \times & B & = & C \\ \text{Order:} \quad 2 \times 3 & & 3 \times 2 & & 2 \times 2 \end{array}$

$(1 \times 10) + (2 \times 11) + (3 \times 12) = 68$

$\begin{bmatrix} 1 & 2 & 3 \\ 4 & 5 & 6 \end{bmatrix} \times \begin{bmatrix} 7 & 10 \\ 8 & 11 \\ 9 & 12 \end{bmatrix} = \begin{bmatrix} 50 & 68 \end{bmatrix}$

$\begin{array}{ccccc} A & \times & B & = & C \\ \text{Order:} \quad 2 \times 3 & & 3 \times 2 & & 2 \times 2 \end{array}$

$\begin{bmatrix} 1 & 2 & 3 \\ 4 & 5 & 6 \end{bmatrix} \times \begin{bmatrix} 7 & 10 \\ 8 & 11 \\ 9 & 12 \end{bmatrix} = \begin{bmatrix} 50 & 68 \\ 122 & \end{bmatrix}$

$(4 \times 7) + (5 \times 8) + (6 \times 9) = 122$

$\begin{bmatrix} 1 & 2 & 3 \\ 4 & 5 & 6 \end{bmatrix} \times \begin{bmatrix} 7 & 10 \\ 8 & 11 \\ 9 & 12 \end{bmatrix} = \begin{bmatrix} 50 & 68 \\ 122 & 167 \end{bmatrix}$

$(4 \times 10) + (5 \times 11) + (6 \times 12) = 167$

WORKED EXAMPLE 15 — Multiplying matrices

Consider the matrices $A = \begin{bmatrix} 2 & 1 & 3 \end{bmatrix}$, $B = \begin{bmatrix} 3 & 2 \\ 1 & 2 \\ 4 & -3 \end{bmatrix}$ and $C = \begin{bmatrix} 1 \\ -2 \end{bmatrix}$.

Determine whether the following matrix products are defined. If so, carry out the matrix multiplication, stating the dimensions of the resulting matrix. If not, justify why.

a AB **b** BA **c** BC

Steps	Working
For each of the products: 1. Write the dimensions next to each other in the order in which the multiplication occurs. 2. Check to see if the number of columns in the first matrix matches the number of rows in the second. 3. If so, state the result dimension and carry out the row-by-column multiplication. 4. If not, state that the matrix multiplication is not defined.	**a** $A \times B$ $(1 \times 3) \times (3 \times 2)$ Defined. $\dim(AB) = 1 \times 2$ $\begin{bmatrix} 2 & 1 & 3 \end{bmatrix} \times \begin{bmatrix} 3 & 2 \\ 1 & 2 \\ 4 & -3 \end{bmatrix}$ $= \begin{bmatrix} 2(3)+1(1)+3(4) & 2(2)+1(2)+3(-3) \end{bmatrix}$ $= \begin{bmatrix} 19 & -3 \end{bmatrix}$ **b** $B \times A$ $(3 \times 2) \times (1 \times 3)$ Not defined, as B has two columns and A has one row. **c** $B \times C$ $(3 \times 2) \times (2 \times 1)$ Defined. $\dim(BC) = 3 \times 1$ $\begin{bmatrix} 3 & 2 \\ 1 & 2 \\ 4 & -3 \end{bmatrix} \times \begin{bmatrix} 1 \\ -2 \end{bmatrix}$ $= \begin{bmatrix} 3(1)+2(-2) \\ 1(1)+2(-2) \\ 4(1)+(-3)(-2) \end{bmatrix}$ $= \begin{bmatrix} -1 \\ -3 \\ 10 \end{bmatrix}$

 Exam hack

Once you are confident with the row-by-column multiplication technique, you can skip the step showing the expanded multiplication. Be careful with your positives and negatives though!

The order of multiplication

Notice that in Worked example 15, the matrix multiplication AB was defined but the multiplication BA was not. This means the order in which matrix multiplication is carried out is important, because the resulting matrix may not be defined.

> **Matrix multiplication and commutativity**
>
> For two matrices A and B, it is said that matrix multiplication is not necessarily commutative. That is, $AB \neq BA$ in most cases. This can be because of two reasons:
>
> 1 If AB is defined, BA may not necessarily be defined.
>   For example,
> $(1 \times 3) \times (3 \times 2)$ will result in a 1×2 matrix, but
> $(3 \times 2) \times (1 \times 3)$ is not possible.
>
> 2 If both AB and BA are defined, they may not necessarily have the same dimensions.
> $(1 \times 3) \times (3 \times 1)$ will result in a 1×1 matrix, but
> $(3 \times 1) \times (1 \times 3)$ will result in a 3×3 matrix.
>
> Order is important!

Because $AB \neq BA$, we can use the language of **pre-multiplication** and **post-multiplication** to describe which of the two matrix products are defined. For example, in Worked example 15, when A was post-multiplied by B (i.e. AB), the matrix product was defined. However, when A was pre-multiplied by B (i.e. BA), the product was not defined.

Recall that the identity matrix I_n is also referred to as the multiplicative identity matrix. This is because when I_n is pre-multiplied by another $m \times n$ matrix, say A, the resulting matrix is just A. Much like the number 1 is the multiplicative identity for regular numbers, because $x \times 1 = x$. The term multiplicative identity means that when multiplied to another, it does not change the result.

For example, let $A = \begin{bmatrix} 1 & 3 \\ 2 & 1 \\ -1 & 0 \end{bmatrix}$ and $I = \begin{bmatrix} 1 & 0 \\ 0 & 1 \end{bmatrix}$, then IA will not be defined, if I is a 2×2 matrix, but AI will be defined.

$$AI = \begin{bmatrix} 1 & 3 \\ 2 & 1 \\ -1 & 0 \end{bmatrix} \times \begin{bmatrix} 1 & 0 \\ 0 & 1 \end{bmatrix}$$

$$= \begin{bmatrix} 1(1) + 3(0) & 1(0) + 3(1) \\ 2(1) + 1(0) & 2(0) + 1(1) \\ -1(1) + 0(0) & -1(0) + 0(1) \end{bmatrix}$$

$$= \begin{bmatrix} 1 & 3 \\ 2 & 1 \\ -1 & 0 \end{bmatrix} = A$$

For IA to be defined, I would need to be the 3×3 identity matrix.

> **Multiplying by the identity matrix**
>
> When a matrix A with dimensions $m \times n$ is multiplied by an identity matrix I, the resulting matrix is A. The dimensions of I depend on whether it is being pre- or post-multiplied with A.

Powers of matrices

One of the few cases when $AB = BA$ is when A and B are the same matrix (i.e. $A = B$) and are both square matrices. For example, if

$$A = \begin{bmatrix} -2 & 1 \\ 1 & 2 \end{bmatrix}$$

then

$$A \times A = \begin{bmatrix} -2 & 1 \\ 1 & 2 \end{bmatrix} \times \begin{bmatrix} -2 & 1 \\ 1 & 2 \end{bmatrix} = \begin{bmatrix} 4+1 & -2+2 \\ -2+2 & 1+4 \end{bmatrix} = \begin{bmatrix} 5 & 0 \\ 0 & 5 \end{bmatrix}$$

which would be the same result regardless of the order of the multiplication.

As a result, we can say that a square matrix can be raised to a power, because the number of rows is always equal to the number of columns, and so the matrix multiplication will always be defined, and the order will not be important. This product, $A \times A$ can be written as A^2.

However, for a non-square matrix, say

$$B = \begin{bmatrix} -2 & 3 \\ 5 & 2 \\ 1 & 4 \end{bmatrix}$$

the product $B \times B$ is not defined as the number of columns is 2 and the number of rows is 3.

Powers of matrices

Only square matrices can be raised to a power.

$$AA = A^2$$

Order: $(m \times m)(m \times m) = (m \times m)$

Product is defined for square matrices.

If A is multiplied by itself k times, then the notation is A^k and the dimensions of A^k will be the same as the dimensions of A.

For non-square matrices, the product is not defined.

$$AA$$

Order: $(m \times n)(m \times n)$

Product is not defined for non-square matrices.

WORKED EXAMPLE 16 — Mixed operations with matrices

Consider the matrices

$$A = \begin{bmatrix} 3 & 6 \\ 1 & 0 \end{bmatrix} \quad B = \begin{bmatrix} 2 & 5 & 1 \\ 0 & 4 & 7 \end{bmatrix} \quad C = \begin{bmatrix} 1 & 6 & 2 \end{bmatrix} \quad D = \begin{bmatrix} 4 \\ 9 \\ 7 \end{bmatrix}$$

For each of the following matrix expressions, state whether the resulting matrix is defined. If not defined, give a reason. If defined, state the dimensions of the resulting matrix before evaluating the matrix.

a AB
b BA
c BC
d BD
e CD
f B^3
g $A^2 + 5A$

Steps	Working
1 Do the number of columns in the first matrix equal the number of rows in the second matrix? 2 If so, the matrix product is defined. If not, the product is not. 3 If defined, carry out the row-by-column multiplication.	**a** $\dim(A) = 2 \times 2 \quad \dim(B) = 2 \times 3$ AB is defined. $\dim(AB) = 2 \times 3$ $AB = \begin{bmatrix} 3 & 6 \\ 1 & 0 \end{bmatrix} \begin{bmatrix} 2 & 5 & 1 \\ 0 & 4 & 7 \end{bmatrix} = \begin{bmatrix} 6 & 39 & 45 \\ 2 & 5 & 1 \end{bmatrix}$ **b** $\dim(B) = 2 \times 3 \quad \dim(A) = 2 \times 2$ BA is not defined, as the number of columns of B (3) is not equal to the number of rows of A (2). **c** $\dim(B) = 2 \times 3 \quad \dim(C) = 1 \times 3$ BC is not defined, as number of columns of B (3) is not equal to the number of rows of C (1). **d** $\dim(B) = 2 \times 3 \quad \dim(D) = 3 \times 1$ BD is defined. $\dim(BD) = 2 \times 1$ $BD = \begin{bmatrix} 2 & 5 & 1 \\ 0 & 4 & 7 \end{bmatrix} \begin{bmatrix} 4 \\ 9 \\ 7 \end{bmatrix} = \begin{bmatrix} 60 \\ 85 \end{bmatrix}$ **e** $\dim(C) = 1 \times 3 \quad \dim(D) = 3 \times 1$ CD is defined. $\dim(CD) = 1 \times 1$ $CD = \begin{bmatrix} 1 & 6 & 2 \end{bmatrix} \begin{bmatrix} 4 \\ 9 \\ 7 \end{bmatrix} = [72]$
f Is B a square matrix? If yes, the power can be defined. If not, then it is not possible.	B is a not square matrix. Only powers of square matrices are defined, so B^3 is not defined.
g Is A a square matrix? If yes, the power can be defined. If not, then it is not possible.	A is a square matrix. Powers of square matrices are always defined, so A^2 is defined. $\dim(A^2) = 2 \times 2 \quad \dim(5A) = 2 \times 2$ Matrices must have the same order to be added, so $A^2 + 5A$ is defined. $A^2 = \begin{bmatrix} 3 & 6 \\ 1 & 0 \end{bmatrix} \begin{bmatrix} 3 & 6 \\ 1 & 0 \end{bmatrix}$ $= \begin{bmatrix} (3 \times 3) + (6 \times 1) & (3 \times 6) + (6 \times 0) \\ (1 \times 3) + (0 \times 1) & (1 \times 6) + (0 \times 0) \end{bmatrix}$ $= \begin{bmatrix} 15 & 18 \\ 3 & 6 \end{bmatrix}$ $A^2 + 5A = \begin{bmatrix} 15 & 18 \\ 3 & 6 \end{bmatrix} + 5 \begin{bmatrix} 3 & 6 \\ 1 & 0 \end{bmatrix}$ $= \begin{bmatrix} 15 & 18 \\ 3 & 6 \end{bmatrix} + \begin{bmatrix} 15 & 30 \\ 5 & 0 \end{bmatrix}$ $= \begin{bmatrix} 30 & 48 \\ 8 & 6 \end{bmatrix}$

When complex multiplications or higher powers (i.e. a power greater than 2) of matrices need to be evaluated, we can use CAS to help with the calculations.

USING CAS 5 — Multiplying matrices

If $A = \begin{bmatrix} 5 & 3 \\ 2 & 6 \end{bmatrix}$ and $B = \begin{bmatrix} 1 & -2 \\ 0 & 3 \end{bmatrix}$, find

a AB **b** B^8 **c** $3A^2 - B^3 A$

ClassPad

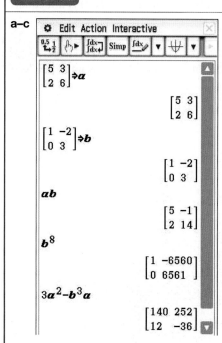

1 Create the two matrices and store them as variables a and b.

2 Perform the calculations as shown above. When you use variables instead of letters, there is no need to insert multiplication signs.

TI-Nspire

1 Create the two matrices and store them as variables a and b.

2 Perform the calculations as shown above.

3 For part **a**, remember to insert a multiplication sign between the a and b.

EXERCISE 2.5 Matrix multiplication and powers

ANSWERS p. 410

Recap

1 Consider the following four matrix expressions.

$$\begin{bmatrix} 8 \\ 12 \end{bmatrix} + \begin{bmatrix} 4 \\ 2 \end{bmatrix} \qquad \begin{bmatrix} 8 \\ 12 \end{bmatrix} + \begin{bmatrix} 4 & 0 \\ 0 & 2 \end{bmatrix} \qquad \begin{bmatrix} 8 & 0 \\ 12 & 0 \end{bmatrix} + \begin{bmatrix} 4 \\ 2 \end{bmatrix} \qquad \begin{bmatrix} 8 & 0 \\ 12 & 0 \end{bmatrix} + \begin{bmatrix} 4 & 0 \\ 0 & 2 \end{bmatrix}$$

How many of these four matrix expressions are defined?

A 0   **B** 1   **C** 2   **D** 3   **E** 4

2 The matrix sum $\begin{bmatrix} 0 & -4 \\ 2 & 5 \end{bmatrix} + \begin{bmatrix} 5 & 4 \\ -2 & 2 \end{bmatrix}$ is equal to

A $\begin{bmatrix} 5 & 0 \\ 0 & 7 \end{bmatrix}$
B $\begin{bmatrix} 0 & 0 \\ 0 & 7 \end{bmatrix}$
C $\begin{bmatrix} 5 & -4 \\ 0 & 7 \end{bmatrix}$

D $\begin{bmatrix} 0 & 5 & -4 & 4 \\ 2 & -2 & 5 & 2 \end{bmatrix}$
E $\begin{bmatrix} 0 & -4 & 5 & 4 \\ 2 & 5 & -2 & 2 \end{bmatrix}$

Mastery

3 WORKED EXAMPLE 15 Consider the matrices $A = \begin{bmatrix} -1 \\ 2 \\ 3 \end{bmatrix}$, $B = \begin{bmatrix} 3 & 10 & -2 \end{bmatrix}$ and $C = \begin{bmatrix} 5 \\ 2 \end{bmatrix}$.

Determine whether the following matrix products are defined. If so, carry out the matrix multiplication, stating the dimensions of the resulting matrix. If not, justify why.

a AB **b** BA **c** BC **d** CB

4 WORKED EXAMPLE 16 Consider the matrices:

$A = \begin{bmatrix} 3 & 4 & 10 \end{bmatrix}$ $B = \begin{bmatrix} 0 \\ 1 \\ 5 \end{bmatrix}$ $C = \begin{bmatrix} -1 & 0 \\ 3 & 1 \end{bmatrix}$ $D = \begin{bmatrix} 5 & 3 \\ 0 & 2 \\ 4 & 1 \end{bmatrix}$

For each of the following matrix expressions, state whether the resulting matrix is defined. If not defined, give a reason. If defined, state the dimensions of the resulting matrix before evaluating the matrix.

a AB **b** BA **c** BC **d** BD
e DC **f** $C^2 - 2C$ **g** D^4

5 Using CAS 5 If $M = \begin{bmatrix} 2 & 6 \\ 8 & 3 \end{bmatrix}$ and $N = \begin{bmatrix} 4.5 & 5.2 \\ 2.8 & 3.6 \end{bmatrix}$, find

a MN **b** M^4 **c** $3M - NM^2$

Calculator-free

6 (6 marks) If $A = \begin{bmatrix} 0 & 1 \\ 1 & 0 \end{bmatrix}$, $B = \begin{bmatrix} 1 \\ 0 \end{bmatrix}$ and $C = \begin{bmatrix} 0 \\ 1 \end{bmatrix}$, determine the following.

a $AB + 2C$ (3 marks)

b $AC - 5B$ (3 marks)

7 (5 marks) Evaluate the following matrix expressions.

a $\begin{bmatrix} 0 & 0 & 0 & 0 \\ 2 & 1 & 1 & 3 \\ 0 & 0 & 0 & 0 \\ 0 & 0 & 0 & 0 \end{bmatrix} \begin{bmatrix} 0 & 0 & 2 & 0 \\ 0 & 0 & 1 & 0 \\ 0 & 0 & 1 & 0 \\ 0 & 0 & 3 & 0 \end{bmatrix}$ (2 marks)

b $\begin{bmatrix} 1 & 0 & 0 & 0 \\ 0 & 1 & 0 & 0 \\ 0 & 0 & 1 & 0 \\ 0 & 0 & 0 & 1 \end{bmatrix} \begin{bmatrix} 2 \\ 0 \\ 0 \\ 2 \end{bmatrix} - 2 \times \begin{bmatrix} 0 \\ -1 \\ -1 \\ 0 \end{bmatrix}$ (3 marks)

8 (6 marks)

   a Matrix A has three rows and two columns. Matrix B has four rows and three columns. State the dimensions of matrix BA. (2 marks)

   b Matrix C has two rows and two columns. Matrix D has three rows and two columns. State the dimensions of matrix DC. (2 marks)

   c Matrix E has four rows and four columns. State the dimensions of matrix E^4. (2 marks)

9 (2 marks)

   a If M is a 3 × 3 matrix, state the matrix I that will give MI = M. (1 mark)

   b If the matrix $N = \begin{bmatrix} 12 & 15 & 3 \\ -6 & 0 & 24 \end{bmatrix}$, state the matrix I that will give IN = N. (1 mark)

Calculator-assumed

10 (10 marks) State whether each of the following matrix products are possible. If possible, determine the dimensions of the resulting matrix. If not, justify why.

   a $\begin{bmatrix} 2 & 7 & 5 \end{bmatrix} \begin{bmatrix} 3 \\ 6 \\ 2 \end{bmatrix}$ (2 marks)

   b $\begin{bmatrix} 1 & 0 \\ 0 & 1 \end{bmatrix} \begin{bmatrix} 3 & 7 \\ 20 & 12 \end{bmatrix}$ (2 marks)

   c $\begin{bmatrix} 6 \\ 3 \\ 7 \end{bmatrix} \begin{bmatrix} 1 & 2 & 5 \\ 7 & 6 & 8 \\ 3 & 5 & 9 \end{bmatrix}$ (2 marks)

   d $\begin{bmatrix} 7 & 15 \\ 3 & 6 \\ 72 & 12 \\ 11 & 1 \end{bmatrix} \begin{bmatrix} 2 & 0 \\ 8 & 13 \end{bmatrix}$ (2 marks)

   e $\begin{bmatrix} 0.5 & 16 \\ 3.8 & 2 \\ 14 & 2.1 \end{bmatrix} \begin{bmatrix} 7 & 6 & 12 & 44 \\ 8 & 52 & 1 & 6 \end{bmatrix}$ (2 marks)

11 (7 marks) Consider the four matrices

$$A = \begin{bmatrix} 5 & 4 \\ 11 & 8 \\ 7 & 6 \end{bmatrix} \quad B = \begin{bmatrix} 2 & 9 & 12 \\ 10 & 5 & 6 \end{bmatrix} \quad C = \begin{bmatrix} 3 & 8 \\ 1 & 0 \end{bmatrix} \quad D = \begin{bmatrix} 1 & 7 \\ 3 & 8 \\ 9 & 2 \end{bmatrix}$$

   a Identify the pair of matrices that can be added. (1 mark)

   b Justify whether the matrix product AC exists. If so, evaluate the product. (2 marks)

   c There are seven possible matrix products that can be formed by pairs of the above matrices. State the two matrix products that gives a resulting matrix with dimensions 3 × 3. (2 marks)

   d Identify which of the above matrices can be raised to a power. Justify your answer. (2 marks)

2.6 Modelling with matrices

In Section 2.3, we saw that matrices can be used to store and display information. Now that we have also learnt matrix operations, we can use the operations within different contexts to solve practical problems.

WORKED EXAMPLE 17 — Solving problems involving matrix addition, subtraction and scalar multiplication

The cost prices of four different laptops in a store are $890, $999, $1300 and $1950. The selling price of each of these four laptops is 140% of the cost price.

a Express the cost prices of the laptops as a column matrix. Let this matrix be C.

b Show a matrix operation involving C that will give the selling price of each laptop. Let this matrix be S.

c On top of the selling price, a $60 commission for the sales representative needs to be included in the final retail price. Show a matrix operation involving S that will give the final retail price of each laptop. Let this matrix be R.

d The store has a sale where laptops with an original cost price less than $1000 have their final retail price reduced by $30, and laptops with an original cost price greater than $1000 have their final retail price reduced by $90. Show a matrix operation involving R that will give the discounted price of each laptop. Let this matrix be D.

Steps	Working
a Construct a 4 × 1 column matrix containing the cost prices and name it C.	$C = \begin{bmatrix} 890 \\ 999 \\ 1300 \\ 1950 \end{bmatrix}$
b 1 Use scalar multiplication to represent a 40% increase. **2** Apply the scalar multiplication of 1.4 to matrix C and name it S.	$S = 1.4C = 1.4 \begin{bmatrix} 890 \\ 999 \\ 1300 \\ 1950 \end{bmatrix} = \begin{bmatrix} 1246 \\ 1398.6 \\ 1820 \\ 2730 \end{bmatrix}$
c Use matrix addition to increase each value by $60 and name it R.	$R = S + \begin{bmatrix} 60 \\ 60 \\ 60 \\ 60 \end{bmatrix} = \begin{bmatrix} 1246 \\ 1398.6 \\ 1820 \\ 2730 \end{bmatrix} + \begin{bmatrix} 60 \\ 60 \\ 60 \\ 60 \end{bmatrix}$ $= \begin{bmatrix} 1306 \\ 1458.6 \\ 1880 \\ 2790 \end{bmatrix}$
d Use matrix subtraction to decrease r_{11} and r_{21} by $30 and r_{31} and r_{41} by $90. Name it D.	$D = R - \begin{bmatrix} 30 \\ 30 \\ 90 \\ 90 \end{bmatrix} = \begin{bmatrix} 1306 \\ 1458.6 \\ 1880 \\ 2790 \end{bmatrix} - \begin{bmatrix} 30 \\ 30 \\ 90 \\ 90 \end{bmatrix}$ $= \begin{bmatrix} 1276 \\ 1428.6 \\ 1790 \\ 2700 \end{bmatrix}$

Video playlist
Modelling with matrices

Costing and pricing problems

The previous worked example was given in the context of consumer arithmetic. This is a very common application of matrices. You may also recall the initial example that was given in Section 2.5 on the number of milkshakes sold at a school canteen and each of their corresponding prices. This milkshake example was an example of a costing and pricing problem, whereby different product categories are given, each with their own associated costs or selling price.

In applied contexts such as costing and pricing, unit row and unit column matrices may be useful to sum the entries of a column or row matrix. For example, suppose the values in the matrix $\begin{bmatrix} 1276 \\ 1428.6 \\ 1790 \\ 2700 \end{bmatrix}$ needed to be added together. Then this matrix could be pre-multiplied by a 1 × 4 unit row matrix to perform this calculation.

$$\begin{bmatrix} 1 & 1 & 1 & 1 \end{bmatrix} \times \begin{bmatrix} 1276 \\ 1428.6 \\ 1790 \\ 2700 \end{bmatrix} = \begin{bmatrix} 7194.6 \end{bmatrix}$$

Similarly, suppose that the values in the row matrix needed to be added together. Then this matrix could be post-multiplied by a 3 × 1 unit column matrix to perform this calculation.

$$\begin{bmatrix} 8 & 15 & 20 \end{bmatrix} \times \begin{bmatrix} 1 \\ 1 \\ 1 \end{bmatrix} = \begin{bmatrix} 43 \end{bmatrix}$$

WORKED EXAMPLE 18 Solving costing and pricing problems

A health food store owner purchases small packets of peppermint tea for $2.50 each and large packets of peppermint tea for $3.50 each. In the last two weeks, she has purchased the number of tea packets shown.

	Small peppermint tea packets	Large peppermint tea packets
Week 1	86	70
Week 2	64	52

a Construct a 2 × 2 matrix N representing the number of different sized packets of peppermint tea that were purchased over the last two weeks.

b Construct a 2 × 1 matrix C representing the cost price of each type of peppermint tea packet.

c Show the use of an appropriate matrix multiplication to find the total cost of purchasing peppermint tea in each of the two weeks. State each cost.

d The health food store owner sells goods at 175% of the cost price. She recorded her purchase costs over the last two weeks for peppermint tea and three other teas in the following table.

	Week 1	Week 2
Peppermint tea		
Apple tea	$473	$542
Chamomile tea	$628	$745
Ginger tea	$263	$220

    i Represent the costs of all four teas over the two-week period in a 4 × 2 cost matrix, T.

    ii Using scalar multiplication, represent the total amount of money that could be made from these goods (i.e. revenue) over the two week period in a 4 × 2 matrix, S.

e If profit is defined as *revenue − cost price*, construct a profit matrix, P.

f Show the use of unit matrices to calculate the total profit to be made if all of the goods purchased over these two weeks are sold.

Steps	Working
a 1 Construct a 2 × 2 matrix called N. 2 Enter the data for Week 1 in row 1 and the data for Week 2 in row 2. The first and second columns should represent the small and large packets of tea, respectively.	$N = \begin{bmatrix} 86 & 70 \\ 64 & 52 \end{bmatrix}$
b 1 Construct a 2 × 1 matrix called C. 2 Enter the cost price of a small packet in row 1 and the cost price of a large packet in row 2.	$C = \begin{bmatrix} 2.5 \\ 3.5 \end{bmatrix}$
c 1 Order the matrix multiplication so that the *number of small packets* is multiplied by the *cost price of small packets*, and the *number of large packets* is multiplied by the *cost price of large packets*. 2 Check to see that the dimensions of the matrix multiplication work and note the dimensions of the resulting matrix product. 3 Use CAS, if needed, to carry out the multiplication. 4 Interpret the information in the matrix.	$NC = \begin{bmatrix} 86 & 70 \\ 64 & 52 \end{bmatrix} \times \begin{bmatrix} 2.5 \\ 3.5 \end{bmatrix}$ $(2 \times 2) \times (2 \times 1)$ $\dim(NC) = 2 \times 1$ $NC = \begin{bmatrix} 460 \\ 342 \end{bmatrix}$ Total cost of peppermint tea in Week 1 is \$460 and total cost of peppermint tea in Week 2 is \$342.
d i 1 Construct a 4 × 2 matrix called T. 2 Let the rows represent the different teas and the columns represent the two weeks. **ii** 1 Represent 175% as a scalar multiple and define a new matrix S in terms of T. 2 Use CAS, if needed, to carry out the scalar multiplication.	$T = \begin{bmatrix} 460 & 342 \\ 473 & 542 \\ 628 & 745 \\ 263 & 220 \end{bmatrix}$ $S = 1.75T = 1.75 \begin{bmatrix} 460 & 342 \\ 473 & 542 \\ 628 & 745 \\ 263 & 220 \end{bmatrix}$ $S = \begin{bmatrix} 805 & 598.5 \\ 827.75 & 948.5 \\ 1099 & 1303.75 \\ 460.25 & 385 \end{bmatrix}$
e 1 Show the matrix subtraction of revenue per tea type per week, S, and total cost price per tea type per week, T. 2 Calculate the 4 × 2 matrix, P.	$P = S - T$ $= \begin{bmatrix} 805 & 598.5 \\ 827.75 & 948.5 \\ 1099 & 1303.75 \\ 460.25 & 385 \end{bmatrix} - \begin{bmatrix} 460 & 342 \\ 473 & 542 \\ 628 & 745 \\ 263 & 220 \end{bmatrix}$ $P = \begin{bmatrix} 345 & 256.5 \\ 354.75 & 406.5 \\ 471 & 558.75 \\ 197.25 & 165 \end{bmatrix}$

2.6

f 1 Construct a 1 × 4 unit row matrix and pre-multiply it to *P* to add the total profit per week.

$$[1 \ 1 \ 1 \ 1] \begin{bmatrix} 345 & 256.5 \\ 354.75 & 406.5 \\ 471 & 558.75 \\ 197.25 & 165 \end{bmatrix}$$

$$= [1368 \ 1368.75]$$

2 Construct a 2 × 1 unit column matrix and post-multiply it to the resulting matrix to add the total profit across both weeks.

$$[1368 \ 1368.75] \begin{bmatrix} 1 \\ 1 \end{bmatrix}$$

$$= [2754.75]$$

3 Interpret the result.

The total profit made across the two weeks if all goods are sold is $2754.75.

> 🔓 **Exam hack**
>
> When using matrix multiplication in an applied context, the labels of the columns of the first matrix should align with the labels of the rows of the second matrix.

Communication networks

Matrices can also be useful to represent how people and systems communicate with one another. Communication links can be shown in a diagram, called a **communication network**, whereby the vertices *A*, *B*, *C* and *D* represent the people or systems, and the directed line segments represent the possibility of a communication link between the two vertices.

These links can also be shown in matrix form, whereby the rows represent the Sender (i.e. the 'from') and the columns represent the Receiver (i.e. the 'to'). **Communication matrices** are square matrices that show the direct links with '1's and '0's, whereby a 1 represents that the communication is possible and 0 represents that the communication is not. Links with the same sender and receiver are called **redundant links**, and are found in the leading diagonal of the communication matrix and will always be '0's. The communication matrix for the network is shown in the below matrix.

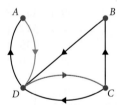

$$M = \begin{array}{c} \\ A \\ B \\ C \\ D \end{array} \begin{array}{c} A \ B \ C \ D \end{array} \begin{bmatrix} 0 & 0 & 0 & 1 \\ 0 & 0 & 0 & 1 \\ 0 & 1 & 0 & 1 \\ 1 & 0 & 1 & 0 \end{bmatrix}$$

The above matrix *M* shows all the possible one-step communications that can occur between *A*, *B*, *C* and *D*.

Now, suppose the one-step communication matrix *M* was raised to the power of 2.

$$M^2 = \begin{bmatrix} 1 & 0 & 1 & 0 \\ 1 & 0 & 1 & 0 \\ 1 & 0 & 1 & 0 \\ 0 & 0 & 0 & 2 \end{bmatrix}$$

Let's focus on the entry $m^2_{44} = 2$. This value corresponds to both the sender and receiver being *D*, which may not necessarily be redundant any more, as a message may need to be sent from *D* to another person, and then back to *D* again. Looking back to the network diagram, *what does this value of 2 represent?*

We can see that there are two distinct ways that a communication can start with *D* being the sender, go to another vertex (i.e. a one-step communication) and then return to *D* (i.e. a two-step communication). This can occur via *D* – *A* – *D* or *D* – *C* – *D*.

Similarly, if matrix M was raised to the power of 3, then

$$M^3 = \begin{bmatrix} 0 & 0 & 0 & 2 \\ 0 & 0 & 0 & 2 \\ 0 & 0 & 0 & 2 \\ 2 & 0 & 2 & 0 \end{bmatrix}$$

Focusing on the entry $m_{14}^3 = 2$ suggests that there are two possible three-step communications from D to A. Observing the network diagram, we can see that its possible via $D - A - D - A$ or $D - C - D - A$.

WORKED EXAMPLE 19	Solving problems involving communication networks

The following communication network shows the way in which six systems interact with one another.

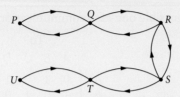

Construct a 6 × 6 communication matrix C representing this network, whereby a '1' represents a possible interaction between two systems and a '0' represents no possible interaction between two systems.

Steps	Working
1 Construct a square matrix, with labels P, Q, R, S, T and U. 2 Complete the matrix with 1s and 0s representing interactions that are possible and not possible, respectively.	Receiver Sender $\begin{array}{c} P \\ Q \\ R \\ S \\ T \\ U \end{array} \begin{bmatrix} 0 & 1 & 0 & 0 & 0 & 0 \\ 1 & 0 & 1 & 0 & 0 & 0 \\ 0 & 1 & 0 & 1 & 0 & 0 \\ 0 & 0 & 1 & 0 & 1 & 0 \\ 0 & 0 & 0 & 1 & 0 & 1 \\ 0 & 0 & 0 & 0 & 1 & 0 \end{bmatrix}$ with columns $P\ Q\ R\ S\ T\ U$

WORKED EXAMPLE 20	Solving problems involving communication matrices

The communication matrix M shows how direct messages can be sent between four people: Arnie (A), Billie (B), Cathy (C) and Detlev (D).

$$M = \text{Sender} \begin{array}{c} A \\ B \\ C \\ D \end{array} \begin{bmatrix} 0 & 1 & 0 & 0 \\ 0 & 0 & 1 & 1 \\ 0 & 1 & 0 & 1 \\ 1 & 0 & 1 & 0 \end{bmatrix}$$

with Receiver columns $A\ B\ C\ D$

a List who each person can send direct messages to.
b Explain why the diagonal from the top left to the bottom right is all zeros.
c Draw a communication network showing the communication links given in the matrix.
d Use the network to explain how Detlev could get a message to Billie via one other person.

Steps	Working
a Starting at the first row, interpret the value of 1 as a possible communication and state who each sender can communicate with.	Arnie can send direct messages to Billie. Billie can send direct messages to Cathy and Detlev. Cathy can send direct messages to Billie and Detlev. Detlev can send direct messages to Arnie and Cathy.
b Refer to the idea of a redundant communication link in the context of the given question.	A person cannot send a direct message to themselves. It is a redundant communication link.

c **1** Draw 4 vertices, one per person.

 2 Use a directed line to represent a possible communication between two people, starting with *A* and working through to *D*.

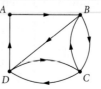

d Observe the network for a possible two-step communication from Detlev to Billie.

Detlev → Cathy → Billie

or

Detlev → Arnie → Billie

USING CAS 6 — Solving problems involving multi-step communication matrices

The matrix below represents the possible one-step communications between four computer systems.

$$M = \text{From} \begin{array}{c} \\ A \\ B \\ C \\ D \end{array} \overset{\displaystyle \text{To}}{\begin{array}{c} A\ B\ C\ D \end{array}} \left[\begin{array}{cccc} 0 & 1 & 0 & 1 \\ 1 & 0 & 1 & 0 \\ 1 & 1 & 0 & 1 \\ 1 & 0 & 1 & 0 \end{array} \right]$$

Show the use of appropriate matrix operations to find the two computer systems with the largest number of possible four-step communications, stating this number.

ClassPad

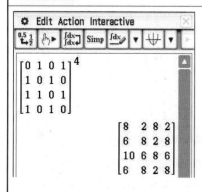

1 In **Main**, enter the **4×4** matrix.
2 Raise it to the power of **4** to find the matrix representing the possible four-step communications.
3 Observe the largest entry and interpret the result in the context of the question.

TI-Nspire

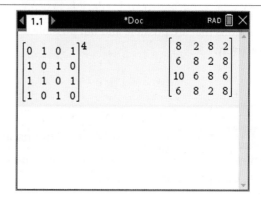

1 In a **Calculator** page, enter the **4×4** matrix.
2 Raise it to the power of **4** to find the matrix representing the possible four-step communications.
3 Observe the largest entry and interpret the result in the context of the question.

The largest number of four-step communications is 10 and this occurs in m_{31}^4. This means that the message is sent from computer *C* to computer *A*.

EXAMINATION QUESTION ANALYSIS

Calculator-free (13 marks)

Four matrices, A, B, C and D, are defined below.

$$A = \begin{bmatrix} 1 & 3 & 6 \end{bmatrix} \qquad B = \begin{bmatrix} 2 & 1 & 5 \\ -1 & x & 1 \end{bmatrix} \qquad C = \begin{bmatrix} 2 \\ 3 \end{bmatrix} \qquad D = \begin{bmatrix} y & 0 \\ 0 & 4 \end{bmatrix}$$

a State the order of B. (1 mark)

b Explain why the matrix product AB is not defined. (2 marks)

c Determine CA, stating its order. (3 marks)

d Which of the following matrix expressions is not defined? Justify your answer. (2 marks)

$CA + B \qquad B - CA \qquad A + C - B \qquad 2B + 3CA$

e If $2a_{12} + 3b_{22} - 3c_{21} = 0$, determine the value of x. (2 marks)

f If the sum of the entries in DC is 0, determine the value of y. (3 marks)

Reading the question

- Highlight the key, high-order command words, for example, *explain, justify*.
- Highlight the relevant matrix terminology, for example, *order, not defined, sum of the entries*.

Thinking about the question

- When is it possible to add and subtract matrices?
- When is it possible to multiply matrices?
- Recall how to solve simple linear equations.

Worked solution (✓ = 1 mark)

a $\dim(B) = 2 \times 3$ ✓

b The order of A is 1×3. ✓

The number of columns in A (3) does not equal the number of rows in B (2). ✓

c $CA = \begin{bmatrix} 2 \\ 3 \end{bmatrix} \times \begin{bmatrix} 1 & 3 & 6 \end{bmatrix}$ ✓

$= \begin{bmatrix} 2 & 6 & 12 \\ 3 & 9 & 18 \end{bmatrix}$ ✓

$\dim(CA) = 2 \times 3$ ✓

d $A + C - B$ is not defined. ✓

All three matrices have different sizes and so the matrix addition/subtraction is not possible. ✓

e $2(3) + 3x - 3(3) = 0$ ✓

$3x - 3 = 0$

$x = 1$ ✓

f $DC = \begin{bmatrix} y & 0 \\ 0 & 4 \end{bmatrix} \times \begin{bmatrix} 2 \\ 3 \end{bmatrix}$

$= \begin{bmatrix} 2y \\ 12 \end{bmatrix}$ ✓

$2y + 12 = 0$ ✓

$y = -6$ ✓

> **Exam hack**
>
> Always write the matrix product first before evaluating it and ensure you haven't copied down numbers incorrectly to avoid making a mistake.

EXERCISE 2.6 Modelling with matrices

ANSWERS p. 410

Recap

1 Let $A = \begin{bmatrix} -2 \\ 0 \end{bmatrix}$, $B = \begin{bmatrix} 0 & 9 \end{bmatrix}$ and $C = \begin{bmatrix} 2 \end{bmatrix}$.

Using these matrices, the matrix product that is **not** defined is

A AB **B** AC **C** BA **D** BC **E** CB

2 If $A = \begin{bmatrix} 8 & 1 \\ 4 & 2 \end{bmatrix}$ and $B = \begin{bmatrix} 3 & 12 \\ 6 & 0 \end{bmatrix}$, then matrix $AB = \begin{bmatrix} 30 & 96 \\ 24 & 48 \end{bmatrix}$.

The element '24' in the matrix AB is correctly obtained using the calculation

A $4 \times 6 + 2 \times 0$ **B** $4 \times 3 + 2 \times 6$ **C** $3 \times 4 + 12 \times 1$

D $4 \times 2 + 8 \times 2$ **E** $8 \times 3 + 1 \times 0$

Mastery

3 WORKED EXAMPLE 17 The cost prices of three different wireless earbuds in a store are $79, $199 and $399. The selling price of each pair of these wireless earbuds is 130% of the cost price.

   a Express the cost prices of the earbuds as a column matrix. Let this matrix be C.

   b Show a matrix operation involving C that will give the selling price of each pair of earbuds. Let this matrix be S.

   c On top of the selling price, a $15 commission for the sales representative needs to be included in the final retail price. Show a matrix operation involving S that will give the final retail price of each pair of earbuds. Let this matrix be R.

   d The store has a sale where earbuds with an original cost price less than $100 have their final retail price reduced by $20, and earbuds with an original cost price greater than $100 have their final retail price reduced by $45. Show a matrix operation involving R that will give the discounted price of each pair of earbuds. Let this matrix be D.

4 WORKED EXAMPLE 18 The manager of a garden supply store purchases small solar garden lights for $3 each and regular solar garden lights for $5. In the last two weeks, he purchased the number of solar garden lights shown.

	Small solar garden lights	Regular solar garden lights
Week 1	133	98
Week 2	75	62

   a Construct a 2 × 2 matrix N representing the number of different sized garden lights that were purchased over the last two weeks.

   b Construct a 2 × 1 matrix C representing the cost price of each type of garden light.

   c Show the use of an appropriate matrix multiplication to find the total cost of purchasing garden lights in each of the two weeks. State each cost.

d The manager sells goods at 155% of the cost price. He recorded his purchase costs over the last two weeks for solar garden lights and three other items in the following table.

	Week 1	Week 2
Pots	$1060	$1555
Gardening tools	$3029	$1124
Solar lights		
Wheelbarrows	$896	$2130

  i Represent the costs of all four items over the two-week period in a 4 × 2 cost matrix, T.

  ii Using scalar multiplication, represent the total amount of money that could be made from these goods (i.e. revenue) in a 4 × 2 matrix, S.

e If profit is defined as *revenue – cost price*, construct a profit matrix, P.

f Show the use of unit matrices to calculate the total profit to be made if all of the goods purchased over these two weeks are sold.

5 **WORKED EXAMPLE 19** The following communication network shows the way in which six systems interact with one another.

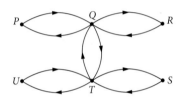

Construct a 6 × 6 communication matrix C representing this network, whereby a '1' represents a possible interaction between two systems and a '0' represents no possible interaction between two systems.

6 **WORKED EXAMPLE 20** The communication matrix M shows how direct messages can be sent between four people: Ahmed (A), Beth (B), Crystal (C) and Daniella (D).

$$M = \text{Sender} \begin{array}{c} \\ A \\ B \\ C \\ D \end{array} \overset{\text{Receiver}}{\begin{bmatrix} A & B & C & D \\ 0 & 1 & 0 & 0 \\ 1 & 0 & 1 & 1 \\ 0 & 0 & 0 & 1 \\ 1 & 0 & 1 & 0 \end{bmatrix}}$$

a List who each person can send direct messages to.
b Explain why the diagonal from the top left to the bottom right is all zeros.
c Draw a communication network showing the communication links given in the matrix.
d Use the network to explain how Crystal could get a message to Ahmed via one other person.

7 **Using CAS 6** The matrix below represents the possible one-step communications between four computer systems.

$$M = \text{From} \begin{array}{c} \\ A \\ B \\ C \\ D \end{array} \overset{\text{To}}{\begin{bmatrix} A & B & C & D \\ 0 & 0 & 1 & 0 \\ 0 & 0 & 1 & 1 \\ 1 & 1 & 0 & 1 \\ 1 & 0 & 0 & 0 \end{bmatrix}}$$

Show the use of appropriate matrix operations to find the two **different** computer systems with the largest number of possible four-step communications between them, stating this number.

Calculator-free

8 (2 marks) Give one reason why each of the following matrices could not be one-step communication matrices.

a $\begin{bmatrix} 0 & 1 & 0 \\ 1 & 0 & 0 \\ 1 & 0 & 1 \end{bmatrix}$ (1 mark)

b $\begin{bmatrix} 0 & 1 \\ 1 & 0 \\ 1 & 0 \end{bmatrix}$ (1 mark)

9 (6 marks) Consider the following two-step communication matrix between four friends, Ariana (A), Beth (B), Casper (C) and Dodge (D).

$$M^2 = \text{From} \begin{array}{c} \\ A \\ B \\ C \\ D \end{array} \begin{array}{c} \text{To} \\ \begin{array}{cccc} A & B & C & D \end{array} \\ \begin{bmatrix} 2 & 0 & 0 & 1 \\ 2 & 1 & 0 & 0 \\ 1 & 0 & 2 & 1 \\ 1 & 1 & 1 & 0 \end{bmatrix} \end{array}$$

Interpret the significance of the following entries in context of the situation:

a m^2_{23} (2 marks)

b m^2_{14} (2 marks)

c m^2_{33} (2 marks)

10 (6 marks) The diagram shows the direct communication links that exist between Sam (S), Tai (T), Umi (U) and Vera (V). For example, the arrow from Umi to Vera indicates that Umi can communicate directly with Vera.

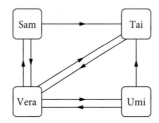

A communication matrix can be used to convey the same information.

In this matrix:

- a '1' indicates that a direct communication link exists between a sender and a receiver
- a '0' indicates that a direct communication link does not exist between a sender and a receiver.

a State the order of the communication matrix for this network. (1 mark)

b Construct the communication matrix, C, for this network. (2 marks)

c Determine $C^2 + C$ and explain its significance in the context of the situation. (3 marks)

Calculator-assumed

11 (5 marks) Peter bought only apples and bananas from his local fruit shop.

The matrix $N = \begin{bmatrix} \overset{A}{3} & \overset{B}{4} \end{bmatrix}$ lists the number of apples (A) and bananas (B) that Peter bought.

The matrix $C = \begin{bmatrix} 0.37 \\ 0.43 \end{bmatrix} \begin{matrix} A \\ B \end{matrix}$ lists the cost (in dollars) of one apple and one banana respectively.

  a Interpret the significance of the matrix product, NC, in the context of the question. (2 marks)

  b Determine NC. (1 mark)

  c Let U be a 2 × 1 unit matrix. Determine NU and state its significance. (2 marks)

12 (5 marks) A company makes Regular (R), Queen (Q) and King (K) size beds. Each bed comes in either the Classic style or the more expensive Deluxe style. The price of each style of bed, in dollars, is listed in a price matrix, P, where

$$P = \begin{bmatrix} \overset{R}{145} & \overset{Q}{210} & \overset{K}{350} \\ 185 & 270 & 410 \end{bmatrix} \begin{matrix} \text{Classic} \\ \text{Deluxe} \end{matrix}$$

The company wants to increase the price of all beds such that the price of the Classic beds are increased by 20% and the price of the Deluxe beds are increased by 35%. A new price matrix, listing the increased prices of the beds can be generated from P by forming a matrix product with the matrix M, where

$$M = \begin{bmatrix} 1.2 & 0 \\ 0 & 1.35 \end{bmatrix}$$

  a State the order in which the matrix product should be performed. Justify your answer. (2 marks)

  b Perform the matrix multiplication from part **a**. (2 marks)

  c State the new price of a Deluxe King size bed. (1 mark)

13 (6 marks) Three types of cheese, Cheddar (C), Gouda (G) and Blue (B), will be bought for a corporate function. The cost matrix P lists the prices of these cheeses, in dollars, at two stores, Idla and AGI respectively.

$$P = \begin{bmatrix} 6.80 & 5.30 & 6.20 \\ 7.30 & 4.90 & 6.15 \end{bmatrix}$$

  a State the order of matrix P. (1 mark)

  b Based on the order of P, outline what the rows and columns represent. (2 marks)

The number of packets of each type of cheese needed is listed in the quantity matrix Q.

$$Q = \begin{matrix} C \\ G \\ B \end{matrix} \begin{bmatrix} 8 \\ 11 \\ 3 \end{bmatrix}$$

  c Evaluate the matrix W = PQ and interpret the entries of the matrix in the context of the question. (2 marks)

  d Hence, justify from which of the two stores the cheese should be bought. (1 mark)

14 (3 marks) A travel company has five employees, Amara (*A*), Ben (*B*), Cheng (*C*), Dana (*D*) and Elka (*E*). The company allows each employee to send a direct message to another employee only as shown in the communication matrix *G*.

$$G = \text{Sender} \begin{array}{c} \\ A \\ B \\ C \\ D \\ E \end{array} \begin{array}{c} \text{Receiver} \\ \begin{array}{ccccc} A & B & C & D & E \end{array} \\ \begin{bmatrix} 0 & 1 & 1 & 1 & 1 \\ 1 & 0 & 1 & 0 & 0 \\ 1 & 1 & 0 & 1 & 0 \\ 0 & 1 & 0 & 0 & 1 \\ 0 & 0 & 0 & 1 & 0 \end{bmatrix} \end{array}$$

The '1' in row *E*, column *D* of matrix *G* indicates that Elka (*sender*) can send a direct message to Dana (*receiver*). The '0' in row *E*, column *C* of matrix *G* indicates that Elka cannot send a direct message to Cheng.

a To whom can Dana send a direct message? (1 mark)

b Cheng needs to send a message to Elka, but cannot do this directly due to a system failure. List the names of the employees who can send the message from Cheng directly to Elka. (2 marks)

15 (4 marks) The communication matrix below shows the direct paths by which messages can be sent between two people in a group of six people, *U* to *Z*.

$$\text{Sender} \begin{array}{c} \\ U \\ V \\ W \\ X \\ Y \\ Z \end{array} \begin{array}{c} \text{Receiver} \\ \begin{array}{cccccc} U & V & W & X & Y & Z \end{array} \\ \begin{bmatrix} 0 & 1 & 1 & 0 & 1 & 1 \\ 1 & 0 & 1 & 0 & 1 & 0 \\ 1 & 1 & 0 & 1 & 0 & 1 \\ 0 & 1 & 0 & 0 & 1 & 1 \\ 0 & 0 & 1 & 1 & 0 & 1 \\ 1 & 1 & 0 & 1 & 1 & 0 \end{bmatrix} \end{array}$$

A '1' in the matrix shows that the person named in that row can send a message directly to the person named in that column. For example, the '1' in row 4, column 2 shows that *X* can send a message directly to *V*. In how many ways can *Y* get a message to *W* by sending it directly to

a one other person? (2 marks)

b two other people? (2 marks)

16 (7 marks) The network diagram shows the possible routes that can be taken between five neighbouring towns, *A*, *B*, *C*, *D* and *E*. All routes between towns are two-way roads, meaning that the route from *A* to *B* can also be travelled from *B* to *A*, and the route from *A* to *A* can also be travelled in the opposite direction.

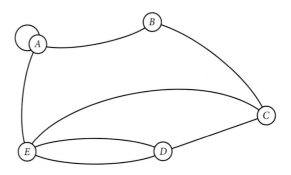

a Construct a labelled 5 × 5 matrix *R* showing the number of possible direct routes between any two towns. (3 marks)

b The matrix *R* should be symmetrical about the leading diagonal. Explain why this is the case. (1 mark)

c Evaluate R^2 and interpret the significance of r_{11}^2. (3 marks)

Chapter summary

Algebraic expressions, equations and formulas

- **Expressions** are collections of terms, either numerical or algebraic. They can be linear (e.g. $2x + 3$) or non-linear (e.g. $2x^2 + \frac{3}{x}$).

- **Equations** have two numerical or algebraic expressions equal. They can be linear, (e.g. $2x + 3 = 5x$) or non-linear, (e.g. $2x^2 + \frac{3}{x} = 1$).

- **Formulas** are equations that have a purpose. They can be single-variable or multi-variable. The variable that the formula is used to calculate is called the **subject**. For example, in $A = \frac{1}{2}bh$, A is the subject. Formulas can be expressed algebraically or in a table of values.
Spreadsheet notation can be used to write formulas, for example, $A2 = 2 \times A1 + B1$.

Introduction to matrices

- **Matrices** are rectangular arrays, using square brackets, containing information in m rows and n columns. Matrices are named using capital letters. For example, A.

  The **size**, **order**, or **dimensions** of a matrix is written as $m \times n$, and can be written as $\dim(A) = m \times n$.

  The entries of a matrix are called **elements**. A specific element in the ith row and jth column of matrix A can be identified using the notation a_{ij}.

Type	Description	Example and order
Row matrix	A matrix with one row and n columns.	$A = \begin{bmatrix} 0 & 1 & 4 \end{bmatrix}$ $\dim(A) = 1 \times 3$
Unit row matrix	A row matrix where all entries are 1s.	$B = \begin{bmatrix} 1 & 1 & 1 & 1 \end{bmatrix}$ $\dim(B) = 1 \times 4$
Column matrix	A matrix with m rows and one column.	$C = \begin{bmatrix} -2 \\ 5 \\ 1 \end{bmatrix}$ $\dim(C) = 3 \times 1$
Unit column matrix	A column matrix where all entries are 1s.	$D = \begin{bmatrix} 1 \\ 1 \end{bmatrix}$ $\dim(D) = 2 \times 1$
Square matrix	A matrix that has the same number of rows and columns. That is, an $n \times n$ matrix.	$E = \begin{bmatrix} 3 & 1 \\ 4 & 2 \end{bmatrix}$ $\dim(E) = 2 \times 2$
Zero matrix	A matrix where all entries are 0s. This is also called the **additive identity** matrix.	$F = \begin{bmatrix} 0 & 0 & 0 \\ 0 & 0 & 0 \end{bmatrix}$ $\dim(F) = 2 \times 3$
Identity matrix	A square matrix that has all entries in its **leading diagonal** as 1s, and the rest 0s. The leading diagonal of a square matrix starts from the top left entry and runs to the bottom right entry. This is also called the **multiplicative identity** matrix and is typically written as I_n where n is the number of rows (or columns).	$I_2 = \begin{bmatrix} 1 & 0 \\ 0 & 1 \end{bmatrix}$ $I_4 = \begin{bmatrix} 1 & 0 & 0 & 0 \\ 0 & 1 & 0 & 0 \\ 0 & 0 & 1 & 0 \\ 0 & 0 & 0 & 1 \end{bmatrix}$

Matrices addition, subtraction and scalar multiplication

- Matrix addition and subtraction is only defined for matrices of the same order.
  - To add and subtract matrices, add the corresponding elements together.
  - Matrix addition is commutative but subtraction is not. $A + B = B + A$

  $A - B \neq B - A$
- **Scalar multiplication** of a matrix is when a matrix is multiplied by a numerical value.
  - Multiply each element by the scalar multiple.
- A **linear combination** is a matrix expression that involves the operations of addition, subtraction or scalar multiplication.
- If two matrices A and B are equal, then $a_{ij} = b_{ij}$.

Matrix multiplication and powers

- **Matrix multiplication** is a new operation called row-by-column multiplication.
  - For the multiplication AB, multiply the elements of each row in A by the elements of each column in B and add them.

$$(1 \times 7) + (2 \times 8) + (3 \times 9) = 50$$

$$\begin{bmatrix} 1 & 2 & 3 \\ 4 & 5 & 6 \end{bmatrix} \times \begin{bmatrix} 7 & 10 \\ 8 & 11 \\ 9 & 12 \end{bmatrix} = \begin{bmatrix} 50 & \\ & \end{bmatrix}$$

$$\begin{array}{ccccc} A & \times & B & = & C \\ \text{Order:} \quad 2 \times 3 & & 3 \times 2 & & 2 \times 2 \end{array}$$

  - If A has dimensions $m \times n$ and B has dimensions $n \times p$, the AB has dimensions $m \times p$.

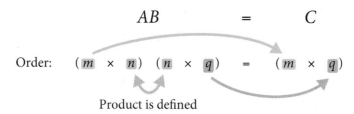

Product is defined

  - Matrix multiplication is not necessarily commutative.
- Matrices can be **pre-multiplied** or **post-multiplied**. The matrix product AB can be interpreted as:
  - B is pre-multiplied by A.
  - A is post-multiplied by B.
- Multiplying the identity matrix I by any other matrix A, results in the same matrix A.
- Square matrices can be raised to a power.
- Costing and pricing problems
  - Multiplying by unit row or unit column matrices can be used to add the entries in matrices.

Modelling with matrices

- A **communication network** diagram can be used to represent possible communication links between people or systems.
- A **communication matrix**, M, is a square matrix that shows all possible one-step communications between people or systems.
  - An entry of '1' represents a possible communication link.
  - An entry of '0' represents no possible communication.
- The M^k communication matrix represents all possible k-step communications. For example, M^2 represents all possible two-step communications.

Cumulative examination: Calculator-free

Total number of marks: 20   Reading time: 2 minutes   Working time: 20 minutes

1 (4 marks) Evaluate the following algebraic expressions for the given values.

   a $x - 2y + 5$ for $x = -1$ and $y = 2$. (2 marks)

   b $\dfrac{4p}{\sqrt{q-4}}$ for $p = 25$ and $q = 20$. (2 marks)

2 (2 marks) Evaluate the formula $V = \dfrac{2}{3}\pi r^2$ when $r = 9$, leaving your answer exact in terms of π.

3 (7 marks) Three matrices, A, B and C, are defined as follows:

$$A = \begin{bmatrix} 1 & 0 & 1 \\ 0 & x & 0 \\ 1 & 0 & 1 \end{bmatrix} \qquad B = \begin{bmatrix} 2 \\ 1 \\ 3 \end{bmatrix} \qquad C = \begin{bmatrix} 1 & 0 & 0 \\ 0 & 1 & 0 \\ 0 & 0 & y \end{bmatrix}$$

   a   State the value of y if $CB = B$. (1 mark)

   b   Determine the matrix product AB, leaving your answer in terms of x. (2 marks)

   c   If $AB = \dfrac{1}{2}\begin{bmatrix} 10 \\ 60 \\ 10 \end{bmatrix}$, determine the value of x. (1 mark)

   d   If $A + C = \begin{bmatrix} 2 & 0 & 1 \\ 0 & 12 & 0 \\ 1 & 0 & -6 \end{bmatrix}$, determine the values of x and y. (3 marks)

4 (7 marks) Apples cost \$3.50 per kg, bananas cost \$4.20 per kg and carrots cost \$1.90 per kg. Ashley buys 3 kg of apples, 2 kg of bananas and 1 kg of carrots.

   a   Represent the cost per kg of each type of fruit in a row matrix C. (2 marks)

   b   Represent the quantity of each type of fruit that Ashley buys in a column matrix, N. (2 marks)

   c   Perform an appropriate matrix multiplication to calculate the total cost of all fruit purchased by Ashley. (3 marks)

Cumulative examination: Calculator-assumed

Total number of marks: 16   Reading time: 2 minutes   Working time: 16 minutes

1 (4 marks) Raina has a job driving disabled children to school. She is paid $16.20 per hour plus $3.65 per day for assisting children. In addition, she receives 65 cents for every work-related kilometre she drives in her car. Calculate Raina's pay for a week when she worked 4 hours each day from Monday to Friday and she used her car for 360 work-related kilometres.

2 (4 marks) The values of a formula $X = \dfrac{-4 + \sqrt{16 - 4AC}}{2A}$ are shown in the table provided for specific values of A and C.

	X	\multicolumn{4}{c}{C}
		−1	0	1	2
A	−1		0		
	1		0		
	2		0		

  a Use the formula to explain why $X = 0$ whenever $C = 0$. (1 mark)

  b Copy and complete the table, correctly rounding your answers to two decimal places where appropriate. (3 marks)

3 (4 marks) Matrix F represents the social media followings on one particular platform amongst four high school students, Anton (A), Ben (B), Cassie (C) and Dana (D).

In this matrix, a '1' represents that one student follows another on the social media platform, whereas a '0' means that a student does not follow another. For example, $f_{21} = 1$ represents that Ben follows Anton on the social media platform, but $f_{23} = 0$ represents that Ben does not follow Cassie.

$$F = \text{From} \begin{array}{c} \\ A \\ B \\ C \\ D \end{array} \overset{\begin{array}{cccc} \text{To} \\ A & B & C & D \end{array}}{\begin{bmatrix} 0 & 1 & 0 & 0 \\ 1 & 0 & 0 & 1 \\ 0 & 1 & 0 & 0 \\ 0 & 0 & 1 & 0 \end{bmatrix}}$$

  a Explain, in context of the question, why the leading diagonal of the matrix F contains 0s. (1 mark)

  b Interpret the meaning of f_{43} in context of the question. (1 mark)

  c Let $U = \begin{bmatrix} 1 & 1 & 1 & 1 \end{bmatrix}$. Evaluate the matrix product UF and describe the significance of this resulting matrix in context of the question. (2 marks)

4 (4 marks) A manufacturer sells three products, A, B and C, through outlets at two shopping centres, Eaton Fair (E) and Newman Outlet (N). The number of units of each product sold per month through each shop is given by the matrix Q, where

$$Q = \begin{bmatrix} 2500 & 3400 & 1890 \\ 1765 & 4588 & 2456 \end{bmatrix} \begin{array}{c} E \\ N \end{array}$$

with columns labelled A, B, C.

  a State the dimensions of Q. (1 mark)

The matrix P shown gives the selling price, in dollars, of products A, B and C.

$$P = \begin{bmatrix} 14.50 \\ 21.60 \\ 19.20 \end{bmatrix} \begin{array}{c} A \\ B \\ C \end{array}$$

  b Explain why the matrix product PQ is not defined. (1 mark)

  c Evaluate $M = QP$ and interpret the significance of the matrix product in context of the question. (2 marks)

PYTHAGORAS' THEOREM AND MENSURATION

CHAPTER  3

Syllabus coverage
Nelson MindTap chapter resources

3.1 Measurement
Units of measurement

3.2 Pythagoras' theorem
Pythagoras' theorem in two dimensions
Pythagoras' theorem in three dimensions

3.3 Perimeter and area
Perimeter and area of quadrilaterals, triangles and circles
Arcs and sectors
Perimeter and area of composite shapes

3.4 Volume
Volume and capacity
Volumes of prisms and cylinders
Volumes of pyramids, cones and spheres

3.5 Surface area
Surface area and nets

Examination question analysis
Chapter summary
Cumulative examination: Calculator-free
Cumulative examination: Calculator-assumed

Syllabus coverage

TOPIC 1.3: SHAPE AND MEASUREMENT

Pythagoras' theorem

1.3.1 use Pythagoras' theorem to solve practical problems in two dimensions and for simple applications in three dimensions

Mensuration

1.3.2 solve practical problems requiring the calculation of perimeters and areas of circles, sectors of circles, triangles, rectangles, parallelograms and composites

1.3.3 calculate the volumes of standard three-dimensional objects, such as spheres, rectangular prisms, cylinders, cones, pyramids and composites in practical situations, for example, the volume of water contained in a swimming pool

1.3.4 calculate the surface areas of standard three-dimensional objects, such as spheres, rectangular prisms, cylinders, cones, pyramids and composites in practical situations; for example, the surface area of a cylindrical food container

Mathematics Applications ATAR Course Year 11 syllabus p. 10 © SCSA

Video playlists (6):
3.1 Measurement
3.2 Pythagoras' theorem
3.3 Perimeter and area
3.4 Volume
3.5 Surface area

Examination question analysis Pythagoras' theorem and mensuration

Skillsheets (2):
3.2 Pythagoras' theorem
3.3, 3.5 Solid shapes

Worksheets (20):
3.1 Units of length and perimeter • Length, area and volume conversions
3.2 Pythagoras' theorem time trial • Pythagorean two-step problems • Applications of Pythagoras' theorem • Pythagoras' problems
3.3 Units of length and perimeter • Area ID • Areas of composite shapes • Composite areas • A page of circular shapes • Applications of area
3.4 A page of solid shapes • Volumes of solids • Measurement in the home
3.5 Nets of solids • A page of solid shapes • Surface area of solids • Surface area • Formula matching game

Puzzle (1):
3.5 Surface area riddle

Nelson MindTap

To access resources above, visit
cengage.com.au/nelsonmindtap

3.1 Measurement

Units of measurement

The following units of measurement for length, area, volume, capacity and angles will be covered in this chapter. These diagrams show how to convert between the units of measurement.

Length units

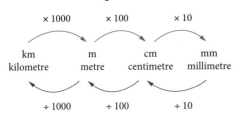

Area units (square units)

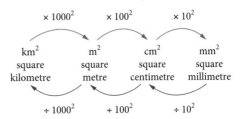

Volume units (cubic units)

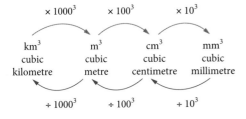

Capacity units

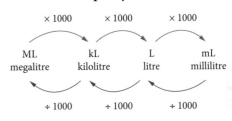

Video playlist
Measurement

Worksheets
Units of length and perimeter

Length, area and volume conversions

WORKED EXAMPLE 1	Converting units of measurement
Convert each of the following units of measurement.	
a 18 645 square centimetres to square metres	**b** 5 metres to millimetres
c 9.62 ML to mL	**d** 4200 m^3 to km^3
Steps	**Working**
a 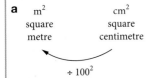	$18\,645\,cm^2 = 18\,645 \div 100^2\,m^2$ $= 1.8645\,m^2$
b 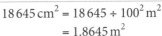	$5\,m = 5 \times 100 \times 10\,mm$ $= 5000\,mm$
c 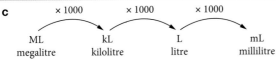	$9.62\,ML = 9.62 \times 1000 \times 1000 \times 1000\,mL$ $= 9\,620\,000\,000\,mL$
d km^3 cubic kilometre ← m^3 cubic metre ÷ 1000^3	$4200\,m^3 = 4200 \div 1000^3\,km^3$ $= 0.000\,0042\,km^3$

EXERCISE 3.1 Measurement

ANSWERS p. 412

Mastery

1 **WORKED EXAMPLE 1** Convert each of the following units of measurement.
  a 43 681 centimetres to kilometres
  b 7 cubic metres to cubic centimetres
  c 5 square metres to square centimetres
  d 600 mL to L
  e 50 m^3 to mm^3
  f 4.2 kL to L
  g 250 000 mm^2 to cm^2
  h 6800 cm^2 to m^2
  i 1.5 km to mm

Calculator-free

2 (3 marks) Convert the following to cm^2.
  a 152 m^2 (1 mark)
  b 300 m^2 (1 mark)
  c 1000 mm^2 (1 mark)

3 (3 marks)
  a How many litres are in a megalitre? (1 mark)
  b How many litres are in a kilolitre? (1 mark)
  c How many litres are in a millilitre? (1 mark)

Calculator-assumed

4 (3 marks) Convert the following to mm^2.
  a 45.3 cm^2 (1 mark)
  b 0.56 m^2 (1 mark)
  c 0.000 0071 km^2 (1 mark)

5 (3 marks) Convert the following to litres.
  a 560 mL (1 mark)
  b 12.67 kL (1 mark)
  c 0.0086 ML (1 mark)

3.2 Pythagoras' theorem

Pythagoras' theorem in two dimensions

A **right-angled triangle** is a triangle where one of the angles is 90°. **Pythagoras' theorem** is a rule for calculating the length of the longest side of a right-angled triangle given the length of the other two sides. For any right-angled triangle, we can write an equation in the form:

longest side squared = one short side squared + the other short side squared

The longest side is called the **hypotenuse**. It is always opposite the right angle.

Example of Pythagoras' theorem

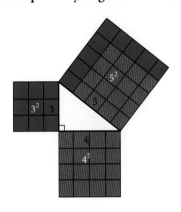

General form of Pythagoras' theorem

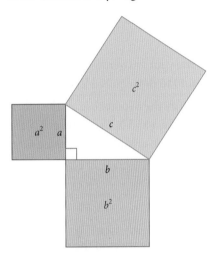

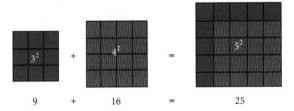

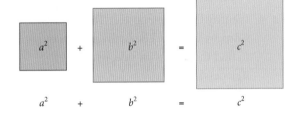

Video playlist
Pythagoras' theorem

Skillsheet
Pythagoras' theorem

Worksheets
Pythagoras' theorem time trial

Pythagorean two-step problems

Applications of Pythagoras' theorem

Pythagoras' problems

Pythagoras' theorem

In a right-angled triangle with side lengths a, b and c, where c is the hypotenuse:

$$c^2 = a^2 + b^2$$

The hypotenuse is always the side opposite the right angle.

This indicates a right angle.

🔓 Exam hack

The following side lengths often appear in Pythagoras' theorem questions:

3, 4, 5   ($3^2 + 4^2 = 5^2$)
6, 8, 10   ($6^2 + 8^2 = 10^2$)
5, 12, 13   ($5^2 + 12^2 = 13^2$)
9, 12, 15   ($9^2 + 12^2 = 15^2$)

Chapter 3 | Pythagoras' theorem and mensuration

WORKED EXAMPLE 2 — Using Pythagoras' theorem to find unknown sides

For each of the following right-angled triangles, find the unknown values, correct to two decimal places.

a 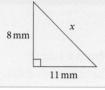   b 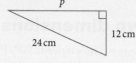   c

Steps	Working
a 1 Identify the hypotenuse, c. The other two sides are a and b. Use Pythagoras' theorem to find the unknown side, using CAS if necessary.	$c = x$, $a = 8$, $b = 11$ $c^2 = a^2 + b^2$ $x^2 = 8^2 + 11^2$ $= 185$ $x = \sqrt{185}$ $= 13.601\ldots$
2 Write your answer in the required units and round to the required level of accuracy.	$x = 13.601\ldots$ $x \approx 13.60$ mm
b 1 Identify the hypotenuse, c. The other two sides are a and b. Use Pythagoras' theorem to find the unknown side, using CAS if necessary.	$c = 24$, $a = 12$, $b = p$ $c^2 = a^2 + b^2$ $24^2 = 12^2 + p^2$ $p^2 = 24^2 - 12^2$ $= 432$ $p = \sqrt{432}$ $= 20.784\ldots$
2 Write your answer in the required units and round to the required level of accuracy.	$p = 20.784\ldots$ $p \approx 20.78$ cm
c 1 Identify the hypotenuse, c. The other two sides are a and b. Use Pythagoras' theorem to find the unknown side, using CAS if necessary.	$c = 30$, $a = x$, $b = x$ $c^2 = a^2 + b^2$ $30^2 = x^2 + x^2$ $2x^2 = 30^2$ $= 900$ $x^2 = 450$ $x = \sqrt{450}$ $= 21.213\ldots$
ClassPad 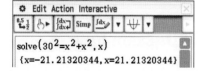	**TI-Nspire** 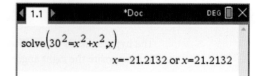
2 Write your answer in the required units and round to the required level of accuracy.	$x = 21.213\ldots$ $x \approx 21.21$ cm

WORKED EXAMPLE 3 — Using Pythagoras' theorem with shapes that contain right-angled triangles

Calculate the value of x, correct to one decimal place.

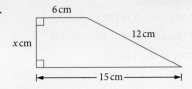

Steps	Working
1 Identify the right-angled triangle.	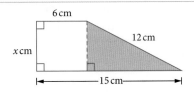
2 Label the sides of the triangle using the information given.	$15 - 6 = 9$ cm 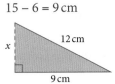
3 Identify the hypotenuse, c. The other two sides are a and b. Use Pythagoras' theorem to find the unknown side, using CAS if necessary.	$c = 12, a = x, b = 9$ $c^2 = a^2 + b^2$ $12^2 = x^2 + 9^2$ $144 = x^2 + 81$ $144 - 81 = x^2$ $x^2 = 63$ $x = \sqrt{63}$ $= 7.937\ldots$
4 Write your answer in the required units and round to the required level of accuracy.	$x = 7.9$ cm

WORKED EXAMPLE 4 — Solving problems using Pythagoras' theorem in two dimensions

A rectangular wooden gate has dimensions 3.8 m by 1.8 m. What is the length of its diagonal plank, to the nearest centimetre?

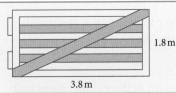

Steps	Working
1 Identify the right-angled triangle and label the unknown side x.	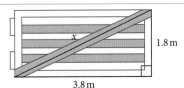
2 Use Pythagoras' theorem to find the unknown side, using CAS if necessary.	$a^2 + b^2 = c^2$ $(1.8)^2 + (3.8)^2 = x^2$ $x^2 = 17.68$ $x = 4.204\ldots$
3 Convert to the required units, round to the required accuracy, and write the answer.	$4.204\ldots$ m $= 4.204\ldots \times 100$ $= 420.4\ldots$ cm The diagonal plank has length 420 cm.

Pythagoras' theorem in three dimensions

WORKED EXAMPLE 5 — Solving problems using Pythagoras' theorem in three dimensions

For this ski jump ramp, what is the distance (in metres) a skier would travel if they skied directly up the centre of the ramp?

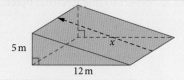

Steps	Working
1 Picture the diagram in three dimensions and find the relevant right-angled triangle with one unknown side.	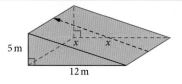
2 Redraw the right-angled triangle separately, labelling the two known sides and the one unknown side.	A right triangle with legs 5 m and 12 m, and hypotenuse x.
3 Use Pythagoras' theorem to find the unknown side, using CAS if necessary.	$c^2 = a^2 + b^2$ $x^2 = 5^2 + 12^2$ $5^2 + 12^2 = x^2$ $x^2 = 169$ $x = 13$
4 Write your answer in the required units and round to the required level of accuracy.	A skier would travel 13 m if they skied directly up the centre of the ramp.

WORKED EXAMPLE 6 — Solving two-step problems using Pythagoras' theorem in three dimensions

This rectangular box has a square base. The lengths of AH and CH are 16 cm and 7 cm respectively.

a Find the length of AC correct to two decimal places.

b Use your answer to part **a** to find x, the length of AD, correct to two decimal places.

Steps	Working
a 1 Picture the diagram in three dimensions and find the relevant right-angled triangle with one unknown side.	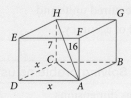
2 Redraw the right-angled triangle separately, labelling the two known sides and one unknown side.	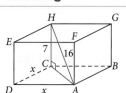
3 Use Pythagoras' theorem to find the unknown side, using CAS if necessary.	$a^2 + b^2 = c^2$ $(AC)^2 + 7^2 = 16^2$ $(AC)^2 = 16^2 - 7^2$ $(AC)^2 = 207$ $AC = 14.387\ldots$
4 Write your answer in the required units and round to the required level of accuracy.	$AC \approx 14.39$ cm

b 1 Picture the diagram in three dimensions and find the relevant right-angled triangle that includes the side whose length we found in part **a**.

2 Redraw the right-angled triangle separately, labelling the sides. Include the unrounded answer from part **a**.

3 Use Pythagoras' theorem and the unrounded answer to part **a** to find x, using CAS if necessary.

$$a^2 + b^2 = c^2$$
$$x^2 + x^2 = (AC)^2$$
$$2x^2 = (14.387\ldots)^2$$
$$2x^2 = 207$$
$$x^2 = 103.5$$
$$x = \sqrt{103.5}$$
$$= 10.173\ldots$$

4 Write your answer in the required units and round to the required level of accuracy.

$AD = 10.17\,\text{cm}$

EXERCISE 3.2 Pythagoras' theorem

ANSWERS p. 412

Recap

1 $2.49\,\text{m}^2 = $ _____ cm^2?

A 2.5 **B** 249 **C** 2490 **D** 24 900 **E** 249 000

2 $6.5\,\text{cm}^3 = $ _____ mm^3?

A 0.65 **B** 65 **C** 650 **D** 6500 **E** 65 000

Mastery

3 WORKED EXAMPLE 2 For each of the following right-angled triangles, find the unknown value, correct to two decimal places.

a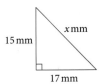

b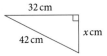

c

d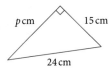

e

f

4 WORKED EXAMPLE 3 Calculate the value of x for each of the following, correct to one decimal place.

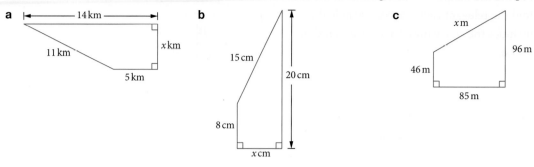

5 WORKED EXAMPLE 4

a Jordan scored a hole-in-one on the hole shown on a golf course. How far did he hit the ball? Give your answer to the nearest metre.

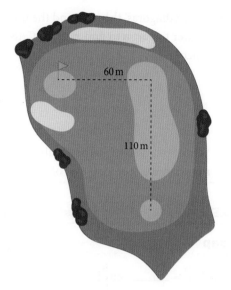

b A 6 m ladder leans against a house so that its base is 2 m from the bottom of the house. How far up the wall of the house does the ladder reach, to the nearest centimetre?

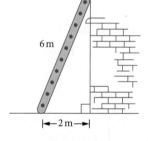

c How many metres does Alice have to walk along Hill Street from her house to get to Mary's house? Give your answer to the nearest metre.

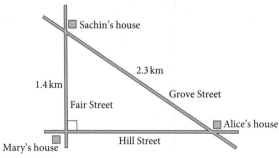

6 **WORKED EXAMPLE 5** For this snowboard jump, what is the distance (in metres) a snowboarder would travel if they snowboarded directly up the centre of the jump?

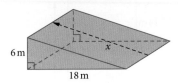

7 **WORKED EXAMPLE 6** This rectangular box has a square base. The lengths of *AH* and *CH* are 26 cm and 14 cm respectively.

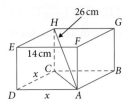

a Find the length of *AC* correct to one decimal places.

b Use your answer to part **a** to find *x*, the length of *AD*, correct to one decimal place.

Calculator-assumed

8 (2 marks) Henry flies a kite attached to a long string, as shown in the diagram.

The horizontal distance of the kite to Henry's hand is 8 m. The vertical distance of the kite above Henry's hand is 15 m. What is the length of the string, in metres?

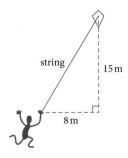

9 (3 marks) Two trees stand on horizontal ground. A 25 m cable connects the two trees at point A and point B, as shown in the diagram.

Point A is 45 m above the ground and point B is 30 m above the ground. Determine the horizontal distance, in metres, between point A and point B.

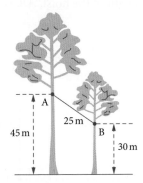

10 (3 marks) Determine the value of d, correct to two decimals places.

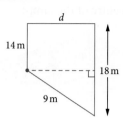

11 (3 marks) Rachel wants to use an old tennis ball container as a pencil case. If the container is a cylinder with a diameter of 7.5 cm and a height of 20 cm, and she has five pencils of the following lengths: 19.5 cm, 20 cm, 21 cm, 21.5 cm and 22 cm, how many of them are too long to fit inside the container?

12 (4 marks) Frank owns a tennis court. A diagram of his tennis court is shown. Assume that all intersecting lines meet at right angles. Frank stands at point A. Another point on the court is labelled point B.

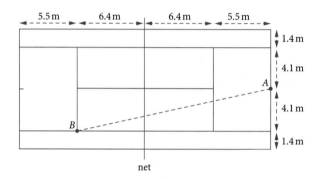

a What is the straight-line distance, in metres, between point A and point B? Round your answer to one decimal place. (2 marks)

b Frank hits a ball when it is at a height of 2.5 m directly above point A. Assume that the ball travels in a straight line to the ground at point B. What is the straight-line distance, in metres, that the ball travels? Round your answer to the nearest whole number. (2 marks)

13 (2 marks) The diagram shows a crane that is used to transfer shipping containers between a port and a cargo ship.

The length of the boom, BC, is 25 m. The length of the hoist, AB, is 15 m. Write a calculation to show that the distance AC is 20 m.

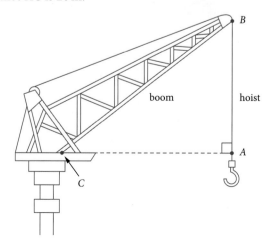

14 (8 marks) The rectangular box has the length, width and height shown. Two right-angled triangles have been drawn.

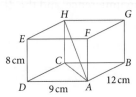

a   Copy the diagram and mark in the right angle in each of the two triangles. (2 marks)
b   Redraw △ACD separately from the three-dimensional figure and label the known side lengths. (1 mark)
c   What is the length of AC, correct to the nearest centimetre? (2 marks)
d   Redraw △ACH separately from the three-dimensional figure and label the known sides. (1 mark)
e   What is the length of AH, correct to the nearest centimetre? (2 marks)

3.3 Perimeter and area

Perimeter and area of quadrilaterals, triangles and circles

The **perimeter** of a **shape** is a measure of the total distance around the outside of the shape. The **area** of a shape is a measure of the amount of space inside the shape. For example:

The perimeter of this shape is $1 + 2 + 4 + 5 + 2 + 2 + 3 + 9 = 28$ m.

The area of this shape is 33 m^2.

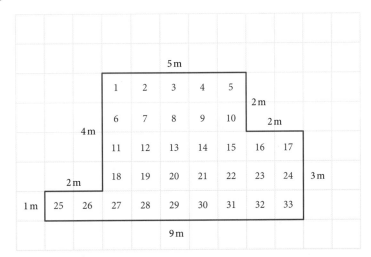

Video playlist
Perimeter and area

Worksheets
Units of length and perimeter

Area ID

We will be looking at the perimeter and area of **quadrilaterals** (shapes with four straight sides), triangles and circles. When dealing with circles we also use the word **circumference** for perimeter. The **radius** of a circle is the distance from its centre to the circumference, and the **diameter** of a circle is the distance from one side to the other through the centre.

We usually use formulas to calculate the perimeter and area of the following shapes.

Shape		Perimeter	Area
Square Four sides All sides are equal All angles are 90°		$P = 4l$	$A = l^2$
Rectangle Four sides Opposite sides are equal in length All angles are 90°		$P = 2l + 2w$	$A = lw$
Parallelogram Four sides Opposite sides are equal in length and parallel		$P = 2a + 2b$	$A = bh$
Rhombus Four sides All sides are equal in length Opposite sides are parallel		$P = 4b$	$A = bh$
Trapezium Four sides Two sides are parallel		Add the lengths of the four sides	$A = \frac{1}{2}(a + b)h$
Triangle Three sides		Add the lengths of the three sides	$A = \frac{1}{2}bh$
Circle Every point is the same distance from the centre		$C = 2\pi r$	$A = \pi r^2$

Exam hack

Sides equal in length are often indicated by dashes.
For example,

is a rectangle.

Parallel sides are often indicated by arrows.
For example,

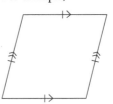

is a rhombus.

WORKED EXAMPLE 7 Calculating the perimeter and area of quadrilaterals, triangles and circles

For each of the following shapes, calculate

  i the perimeter to the nearest centimetre

  ii the area to the nearest square centimetre.

a

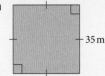

35 m

b

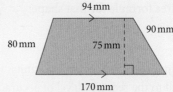

c

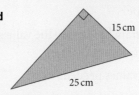

23 cm

d

15 cm

25 cm

Steps	Working
a **i** 1 Identify the shape.	square
2 State the values from the diagram. Use the perimeter formula for that shape.	$l = 35$ $P = 4l$ $= 4 \times 35$ $= 140$
3 Write your answer in the required units and round to the required level of accuracy.	$P = 140\,\text{m}$ $= 140 \times 100\,\text{cm}$ $= 14\,000\,\text{cm}$
ii 1 Use the area formula for that shape.	$A = l^2$ $= 35^2$ $= 1225$
2 Write your answer in the required units and round to the required level of accuracy.	$A = 1225\,\text{m}^2$ $= 1225 \times 100^2\,\text{cm}^2$ $= 12\,250\,000\,\text{cm}^2$
b **i** 1 Identify the shape.	trapezium
2 State the values from the diagram. Use the perimeter formula for that shape.	$a = 94$, $b = 170$, other sides = 80 and 90, $h = 75$ $P = 94 + 170 + 80 + 90$ $= 434$
3 Write your answer in the required units and round to the required level of accuracy.	$P = 434\,\text{mm}$ $= 434 \div 10\,\text{cm}$ $= 43.4\,\text{cm}$ $P = 43\,\text{cm}$

ii 1 Use the area formula for that shape.	$A = \frac{1}{2}(a+b)h$	
	$= \frac{1}{2}(94 + 170) \times 75$	
	$= 9900$	
2 Write your answer in the required units and round to the required level of accuracy.	$A = 9900 \text{ mm}^2$ $= 9900 \div 10^2 \text{ cm}^2$ $A = 99 \text{ cm}^2$	
c i 1 Identify the shape.	circle	
2 State the values from the diagram. Use the formula for that shape.	$r = 23$ $C = 2\pi r$ $= 2\pi \times 23$ $= 144.513...$	
3 Write your answer in the required units and round to the required level of accuracy.	$C = 145 \text{ cm}$	
ii 1 Use the area formula for that shape.	$A = \pi r^2$ $= \pi \times 23^2$ $= 1661.902...$	
2 Write your answer in the required units and round to the required level of accuracy.	$A = 1662 \text{ cm}^2$	
d i 1 Identify the shape.	triangle	
2 Redraw the diagram to identify the values. State the values from the diagram. Calculate other values needed. Use the perimeter formula for that shape.	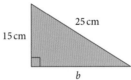 $h = 15, b = ?$ Using Pythagoras' theorem: $a^2 + b^2 = c^2$ $15^2 + b^2 = 25^2$ $b^2 = 25^2 - 15^2$ $= 400$ $b = 20$ $P = 20 + 15 + 25 = 60$	
3 Write your answer in the required units and round to the required level of accuracy.	$P = 60 \text{ cm}$	
ii 1 Use the area formula for that shape.	$A = \frac{1}{2}bh$ $= \frac{1}{2} \times 20 \times 15$ $= 150$	
2 Write your answer in the required units and round to the required level of accuracy.	$A = 150 \text{ cm}^2$	

Exam hack

Always use the π value on CAS when doing circle-related calculations.

Arcs and sectors

An **arc** is a part of the circumference of a circle formed by two radiuses. If the angle between two radiuses is known, we can calculate the **arc length**.

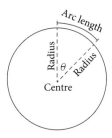

circumference of circle = $2\pi r$

arc length = $\dfrac{\theta}{360}$ of circle

$= \dfrac{\theta}{360} \times 2\pi r = \dfrac{\pi r \theta}{180}$

A **sector** is the part of a circle formed by two radiuses and the arc between them.

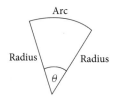

Arcs and sectors

For a sector of a circle of radius r, angle θ, and arc length l:

arc length $l = \dfrac{\pi r \theta}{180}$

sector perimeter $= 2r + l$

sector area $= \dfrac{1}{2} rl$

WORKED EXAMPLE 8 — Calculating the perimeter and area of sectors

Calculate the following correct to two decimal places for each of the sectors shown.

   i arc length   ii perimeter   iii area

a

b

Steps	Working
a i 1 State the values from the diagram. Use the arc length formula.	$r = 14,\ \theta = 30$ $l = \dfrac{\pi r \theta}{180}$ $= \dfrac{\pi \times 14 \times 30}{180}$ $= 7.330\ldots$
2 Write your answer in the required units and round to the required level of accuracy.	arc length = 7.33 cm
ii 1 State the values from the diagram. Use the sector perimeter formula, using unrounded values where necessary.	$r = 14,\ \theta = 30,\ l = 7.330\ldots$ perimeter $= 2r + l$ $= 2 \times 14 + 7.330\ldots$ $= 35.330\ldots$
2 Write your answer in the required units and round to the required level of accuracy.	perimeter = 35.33 cm
iii 1 State the values from the diagram. Use the sector area formula, using unrounded values where necessary.	$r = 14,\ \theta = 30,\ l = 7.330\ldots$ area $= \dfrac{1}{2} rl$ $= \dfrac{1}{2} \times 14 \times 7.330\ldots$ $= 51.312\ldots$
2 Write your answer in the required units and round to the required level of accuracy.	area = 51.31 cm²

b i 1 State the values from the diagram. Use the arc length formula.	$r = 23.6$, $\theta = 360 - 75 = 285$ $l = \dfrac{\pi r \theta}{180}$ $= \dfrac{\pi \times 23.6 \times 285}{180}$ $= 117.390\ldots$	
2 Write your answer in the required units and round to the required level of accuracy.	arc length = 117.39 mm	
ii 1 State the values from the diagram. Use the sector perimeter formula, using unrounded values where necessary.	$r = 23.6$, $\theta = 285$, $l = 117.390\ldots$ perimeter $= 2r + l$ $= 2 \times 23.6 + 117.390\ldots$ $= 164.590\ldots$	
2 Write your answer in the required units and round to the required level of accuracy.	perimeter = 164.59 mm	
iii 1 State the values from the diagram. Use the sector area formula, using unrounded values where necessary.	$r = 23.6$, $\theta = 285$, $l = 117.390\ldots$ area $= \dfrac{1}{2}rl$ $= \dfrac{1}{2} \times 23.6 \times 117.390\ldots$ $= 1385.212\ldots$	
2 Write your answer in the required units and round to the required level of accuracy.	area = 1385.21 mm^2	

Skillsheet
Solid shapes

Worksheets
Areas of composite shapes

Composite areas

A page of circular shapes

Applications of area

Perimeter and area of composite shapes

A **composite shape** is formed by combining two or more shapes. Use the formulas for quadrilaterals, triangles and circles to calculate the perimeter and area of composite shapes.

WORKED EXAMPLE 9 | Calculating the perimeter and area of composite shapes

For each of the following shapes, calculate the
  i perimeter
  ii area

Give your answers using the units stated in each diagram to two decimal places.

a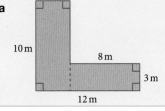

b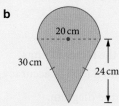

Steps	**Working**
a i 1 Identify the shapes that make up the composite shape. **2** Calculate the missing lengths.	2 rectangles 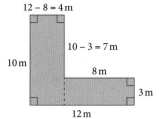
3 Add all the lengths.	$P = 10 + 4 + 7 + 8 + 3 + 12$ $= 44$
4 Write your answer in the required units and round to the required level of accuracy.	$P = 44$ m

ii 1 Separate the shapes that make up the composite shape.	rectangle 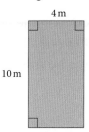	rectangle 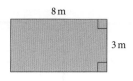
2 Use the area formulas for those shapes.	$l = 10$, $w = 4$ $A = lw$ $ = 4 \times 10$ $ = 40$	$l = 8$, $w = 3$ $A = lw$ $ = 8 \times 3$ $ = 24$
3 Add the areas. Write your answer in the required units and round to the required level of accuracy.	$A = 40 + 24$ $ = 64\,\text{m}^2$	

b **i** **1** Identify the shapes that make up the composite shape.

semicircle and triangle

2 Calculate the missing lengths.

The curve of a semicircle is half the circumference.

$r = 10$
$C = 2\pi r$
$ = 62.831\ldots$

$\dfrac{1}{2}C = \dfrac{1}{2} \times 62.831\ldots$
$\phantom{\dfrac{1}{2}C} = 31.415\ldots$

3 Add all the lengths.

$P = 30 + 31.415\ldots + 30$
$ = 91.415\ldots$

4 Write your answer in the required units and round to the required level of accuracy.

$P = 91.42\,\text{cm}$

ii 1 Separate the shapes that make up the composite shape.	semicircle 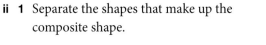	triangle 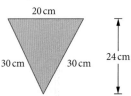
2 Use the area formulas for those shapes.	The area of a semicircle is half the area of a circle. $r = 10$ $A = \pi r^2$ $ = \pi \times 100$ $ = 314.159\ldots$ $\dfrac{1}{2}A = 157.079\ldots$	$b = 20$, $h = 24$ $A = \dfrac{1}{2}bh$ $ = \dfrac{1}{2} \times 20 \times 24$ $ = 240$
3 Add the areas. Write your answer in the required units and round to the required level of accuracy.	$A = 157.079\ldots + 240$ $ = 397.08\,\text{cm}^2$	

Some composite shapes are made by taking a piece away from a shape.

WORKED EXAMPLE 10 — Calculating the area of composite shapes where a shape is removed

Find the following shaded areas. Give your answers using the units stated in each diagram to one decimal place.

a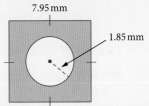

b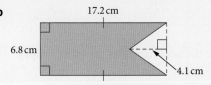

Steps	Working	
a 1 Separate the shapes that make up the composite shape.	square 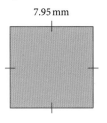	circle 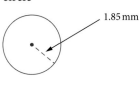
2 Use the area formulas for those shapes.	$l = 7.95$ $A = l^2$ $\quad = (7.95)^2$ $\quad = 63.2025$	$r = 1.85$ $A = \pi r^2$ $\quad = \pi \times (1.85)^2$ $\quad = 10.752\ldots$
3 Subtract the area of the empty space from the other area. Write your answer in the required units and round to the required level of accuracy.	$A = 63.2025 - 10.752$ $\quad = 52.5 \text{ mm}^2$	
b 1 Separate the shapes that make up the composite shape.	rectangle 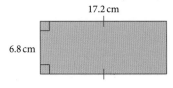	triangle
2 Use the area formulas for those shapes.	$l = 17.2, w = 6.8$ $A = lw$ $\quad = 17.2 \times 6.8$ $\quad = 116.96$	$b = 6.8, h = 4.1$ $A = \dfrac{1}{2}bh$ $\quad = \dfrac{1}{2} \times 6.8 \times 4.1$ $\quad = 13.94$
3 Subtract the area of the empty space from the other area. Write your answer in the required units and round to the required level of accuracy.	$A = 116.96 - 13.94$ $\quad = 103.0 \text{ cm}^2$	

WORKED EXAMPLE 11 — Applying perimeter and area formulas

A sports ground 140 m long and 50 m wide has semi-circular ends.

a A fence is to be built all the way around, with four gates each 2 m wide. The gates cost $150 each and the fencing costs $45 per metre. Calculate the cost of fencing the sports ground, correct to the nearest dollar.

b The sports ground needs to be covered in grass that costs $90 per square metre. Calculate the cost of the grass, correct to the nearest dollar.

Steps	Working
a 1 Decide if it is a perimeter or area problem.	perimeter
2 Sketch the shape, showing the measurements. Calculate the missing lengths.	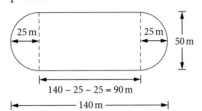
3 Add all the lengths. Add or delete any other given amounts. Write your unrounded answer in the required units.	The two semicircles together make a full circle circumference. $r = 25 \qquad C = 2\pi r$ $\qquad\qquad\quad = 2\pi \times 25$ $\qquad\qquad\quad = 157.079\ldots$ $P = 157.079\ldots + 90 + 90$ $\quad = 337.079\ldots$ m Four 2 m gates $= 4 \times 2$ $\qquad\qquad\qquad\quad = 8$ m Fence required $= 337.079\ldots - 8$ $\qquad\qquad\qquad\; = 329.079\ldots$ m
4 Calculate the cost and write the answer to the required level of accuracy.	Total cost = fence cost + gate cost $\qquad\quad\; = 329.079\ldots \times 45 + 4 \times \$150$ $\qquad\quad\; = \$15\,408.58\ldots$ The cost of fencing is $15 409.
b 1 Decide if it is a perimeter or area problem. Separate the shapes that make up the composite shape.	area 2 semicircles = 1 circle rectangle 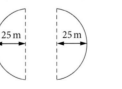 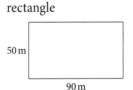
2 Use the area formulas for those shapes.	$r = 25$ $l = 90, w = 50$ $A = \pi r^2$ $A = lw$ $\; = \pi \times 25^2$ $\; = 90 \times 50$ $\; = 1963.495\ldots$ m² $\; = 4500$ m²
3 Add all the areas. Write your unrounded answer in the required units.	Total area $= 1963.495\ldots + 4500$ $\qquad\qquad\; = 6463.495\ldots$ m²
4 Calculate the cost and write the answer to the required level of accuracy.	cost $= 6463.495\ldots \times \$90$ $\qquad = \$581\,714.58\ldots$ The cost of grass is $581 715.

EXERCISE 3.3 Perimeter and area

ANSWERS p. 412

Recap

1 Which of the following is a true statement for the triangle shown?
   - A $b = 53$ (rounded to the nearest whole number)
   - B $(107)^2 + b^2 = (93)^2$
   - C $b = 142$ (rounded to the nearest whole number)
   - D $(107)^2 - b^2 = (93)^2$
   - E The length of the hypotenuse $= (107)^2 + (93)^2$

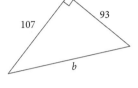

2 Which of the following is the height, h, of this ice-cream cone, correct to the nearest millimetre?
   - A 98 mm
   - B 100 mm
   - C 107 mm
   - D 108 mm
   - E 112 mm

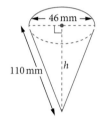

Mastery

3 **WORKED EXAMPLE 7** For each of the following shapes, calculate the
   i perimeter to the nearest centimetre
   ii area to the nearest square centimetre.

a

b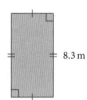

c

d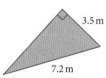

e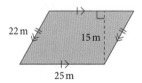

f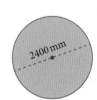

g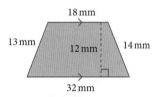

h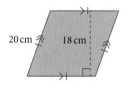

i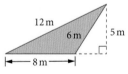

j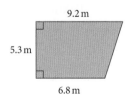

k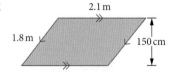

l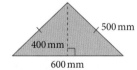

4 **WORKED EXAMPLE 8** Calculate the following, correct to two decimal places, for each of the sectors shown.

    i arc length   ii perimeter   iii area

a 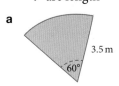   b 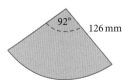   c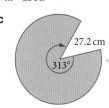

5 **WORKED EXAMPLE 9** For each of the following shapes, calculate the

    i perimeter   ii area

Give your answers using the units stated in each diagram to two decimal places.

a 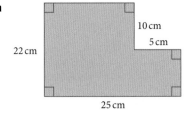   b 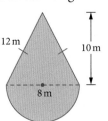   c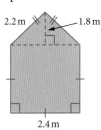

d 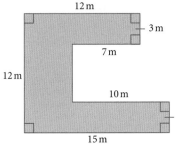   e 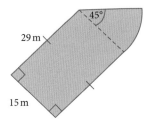   f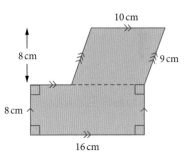

6 **WORKED EXAMPLE 10** Find the following shaded areas of each shape. Give your answers using the units stated in each diagram to one decimal place.

a 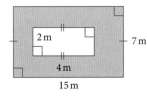   b 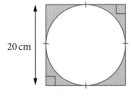   c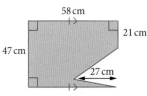

d 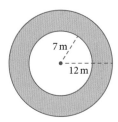   e 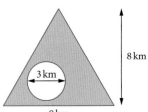   f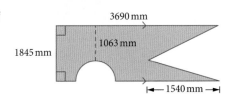

7 **WORKED EXAMPLE 11** The L-shaped island kitchen bench shown is 600 mm wide and has semi-circular ends.

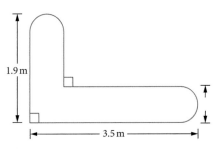

a Copy the diagram and add in the missing measurement in metres.

b The stone surface costs $1000 per square metre. Calculate the cost of the stone, correct to the nearest dollar.

c The edges around the outside of the bench will require polishing at a cost of $55 per metre. Calculate the cost of polishing the edges of the bench, correct to the nearest dollar.

Calculator-free

8 (2 marks) The four bases of a baseball field form four corners of a square of side length 27.43 m, as shown in the diagram.

A player ran from home base to first base, then to second base, then to third base and finally back to home base. Determine the minimum distance, in metres, that the player ran.

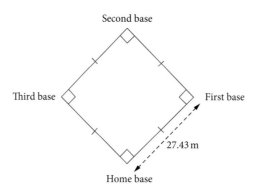

9 (2 marks) The top of a table is in the shape of a trapezium, as shown. Determine the area of the tabletop.

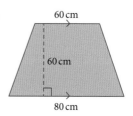

Calculator-assumed

10 (3 marks) Calculate the area of the shaded section in the diagram, in square centimetres.

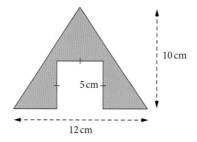

> 🔒 **Exam hack**
>
> Always check the number of marks allocated to a question and ensure you show full working for questions worth 2 or more marks.

11 (4 marks) A flag consists of three different coloured sections: red, white and blue. The flag is 3 m long and 2 m wide, as shown in the diagram.

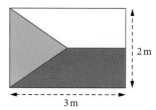

The blue section is an isosceles triangle that extends to half the length of the flag.

    a Determine the area of the blue section. (2 marks)

    b Determine the area of the red section. (2 marks)

12 (4 marks) The floor of a chicken coop is in the shape of a trapezium. The floor, ABCD, and the chicken coop are shown.

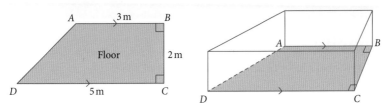

$AB = 3$ m, $BC = 2$ m and $CD = 5$ m

a What is the area of the floor of the chicken coop? Write your answer in square metres. (2 marks)

b What is the perimeter of the floor of the chicken coop? Write your answer in metres, correct to one decimal place. (2 marks)

13 (4 marks) A dairy farm is situated on a large block of land. The shaded area in the diagram represents the block of land.

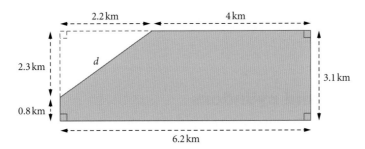

a Show that the length d is 3.2 km, rounded to one decimal place. (1 mark)

b Using $d = 3.2$, calculate the perimeter, in kilometres, of this block of land. (1 mark)

c Calculate the area of this block of land. Round your answer to the nearest square kilometre. (2 marks)

14 (5 marks) A swimming pool is set in a wooden deck, as shown, with quarter circle sections. A pool fence is to be erected around the border, with a 1200 mm wide gate. The cost of materials for a 1200 mm high fence is $42/m, and the gate costs $120. Installation costs are $10/m for the fence and gate.

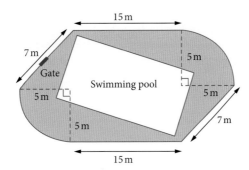

a Calculate the perimeter of the deck in metres, correct to one decimal place. (3 marks)

b Calculate the cost of materials to complete the job to the nearest dollar. (1 mark)

c Calculate the cost of installing the fence and gate to the nearest dollar. (1 mark)

3.4 Volume

Volume and capacity

The **volume** of a three-dimensional object or **solid** is the amount of space it takes up. The **capacity** of a three-dimensional object is the amount of liquid it can hold. We measure volume in cubic units based on metres, and we measure capacity in units based on litres. We can convert between the two.

1 cubic centimetre (cm^3) holds 1 millilitre (mL)

Volume and capacity conversion

When calculating capacity, calculate volume first and then convert.

$1\,cm^3$ = 1 millilitre (mL)

$1000\,cm^3$ = 1 litre (L)

$1\,m^3$ = 1000 litres (L)

Volumes of prisms and cylinders

A **prism** is a three-dimensional object with straight edges that has the same **cross-section** along its full length. A **cylinder** isn't a prism because it has curves, but it does have the same cross-section along its full length. For cylinders, the length is often called the height.

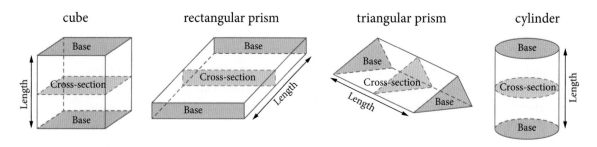

Volume of prisms and cylinders

Prisms and cylinders have the same cross-section from one **base** to the other.

volume of prisms and cylinders = base area × length

Object	Diagram	Area of base	Volume formula
cube		$A = l^2$	$V = l^3$
rectangular prism		$A = wh$	$V = whl$
triangular prism		$A = \dfrac{1}{2}bh$	$V = \dfrac{1}{2}bhl$
cylinder		$A = \pi r^2$	$V = \pi r^2 h$

WORKED EXAMPLE 12 — Calculating the volume and capacity of prisms and cylinders

Calculate the volume (V) and capacity (C) of each of the following, rounding your answer to the nearest whole unit.

a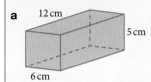

b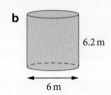

c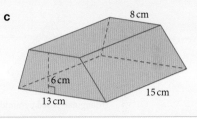

Steps	Working
a 1 Identify the object. Use a formula, or volume = base area × length if it is a prism we have no formula for.	rectangular prism $V = whl$ $w = 6, h = 5, l = 12$ $V = 6 \times 5 \times 12$ $= 360$
2 Write the volume including units and round to the required level of accuracy.	$V = 360\,\text{cm}^3$
3 Convert volume to capacity including units and round to the required level of accuracy.	$1\,\text{cm}^3 = 1$ millilitre (mL) $C = 360\,\text{mL}$

b 1 Identify the object. cylinder
   Use a formula, or $V = \pi r^2 h$
   volume = base area × length $r = 3, h = 6.2$
   if it is a prism we have no formula for. $V = \pi \times 3^2 \times 6.2$
   $= 175.300...$

 2 Write the volume including units and round $V = 175\,m^3$
   to the required level of accuracy.

 3 Convert volume to capacity including units $1\,m^3 = 1000$ litres (L)
   and round to the required level of accuracy. $C = 175\,000\,L$

c 1 Identify the object. prism with a trapezium base
   Use a formula, or $V =$ base area × length $= \frac{1}{2}(a + b)h \times l$
   volume = base area × length
   if it is a prism we have no formula for. $a = 8, b = 13, h = 6, l = 15$

   $V = \frac{1}{2}(8 + 13) \times 6 \times 15$
   $= 945$

 2 Write the volume including units and round $V = 945\,cm^3$
   to the required level of accuracy.

 3 Convert volume to capacity including units $1\,cm^3 = 1$ millilitre (mL)
   and round to the required level of accuracy. $C = 945\,mL$

Volumes of pyramids, cones and spheres

A pyramid has a base and triangular faces meeting at a point called the **apex**. Pyramids take their names from the shape of their base. For example:

triangular-based pyramid

square-based pyramid

hexagonal-based pyramid

A **cone** is similar to a pyramid except its base is a circle.

A **sphere** looks like a ball.

A **hemisphere** is half a sphere.

We can calculate the volume of pyramids and cones if we know the area of the base and the height. To calculate the volume of a sphere or hemisphere, we only need to know the radius.

Volume of pyramids, cones and spheres
Volume of pyramids and cones $= \frac{1}{3} \times$ area of base × height
Volume of a sphere $= \frac{4}{3}\pi r^3$

Object	Diagram	Area of base	Volume formula
square-based pyramid		$A = l^2$	$V = \dfrac{1}{3}l^2 h$
cone		$A = \pi r^2$	$V = \dfrac{1}{3}\pi r^2 h$
sphere		no base	$V = \dfrac{4}{3}\pi r^3$

Exam hack

If the **slant length** of a pyramid or cone is given, use Pythagoras' theorem to find the height.

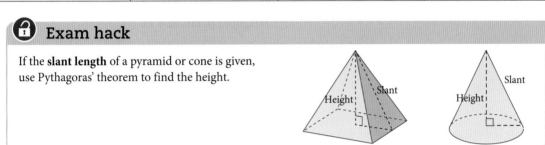

WORKED EXAMPLE 13 Calculating the volume and capacity of pyramids, cones and spheres

Calculate the volume (V) and capacity (C) of each of the following, rounding your answer to the nearest whole unit.

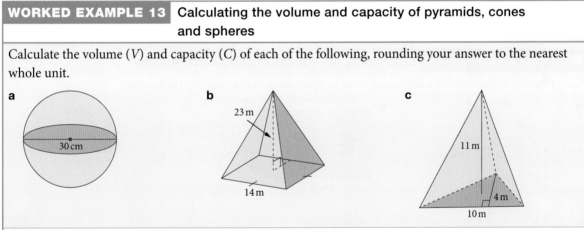

Steps	Working
a 1 Identify the object. Use the required formula to determine the volume.	sphere $V = \dfrac{4}{3}\pi r^3$ $r = 15$ $V = \dfrac{4}{3} \times \pi \times 15^3$ $= 14\,137.16\ldots$
2 Write the volume in the required units and round to the required level of accuracy.	$V = 14\,137\text{ cm}^3$
3 Convert volume to capacity including units and round to the required level of accuracy.	$1\text{ cm}^3 = 1$ millilitre (mL) $C = 14\,137\text{ mL}$

b 1 Identify the object. Use a formula, or

$\text{volume} = \frac{1}{3} \times \text{area of base} \times \text{height}$

if it is a pyramid we have no formula for.

pyramid

$V = \frac{1}{3}l^2h$

$w = 14, l = 14, h = 23$

$V = \frac{1}{3} \times 14 \times 14 \times 23$

$= 1502.666...$

2 Write the volume in the required units and round to the required level of accuracy.

$V = 1503 \, \text{m}^3$

3 Convert volume to capacity including units and round to the required level of accuracy.

$1 \, \text{m}^3 = 1000$ litres (L)

$C = 1\,503\,000 \, \text{L}$

c 1 Identify the object. Use a formula, or

$\text{volume} = \frac{1}{3} \times \text{area of base} \times \text{height}$

if it is a pyramid we have no formula for.

pyramid with a triangular base

$b = 10, h = 4$

$\text{Area of base} = \frac{1}{2}bh$

$= \frac{1}{2} \times 10 \times 4$

$= 20$

Height of pyramid = 11

$V = \frac{1}{3} \times \text{area of base} \times \text{height of pyramid}$

$= \frac{1}{3} \times 20 \times 11$

$= 73.333...$

2 Write the volume in the required units and round to the required level of accuracy.

$V = 73 \, \text{m}^3$

3 Convert volume to capacity including units and round to the required level of accuracy.

$1 \, \text{m}^3 = 1000$ litres (L)

$C = 73\,000 \, \text{L}$

WORKED EXAMPLE 14 — Calculating the volume of composite objects

Calculate the volume of the following, giving your answers to the nearest whole unit.

a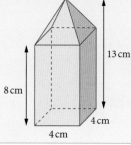
b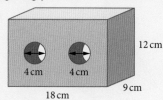
c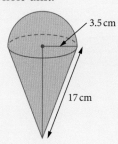

Steps	Working	
a 1 Identify the objects that make up the composite object. Use the formulas needed, calculating any missing values.	rectangular prism $w = 4, h = 8, l = 4$ $V = whl$ $= 4 \times 8 \times 4$ $= 128 \, \text{cm}^3$	pyramid 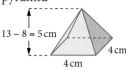 $13 - 8 = 5 \, \text{cm}$ $w = 4, l = 4, h = 5$ $V = \frac{1}{3}l^2h$ $V = \frac{1}{3} \times 4 \times 4 \times 5$ $= 26.666... \, \text{cm}^3$

2 Calculate the total volume by adding or subtracting the volumes. Write your answer in the required units and round to the required level of accuracy.

Total volume = rectangular prism volume + pyramid volume
= 128 + 26.66...
= 154.666...
= 155 cm³

b 1 Identify the objects that make up the composite object.
Use the formulas needed, calculating any missing values.

rectangular prism

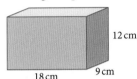

two cylinders

$w = 18, h = 12, l = 9$
$V = whl$
$V = 18 \times 12 \times 9$
$= 1944 \text{ cm}^3$

$r = 2, h = 9$
$V = \pi r^2 h$
$= \pi \times 2^2 \times 9$
$= 113.09... \text{ cm}^3$

2 Calculate the total volume by adding or subtracting the volumes. Write your answer in the required units and round to the required level of accuracy.

Total volume = rectangular prism volume − 2 × cylinder volume
= 1944 − 2 × 113.09...
= 1717.80...
= 1718 cm³

c 1 Identify the objects that make up the composite object.
Use the formulas needed, calculating any missing values.

hemisphere (half sphere)

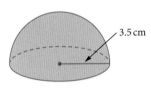

cone

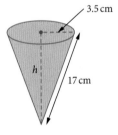

$V = \dfrac{1}{2} \times \dfrac{4}{3} \pi r^3$

$r = 3.5$

$V = \dfrac{1}{2} \times \dfrac{4}{3} \times \pi \times 3.5^3$
$= 89.797... \text{ cm}^3$

Using Pythagoras' theorem:
$17^2 = 3.5^2 + h^2$
$h = \sqrt{17^2 - 3.5^2}$
$= 16.63... \text{ cm}$

$V = \dfrac{1}{3} \pi r^2 h$

$r = 3.5, h = 16.63...$

$V = \dfrac{1}{3} \pi \times 3.5^2 \times 16.636...$
$= 213.407... \text{ cm}^3$

2 Calculate the total volume by adding or subtracting the volumes. Write your answer in the required units and round to the required level of accuracy.

Total volume = hemisphere volume + cone volume
= 89.797... + 213.407...
= 303.204...
= 303 cm³

EXERCISE 3.4 Volume

ANSWERS p. 413

Recap

1 A stove has one circular burner. This burner has a radius of 11 cm. The area of the top of this burner, in square centimetres, is closest to

   A 35   **B** 69   **C** 95   **D** 380   **E** 1521

2 A piece of cardboard is shown in the diagram.

The area of the cardboard, in square centimetres, is

   A 4
   B 5
   C 21
   D 25
   E 29

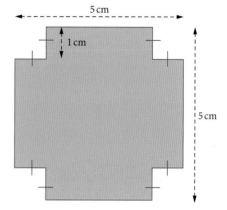

Mastery

3 WORKED EXAMPLE 12 Calculate the volume (V) and capacity (C) of each of the following, rounding your answer to the nearest whole unit.

a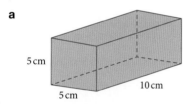

b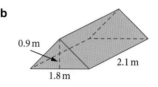

c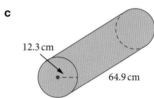

d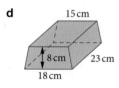

e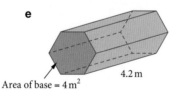
Area of base = $4\,m^2$

f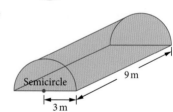

4 WORKED EXAMPLE 13 Calculate the volume (V) and capacity (C) of each of the following, rounding your answer to the nearest whole unit.

a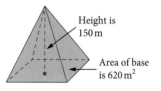
Height is 150 m
Area of base is $620\,m^2$

b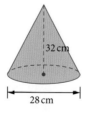

c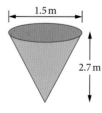

d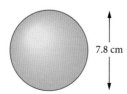

e

f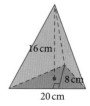

5 **WORKED EXAMPLE 14** Calculate the volume of the following, giving your answers to the nearest whole unit.

a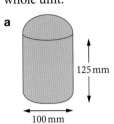

b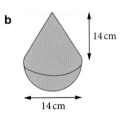

c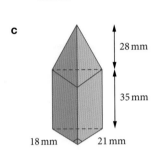

d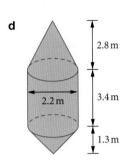

e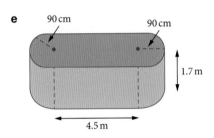

f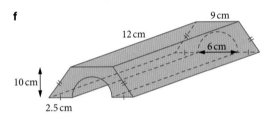

Calculator-free

6 (2 marks) A square-based pyramid is shown in the diagram. The base lengths of this pyramid are 20 cm. The slant edges of this pyramid are 20 cm. Show the calculation that gives the volume of this pyramid in cubic centimetres. (You do not need to evaluate the answer.)

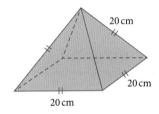

7 (2 marks) A steel beam used for constructing a building has a cross-sectional area of 0.048 m^2 as shown. The beam is 12 m long.

In cubic metres, determine the volume of this steel beam.

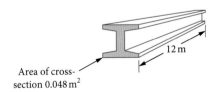

Calculator-assumed

8 (3 marks) A grain storage silo in the shape of a cylinder with a conical top is shown in the diagram.

Determine the volume of this silo.

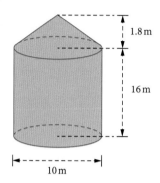

9 (3 marks) A cylindrical block of wood has a diameter of 12 cm and a height of 8 cm. A hemisphere is removed from the top of the cylinder, 1 cm from the edge, as shown.

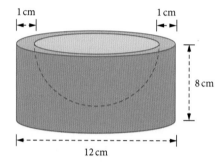

Determine the volume of the block of wood, in cubic centimetres, after the hemisphere has been removed.

10 (3 marks) A cake made using three cylindrical sections is shown in the diagram.

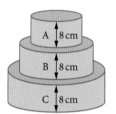

Each section of the cake has a height of 8 cm, as shown in the diagram. The middle section of the cake, B, has twice the volume of the top section of the cake, A. The bottom section of the cake, C, has twice the volume of the middle section of the cake, B. The volume of the top section of the cake, A, is 900 cm³.

Determine the diameter of the bottom section of the cake, C, in centimetres.

11 (2 marks) The building shown in the diagram is 8 m wide and 24 m long. The side walls are 4 m high. The peak of the roof is 6 m vertically above the ground. Determine the volume of the building.

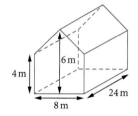

12 (5 marks) A rectangular block of land has width 50 metres and length 85 metres.

   a Calculate the area of this block of land. Write your answer in m². (1 mark)

In order to build a house, the builders dig a hole in the block of land. The hole has the shape of a triangular prism, *ABCDEF*. The width *AD* = 20 m, length *DC* = 25 m and height *EC* = 4 m are shown in the diagram.

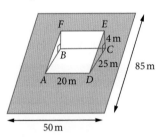

   b Calculate the volume of the triangular prism, *ABCDEF*. Write your answer in m³. (2 marks)

Once the triangular prism shape has been dug, a fence will be placed along the two sloping edges, *AF* and *DE*, and along the edges *AD* and *FE*.

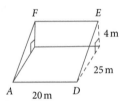

   c Calculate the total length of fencing that will be required. Write your answer, in metres, correct to one decimal place. (2 marks)

13 (4 marks) A concrete staircase leading up to a grandstand has 10 steps. The staircase is 1.6 m high and 3.0 m deep. Its cross-section comprises identical rectangles. One of these rectangles is shaded in the diagram.

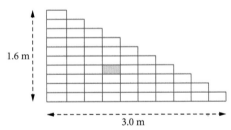

   a Find the area of the shaded rectangle in square metres. (2 marks)

The concrete staircase is 2.5 m wide.

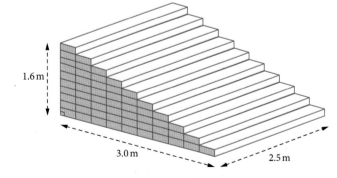

   b Find the volume of the solid concrete staircase in cubic metres. (2 marks)

3.5 Surface area

Surface area and nets

The **surface area** of a three-dimensional object is the area of all of its faces added together. Drawing or visualising the **net** of a three-dimensional object helps us to identify the shapes of all the faces. A net is a two-dimensional shape that can be folded up to form a three-dimensional object.

Three-dimensional object	Net	Surface area (SA) formula
rectangular prism	(net with $A = lw$, $A = wh$, $A = lw$, $A = wh$, $A = lh$, $A = lh$)	3 pairs of rectangles $SA = 2(lw + wh + lh)$
cylinder	(net with two circles of radius r and a rectangle of dimensions $2\pi r$ by h)	2 circles and 1 rectangle $SA = 2\pi r^2 + 2\pi rh$ $= 2\pi r(r + h)$
square-based pyramid s = slant	(net with square of side l and 4 triangles with slant s)	1 square and 4 triangles $SA = l^2 + 4 \times \left(\dfrac{1}{2}ls\right)$ $= l^2 + 2ls$
cone s = slant	(net with circle of radius r and sector with slant s)	1 circle and a sector of a circle $SA = \pi r^2 + \pi rs$ $= \pi r(r + s)$
sphere	The net of a sphere doesn't help us to calculate the surface area.	$SA = 4\pi r^2$

WORKED EXAMPLE 15 — Using surface area formulas

Calculate the surface area (SA) of each of the following, rounding your answer to two decimal places.

a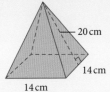

b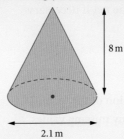

Steps	Working
a 1 Identify the object and the shapes that make up the net. Use the formula needed, calculating any missing values.	square-based pyramid 1 square and 4 triangles $SA = l^2 + 2ls$ $l = 14, s = 20$ $SA = (14)^2 + 2 \times 14 \times 20$ $= 756$
2 Write your answer in the required units and round to the required level of accuracy.	$SA = 756.00 \text{ cm}^2$
b 1 Identify the object and the shapes that make up the net. Use the formula needed, calculating any missing values.	cone 1 circle and a sector of a circle $SA = \pi r(r + s)$ $r = \dfrac{1}{2} \times 2.1 = 1.05, s = ?$ 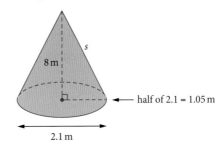 $s^2 = 1.05^2 + 8^2 = 65.1025$ $s = \sqrt{65.1025} = 8.068\ldots$ $SA = \pi r(r + s)$ $= \pi \times 1.05 \times (1.05 + 8.068\ldots)$ $= 30.079\ldots$
2 Write your answer in the required units and round to the required level of accuracy.	$SA = 30.08 \text{ m}^2$

WORKED EXAMPLE 16 — Applying surface area formulas

Calculate the surface area (SA) of each of the following, rounding your answer to two decimal places.

a an open-top cylindrical tin   b a hemisphere salad bowl with a flat lid   c a solid wooden ramp

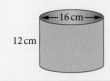 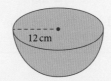 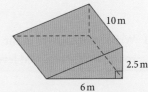

Chapter 3 | Pythagoras' theorem and mensuration

Steps	Working
a 1 Identify the object and the shapes that make up the net, drawing the net if necessary.	cylinder 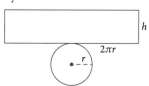
2 Adapt the longer version of the formulas needed, calculating any missing values.	1 circle and 1 rectangle $SA = \pi r^2 + 2\pi rh$ $r = 8, h = 12$ $SA = \pi(8^2) + 2\pi \times 8 \times 12$ $= 804.247\ldots$
3 Write your answer in the required units and round to the required level of accuracy.	$SA = 804.25\,\text{cm}^2$
b 1 Identify the object and the shapes that make up the net, drawing the net if necessary.	half a sphere circle and curved surface of half a sphere
2 Adapt the longer version of the formulas needed, calculating any missing values.	$SA = \pi r^2 + \dfrac{1}{2} \times 4\pi r^2$ $= \pi r^2 + 2\pi r^2$ $= 3\pi r^2$ $r = 12$ $SA = 3 \times \pi \times 12^2$ $= 1357.168\ldots$
3 Write your answer in the required units and round to the required level of accuracy.	$SA = 1357.17\,\text{cm}^2$
c 1 Identify the object and the shapes that make up the net, drawing the net if necessary.	3 rectangles and 2 triangles 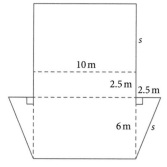 area of rectangles $= lw$ area of triangles $= \dfrac{1}{2}bh$
2 Adapt the longer version of the formulas needed, calculating any missing values.	$s = ?$ $s^2 = 2.5^2 + 6^2$ $= 42.25$ $s = \sqrt{42.25}$ $= 6.5$ $SA = 6.5 \times 10 + 2.5 \times 10 + 6 \times 10 + 2 \times \left(\dfrac{1}{2} \times 2.5 \times 6\right)$ $= 165$
3 Write your answer in the required units and round to the required level of accuracy.	$SA = 165.00\,\text{m}^2$

EXAMINATION QUESTION ANALYSIS

Calculator-assumed (10 marks)

The following diagram shows a cargo ship viewed from above.

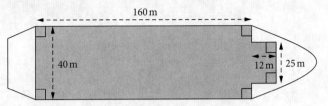

The shaded region illustrates the part of the deck on which shipping containers are stored.

a What is the area, in square metres, of the shaded region? (2 marks)

Each shipping container is in the shape of a rectangular prism.

Each shipping container has a height of 2.6 m, a width of 2.4 m and a length of 6 m, as shown in the diagram.

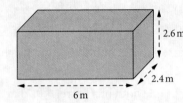

b What is the volume, in cubic metres, of one shipping container? (2 marks)

c What is the surface area, in square metres, of the outside of one shipping container? (2 marks)

One shipping container is used to carry barrels. Each barrel is in the shape of a cylinder.

Each barrel is 1.25 m high and has a diameter of 0.73 m, as shown in the diagram. Each barrel must remain upright in the shipping container.

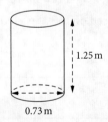

d What is the maximum number of barrels that can fit in one shipping container? (4 marks)

Reading the question

- Identify the shapes and three-dimensional objects involved.
- Be clear on whether you are using an area, volume or surface area formula.

Thinking about the question

- For part **d**, picture how upright barrels will be stacked inside the container.
- What happens in part **d** if you don't get neat whole barrels in your calculation?

> **Worked solution** (✓ = 1 mark)

a Separate the shapes into the two rectangles that make up the composite shape.

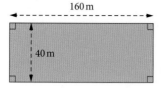

Area = 160 × 40 + 12 × 25 ✓
= 6700 m² ✓

b Use the formula for the volume of a rectangular prism.

$V = whl$
= 2.4 × 2.6 × 6 ✓
= 37.44 m³ ✓

c Use the formula for the surface area of a rectangular prism.

$SA = 2(lh + wh + lw)$
= 2(6 × 2.6 + 2.4 × 2.6 + 6 × 2.4) ✓
= 72.48 m² ✓

d Calculate the number of upright barrels that can fit the width, height and length of the shipping container.

Width of the shipping container = 2.4 m, width of a barrel = 0.73 m

So, the number of barrels that can be placed across the width = 2.4 ÷ 0.73 = 3.287... which is **3 barrels**. ✓

Length of the shipping container = 6 m, length of a barrel = 0.73 m

So, the number of barrels that can be placed along the length = 6 ÷ 0.73 = 8.219... which is **8 barrels**. ✓

Height of the shipping container = 2.6 m, height of a barrel = 1.25 m

So, the number of barrels that can be placed on top of each other = 2.6 ÷ 1.25 = 2.08 which is **2 barrels**. ✓

This means the maximum number of barrels that can fit in one shipping container = 3 × 8 × 2 = **48** ✓

EXERCISE 3.5 Surface area   ANSWERS p. 413

Recap

1 Which of the following calculations would give the volume for the pyramid shown?

A $V = \frac{1}{3} \times 8^2 \times 12$

B $V = \frac{1}{3} \times 8 \times 12^2$

C $V = \frac{1}{3} \times 8 \times 12$

D $V = \frac{1}{3} \times 4^2 \times 12$

E $V = \frac{1}{3} \times 4^2 \times 12.6$

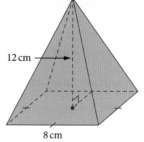

2 Sand is poured out of a truck and forms a pile in the shape of a circular cone. The diameter of the base of the pile of sand is 2.6 m. The height is 1.2 m. The volume (in m³) of sand in the pile is closest to

    A 2.1   **B** 3.1   **C** 6.4   **D** 8.5   **E** 25.5

Mastery

3 WORKED EXAMPLE 15 Calculate the surface area (SA) of each of the following, rounding your answer to two decimal places.

a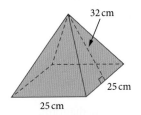

b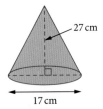

c

d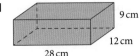

e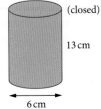

f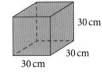

4 WORKED EXAMPLE 16 Calculate the surface area (SA) of each of the following, rounding your answer to two decimal places.

a cylindrical pipe open at both ends

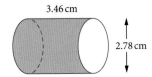

b spherical marble cut in half

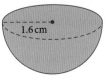

c bookend

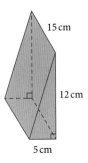

d wooden ramp

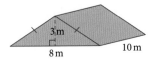

e souvenir Egyptian pyramid

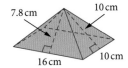

f chocolate formed by cutting a small cone off the end of a large cone

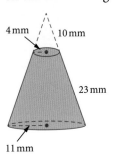

Calculator-free

5 (2 marks) The rectangular box shown in this diagram is closed at the top and at the bottom. It has a volume of $6\,m^3$. The base dimensions are $1.5\,m \times 2\,m$.

Determine the surface area of the box.

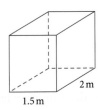

Calculator-assumed

6 (2 marks) Write the correct formula to calculate the surface area of this solid.

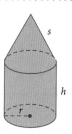

7 (2 marks) A greenhouse is built in the shape of a trapezoidal prism, as shown in the diagram. The cross-section of the greenhouse (shaded) is an isosceles trapezium. The parallel sides of this trapezium are 4 m and 10 m respectively. The two equal sides are each 5 m. The length of the greenhouse is 12 m. The five exterior surfaces of the greenhouse, **not** including the base, are made of glass. Determine the total area, in m^2, of the glass surfaces of the greenhouse.

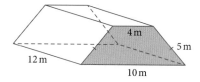

8 (3 marks) A cone with a radius of 2.5 cm is shown in the diagram.

The slant edge, x, of this cone is also shown. The volume of this cone is 36 cm^3. The surface area of this cone, including the base, can be found using the rule: surface area = $\pi r(r + x)$. Determine the surface area of this cone, including the base, in square centimetres.

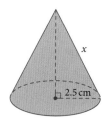

9 (3 marks) Paula has built a model house using a triangular prism on top of a rectangular box. The dimensions of the model house are shown on the diagram.

Paula will paint the outside walls and the roof of the model house. Determine the area that will be painted, in square centimetres.

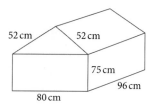

10 (3 marks) A golf ball is spherical in shape and has a radius of 21.4 mm, as shown in the diagram.

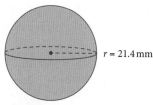

Assume that the surface of the golf ball is smooth.

  a What is the surface area of the golf ball shown? Round your answer to the nearest square millimetre. (2 marks)

  b Golf balls are sold in a rectangular box that contains five identical golf balls, as shown in the diagram.

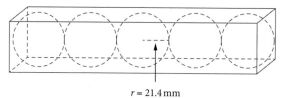

    What is the minimum length, in millimetres, of the box? (1 mark)

11 (3 marks) Shannon is a baker. One of her baking tins has a rectangular base of length 28 cm and width 20 cm. The height of this baking tin is 5 cm, as shown in the diagram.

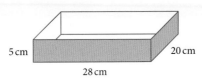

a What is the volume of this tin, in cubic centimetres? (1 mark)

Another baking tin has a circular base with a radius of 12 cm. The height of this baking tin is 8 cm, as shown in the below diagram.

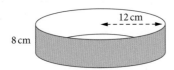

b Shannon needs to cover the inside of both the base and side of this tin with baking paper. What is the area of baking paper required, in square centimetres? Round your answer to one decimal place. (2 marks)

12 (8 marks) A shed has the shape of a prism. Its front face, $AOBCD$, is shaded in the diagram. $ABCD$ is a rectangle and M is the midpoint of AB.

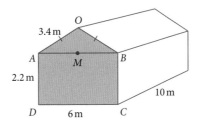

a Show that the length of OM is 1.6 m. (2 marks)
b Show that the area of the front face of the shed, $AOBCD$, is $18\,m^2$. (2 marks)
c Find the volume of the shed in m^3. (1 mark)
d All inside surfaces of the shed, including the floor, will be painted.
   i Find the total area that will be painted in m^2. (2 marks)
   One litre of paint will cover an area of $16\,m^2$.
   ii Determine the number of litres of paint that is required. (1 mark)

3 Chapter summary

Converting units of measurement

Length
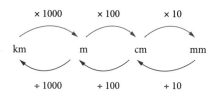

Area
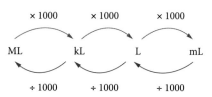

Volume
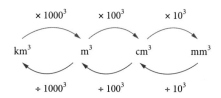

Capacity
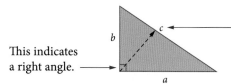

Pythagoras' theorem

- $c^2 = a^2 + b^2$, where c is the **hypotenuse** (longest side).

The hypotenuse is always the side opposite the right angle.

This indicates a right angle.

Perimeter and area

- The **perimeter** of a **shape** is a measure of the total distance around the outside of the shape.
- The **area** of a shape is a measure of the amount of space inside the shape.

Shape		Perimeter	Area
Square Four sides All sides are equal in length All angles are 90°		$P = 4l$	$A = l^2$
Rectangle Four sides Opposite sides are equal in length All angles are 90°		$P = 2l + 2w$	$A = lw$
Parallelogram Four sides Opposite sides are equal in length and parallel		$P = 2a + 2b$	$A = bh$
Rhombus Four sides All sides are equal in length Opposite sides are parallel		$P = 4b$	$A = bh$
Trapezium Four sides Two sides are parallel		Add the lengths of the four sides	$A = \frac{1}{2}(a + b)h$
Triangle Three sides		Add the lengths of the three sides	$A = \frac{1}{2}bh$
Circle Every point is the same distance from the centre		$C = 2\pi r$	$A = \pi r^2$
Sectors Part of a circle formed by two radiuses and the arc between them		$P = 2r + l$ where $l = \frac{\pi r \theta}{180}$	$A = \frac{1}{2}rl$ where $l = \frac{\pi r \theta}{180}$

Volume and capacity

- The **volume** of a three-dimensional object is the amount of space it takes up.
- The **capacity** of a three-dimensional object is the amount of liquid it can hold.
- When calculating capacity, calculate volume first and then convert:
    - $1 \text{ cm}^3 = 1$ millilitre (mL)
    - $1000 \text{ cm}^3 = 1$ litre (L)
    - $1 \text{ m}^3 = 1000$ litres (L)
- **Prisms** and **cylinders** have the same cross-section from one base to the other.
- Volume of prisms and cylinders = base area × length
- Volume of pyramids and cones = $\frac{1}{3}$ × base area × height

Volume formulas

Object	Diagram	Area of base	Volume formula
cube		$A = l^2$	$V = l^3$
rectangular prism		$A = wh$	$V = whl$
triangular prism		$A = \frac{1}{2}bh$	$V = \frac{1}{2}bhl$
cylinder		$A = \pi r^2$	$V = \pi r^2 h$
square-based pyramid		$A = l^2$	$V = \frac{1}{3}l^2 h$
cone		$A = \pi r^2$	$V = \frac{1}{3}\pi r^2 h$
sphere		No base	$V = \frac{4}{3}\pi r^3$

Surface area

- The **surface area** of a three-dimensional object is the sum of the area of all of its faces.
- Drawing or visualising the **net** of a three-dimensional object helps to identify the shapes of all the faces. A net is a two-dimensional shape that can be folded up to form the three-dimensional object.

Three-dimensional object	Net	Surface area (SA) formula
rectangular prism	Faces: $A = lh$, $A = lw$, $A = wh$, $A = lw$, $A = wh$, $A = lh$	3 pairs of rectangles $SA = 2(lw + wh + lh)$
cylinder	2 circles of radius r and a rectangle of dimensions $2\pi r$ by h	2 circles and 1 rectangle $SA = 2\pi r^2 + 2\pi rh$ $ = 2\pi r(r + h)$
square-based pyramid s = slant	Square of side l with 4 triangles of slant s	1 square and 4 triangles $SA = l^2 + 4 \times \left(\dfrac{1}{2}ls\right)$ $ = l^2 + 2ls$
cone s = slant	Circle of radius r and a sector of a circle with slant s	1 circle and a sector of a circle $SA = \pi r^2 + \pi rs$ $ = \pi r(r + s)$
sphere	The net of a sphere doesn't help us calculate the surface area.	$SA = 4\pi r^2$

Cumulative examination: Calculator-free

Total number of marks: 23   Reading time: 3 minutes   Working time: 23 minutes

1 (4 marks) The cost of fruit at a stall, in dollars per kilogram, is shown in the table.

Apples	$2.50
Pears	$3.20
Bananas	$1.90

  a Represent this information in a column matrix, C. (1 mark)

  b Sean wants to buy 2 kg of apples, 1 kg of pears and 3 kg of bananas. Represent this information in a row matrix, W. (1 mark)

  c Perform an appropriate matrix multiplication to calculate the total cost of Sean's fruit purchase. (2 marks)

2 (2 marks) Write a calculation that can be used to convert 2 cubic metres to cubic centimetres.

3 (3 marks) PQRS is a square of side length 4 m, as shown in the diagram. The distance ST is 1 m.

Determine the shaded area PQTS shown in the diagram, in m^2.

4 (2 marks) A right-angled triangle has side lengths of 8 cm and 6 cm. Determine the length of the longest side.

5 (5 marks) A water trough in the shape of a rectangular prism has a length of 1 m, a width of 2 m and a height of 0.75 metres. The outside of the prism, excluding the top, needs to be painted at a cost of $8/$m^2$.

  a Determine the surface area of the prism, excluding the top. (2 marks)

  b Determine the cost to paint the outside of the rectangular prism. (1 mark)

  c How much water in litres can the trough hold? (2 marks)

6 (2 marks) The square based pyramid, shown in the diagram, has a base area of 64 cm^2. Determine its capacity in millilitres.

7 (5 marks) The building shown in the diagram is 8 m wide and 24 m long. The side walls are 4 m high. The peak of the roof is 7 m vertically above the ground.

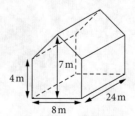

  a Determine the dimensions of the roof. (1 mark)

  b Determine the surface area of the building, not including the base. (3 marks)

One litre of paint will cover an area of 8 m^2.

  c Determine the number of litres of paint that is required to paint the outside of the entire building. (1 mark)

Cumulative examination: Calculator-assumed

Total number of marks: 37   Reading time: 4 minutes   Working time: 37 minutes

1 (14 marks) Sam is a casual junior baker at a hot bread shop. A casual junior baker earns $18.32 per hour on Monday till Friday. On Saturday all bakers receive their normal pay plus 50% and on Sunday, casual bakers receive 98% more than their normal pay per hour.

  a The table shows the times Sam worked last week. Find the missing values **i** to **xii**. (12 marks)

Shift	Starting time	Finishing time	Unpaid breaks	Number of hours worked	Pay per hour	Pay
1	Thursday 10 pm	Friday 6:30 am	30 minutes	i	v	ix
2	Saturday 12:00 am	8 am	1 hour	ii	vi	x
3	Saturday 8 pm	Sunday 12:00 am	0	iii	vii	xi
4	Sunday 6:30 pm	Monday 12:00 am	30 minutes	iv	viii	xii

  b Calculate Sam's total pay. (2 marks)

2 (4 marks) A fence has been constructed around the grassed area in the following shape.

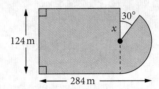

  a What is the length x shown? (1 mark)
  b What is the total length of the fence to the nearest metre? (3 marks)

3 (2 marks) Determine the perimeter of the following shape rounded to two decimal places.

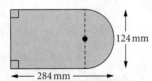

4 (7 marks) Shona is in charge of decorations for an event. She wants to hang the decorations from the ceiling. The ceiling is triangular in shape, as shown in the diagram. All three sides of the ceiling are 23 m long.

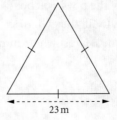

  a What is the perimeter, in metres, of the ceiling? (1 mark)

The decorations are to be hung from a beam *AB* that runs across the centre of the ceiling, as shown in the diagram.

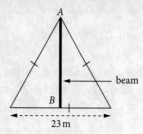

b Write a calculation that shows that the length of this beam, *AB*, rounded to one decimal place, is 19.9 m. (1 mark)

c What is the area, in square metres, of the ceiling? Round your answer to the nearest whole number. (1 mark)

Shona wants to hang spheres from the beam. Each sphere has a radius of 18 cm, as shown in the diagram below.

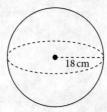

d What is the volume, in cubic centimetres, of one sphere? Round your answer to the nearest whole number. (1 mark)

Shona will place cylindrical bowls on each table at the event as a centrepiece. Each cylindrical bowl has a radius of 6 cm, as shown in the diagram.

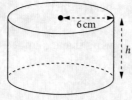

Each bowl has a volume of 1244 cm^3.

e Write a calculation that shows that the height, *h*, of one cylindrical bowl, rounded to the nearest whole number, is 11 cm. (1 mark)

f A candle, also in the shape of a cylinder, is to be placed upright inside each bowl so that it touches the base of the bowl.

The candle has a radius of 3 cm and a height of 18 cm. Once the candle has been placed inside the bowl, the remaining volume of the bowl will be filled with sand. What volume of sand, in cubic centimetres, is required to fill the cylindrical bowl once the candle is placed inside it? Round your answer to the nearest whole number. (2 marks)

5 (3 marks) A cone with a radius of 117 mm is shown in the diagram.

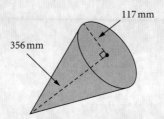

a Determine, correct to two decimal places, the slant length of the cone. (1 mark)

b Determine, correct to two decimal places, the surface area of this cone, including the base. (2 marks)

6 (3 marks) A salad bowl is in the shape of a hemisphere with a radius of 11 cm.

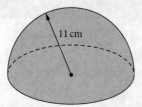

a Calculate the volume of the hemisphere to the nearest whole unit. (2 marks)

b Determine the capacity of the hemisphere in litres, rounded to one decimal place. (1 mark)

7 (4 marks) A cylindrical fuel tank is shown in the diagram.

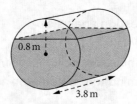

The radius of the fuel tank is 0.8 m. The length of the fuel tank is 3.8 m. The depth of fuel in the tank is 1.2 m. One thousand litres of fuel has a volume of 1 m³. Determine the amount of fuel in this tank to the nearest whole unit.

CHAPTER 4
SIMILAR FIGURES AND SIMILARITY

Syllabus coverage
Nelson MindTap chapter resources

4.1 Scale factor and similarity
Scale factor
Identifying similar figures
Proving that two triangles are similar

4.2 Area and volume of similar figures
Similar three-dimensional objects

4.3 Scale drawings
Examination question analysis
Chapter summary
Cumulative examination: Calculator-free
Cumulative examination: Calculator-assumed

Syllabus coverage

TOPIC 1.3: SHAPE AND MEASUREMENT

Similar figures and scale factors

1.3.5 review the conditions for similarity of two-dimensional figures, including similar triangles
1.3.6 use the scale factor for two similar figures to solve linear scaling problems
1.3.7 obtain measurements from scale drawings, such as maps or building plans, to solve problems
1.3.8 obtain a scale factor and use it to solve scaling problems involving the calculation of the areas of similar figures and surface areas and volumes of similar solids

Mathematics Applications ATAR Course Year 11 syllabus p. 10 © SCSA

Video playlists (4):
4.1 Scale factor and similarity
4.2 Area and volume of similar figures
4.3 Scale drawings
Examination question analysis Similar figures and similarity

Skillsheet (1):
4.1 Finding sides in similar triangles

Worksheet (1):
4.2 Areas and volumes of similar figures

Nelson MindTap

To access resources above, visit
cengage.com.au/nelsonmindtap

Video playlist
Scale factor and similarity

Skillsheet
Finding sides in similar triangles

4.1 Scale factor and similarity

Scale factor

When looking at similarity, we are looking at an increase or decrease in size of the same figure or shape.

A **scale factor** measures how much a shape needs to be enlarged or reduced to produce a **similar figure**. An enlargement or reduction may also be called a **dilation**. For example:

Circles

5 cm

10 cm

scale factor = $\dfrac{10}{5} = 2$

Rectangles

8 m
16 m

2 m
4 m

scale factor = $\dfrac{4}{16} = \dfrac{2}{8} = \dfrac{1}{4} = 0.25$

Triangles

6 cm, 10 cm, 8 cm

60 cm, 100 cm, 80 cm

scale factor = $\dfrac{100}{10} = \dfrac{80}{8} = \dfrac{60}{6} = 10$

Similar figures and scale factors

k = scale factor for similar figures = $\dfrac{\text{any length of second shape}}{\text{matching length of first shape}}$

$k > 1$ gives a larger second shape

$k < 1$ gives a smaller second shape

WORKED EXAMPLE 1 Working with scale factors

For each of these pairs of similar figures, find the

  i scale factor, k

  ii value of x, rounded to one decimal place.

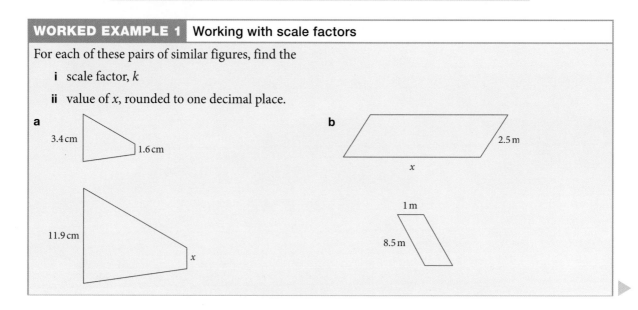

a
3.4 cm
1.6 cm

11.9 cm
x

b
2.5 m
x

1 m
8.5 m

154 Nelson WAmaths Mathematics Applications 11

Steps	Working
a i Find a pair of matching lengths and use $k = \dfrac{\text{any length of second shape}}{\text{matching length of first shape}}$.	$k = \dfrac{11.9}{3.4} = 3.5$
ii Use the scale factor to find and solve an equation for the unknown, using CAS if necessary. Write your answer to the required accuracy.	$\dfrac{x}{1.6} = 3.5$ $x = 3.5 \times 1.6$ $ = 5.6\,\text{cm}$
b i Find a pair of matching lengths and use $\text{scale factor} = \dfrac{\text{any length of second shape}}{\text{matching length of first shape}}$.	$k = \dfrac{1}{2.5} = 0.4$
ii Use the scale factor to find and solve an equation for the unknown, using CAS if necessary. Write your answer to the required accuracy.	$\dfrac{8.5}{x} = 0.4$ $0.4x = 8.5$ $x = \dfrac{8.5}{0.4}$ $ = 21.3\,\text{m}$

  Exam hack

Don't round your answer if you're not asked to round it.

Identifying similar figures

We can't always tell if two figures are similar just by looking at them. It depends on the shape and often the angles within the shape. The symbol for similar figures is ~.

Circles	Squares	Rectangles
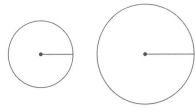		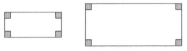
All circles are similar. The radius is the only length measurement needed.	All squares are similar. The scale factors for each matching pair of sides will be the same because all sides in each figure are of equal length.	Rectangles are similar if the two pairs of matching sides have the same scale factor.

All figures with straight sides (polygons)

 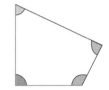   

Polygons are similar if every pair of matching sides have the same scale factor *and* all pairs of matching angles are equal.

WORKED EXAMPLE 2 | Identifying similar figures

State whether the following pairs of figures are similar, giving a reason.

a 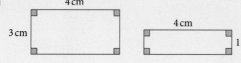 b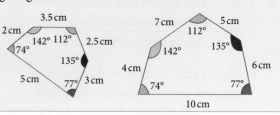

Steps	Working
a 1 Identify the shape.	rectangle
2 Calculate the scale factors of the matching sides in order, starting with the largest values in each shape. If necessary, look at the angles.	$\frac{4}{4} = 1, \frac{1}{3} = 0.333...$ The scale factors are not the same, so the figures aren't similar.
b 1 Identify the shape.	polygon
2 Calculate the scale factors of the matching sides in order, starting with the largest values in each shape. If necessary, consider the angles.	$\frac{10}{5} = \frac{7}{3.5} = \frac{6}{3} = \frac{5}{2.5} = \frac{4}{2} = 2$ The scale factors are the same. The pairs of matching angles are equal in each shape, so the figures are similar.

🔓 Exam hack

Questions sometimes present two similar figures like this where the smaller and larger figures are in the one diagram.

  is the same as

Proving that two triangles are similar

The following tests can be used to show whether two triangles are similar.

Triangles

   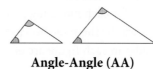   

Side-Side-Side (SSS) or **Angle-Angle (AA)** or **Side-Angle-Side (SAS)**

Triangles are similar if the three pairs of matching sides have the same scale factor. | Triangles are similar if two pairs of matching angles are equal. | Triangles are similar if two pairs of matching sides have the same scale factor and the angles between them are equal.

🔓 Exam hack

Unless the question states that two figures are similar, you cannot assume they are. You may need to prove they are similar before finding a missing length.

By using the tests SSS, AA or SAS, we can show that two triangles are similar. When two triangles are similar, we write a similarity statement such as △ABC ~ △DEF.

> **Exam hack**
>
> When writing a similarity statement, take note of the order of the letters so that the corresponding angles match.

WORKED EXAMPLE 3 Proving two triangles are similar

Show that the following triangles are similar using either SSS, AA or SAS.

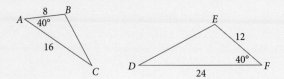

Steps	Working
1 Calculate the scale factors of the matching sides in order, starting with the largest values in each shape.	$\frac{24}{16} = 1.5, \frac{12}{8} = 1.5$ The scale factors are the same.
2 Determine which test proves they are similar. If necessary, look at the angles.	Use SAS. The angle between the two sides is 40° for both triangles and the corresponding sides are in the same ratio, so the triangles are similar.
3 Write the similarity statement.	Write the similarity statement, taking note of the order of the letters so that the corresponding angles match. △ABC ~ △FED

EXERCISE 4.1 Scale factor and similarity ANSWERS p. 414

Mastery

1 **WORKED EXAMPLE 1** For each of these pairs of similar figures, find the

   i scale factor, k

   ii value of x, rounded to one decimal place.

a   b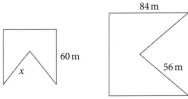

c 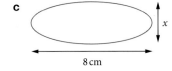   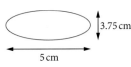   d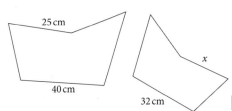

Chapter 4 | Similar figures and similarity 157

2 **WORKED EXAMPLE 2** State whether the following pairs of figures are similar, giving a reason.

a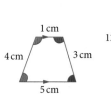

b

c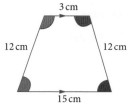

3 **WORKED EXAMPLE 3** Show that the following pairs of triangles are similar.

a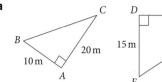

b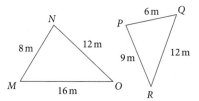

c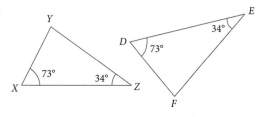

Calculator-free

4 (2 marks) Triangle ABC is similar to triangle DEF.

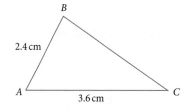

a Determine the scale factor. (1 mark)
b Determine the length of DF. (1 mark)

5 (2 marks) Triangle M, shown here, has side lengths of 3 cm, 4 cm and 5 cm.

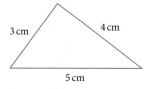

Four other triangles have the following side lengths:
- Triangle N has side lengths of 3 cm, 6 cm and 8 cm.
- Triangle O has side lengths of 4 cm, 8 cm and 12 cm.
- Triangle P has side lengths of 6 cm, 8 cm and 10 cm.
- Triangle Q has side lengths of 9 cm, 12 cm and 15 cm.

Which of the above triangles are similar to triangle M?

Calculator-assumed

6 (4 marks) In the diagram, $AD = 9$ cm, $AC = 24$ cm and $DB = 27$ cm. Line segments AC and DE are parallel.

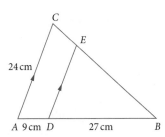

- **a** Show that the triangles are similar. (2 marks)
- **b** Determine the scale factor. (1 mark)
- **c** Determine the length of DE. (1 mark)

7 (2 marks) Two cones, as shown, have the same angle at the base. The larger cone has a slant length of 15 cm and the smaller cone has a slant length of 12 cm. The diameter of the larger cone is 9 cm.

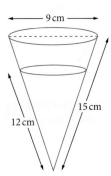

- **a** Determine the scale factor. (1 mark)
- **b** What is the diameter of the smaller cone? (1 mark)

8 (3 marks) A triangle has sides of length 20 cm, 48 cm and 52 cm. A second triangle which is similar to the first triangle has a longest side of 65 cm.

- **a** Determine the scale factor and dimensions of the second triangle. (2 marks)
- **b** What is the perimeter of the second triangle? (1 mark)

9 (2 marks) A torch, which is held horizontally, is shone on to a wall from a distance of 1.2 metres as shown. The circular area of light it creates on the wall has a radius of 8 centimetres.

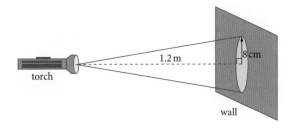

The torch is now moved an additional 2 metres away from the wall.

- **a** What is the scale factor of the two triangles formed by the torch? (1 mark)
- **b** What is the radius of the circular area of light on the wall now, rounded to one decimal place? (1 mark)

4.2 Area and volume of similar figures

Similar three-dimensional objects

Similarity works the same way for three-dimensional objects. Three-dimensional objects are similar if they are exactly the same type and the scale factor of pairs of matching lengths are the same. For example, these pairs are all similar:

Rectangular prisms

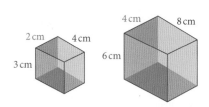

scale factor $= \dfrac{8}{4} = \dfrac{6}{3} = \dfrac{4}{2} = 2$

Cylinders

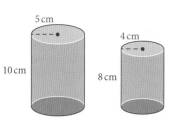

scale factor $= \dfrac{8}{10} = \dfrac{4}{5} = 0.8$

Cones

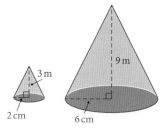

scale factor $= \dfrac{9}{3} = \dfrac{6}{2} = 3$

Scaling areas and volumes

When a shape is scaled by a factor of k:

area of second shape
$= k^2 \times$ area of first shape

When a three-dimensional object is scaled by a factor of k:

surface area of second object
$= k^2 \times$ surface area of first object

volume of second object
$= k^3 \times$ volume of first object

where

$k = \dfrac{\text{any length of second shape/object}}{\text{matching length of first shape/object}}$

WORKED EXAMPLE 4 — Scaling areas and volumes

A square-based pyramid with a base length of 4 cm is enlarged to produce a similar pyramid with a base length of 10 cm.

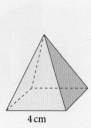 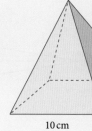

4 cm　　　10 cm

Find the

a scale factor, k

b surface area of the larger pyramid if the surface area of the smaller pyramid is 64 cm².

c volume of the smaller pyramid if the volume of the larger pyramid is 32 cm³.

Steps	Working
a Use $k = \dfrac{\text{any length of second object}}{\text{matching length of first object}}$.	$k = \dfrac{10}{4} = 2.5$
b 1 Use surface area of second object $= k^2 \times$ surface area of first object.	Surface area of larger pyramid $= k^2 \times$ surface area of smaller pyramid $= (2.5)^2 \times 64$ $= 400$
2 Write your answer in the required units and round to the required level of accuracy.	The surface area of the larger pyramid is $400\,\text{cm}^2$.
c 1 Use volume of second object $= k^3 \times$ volume of first object.	Volume of larger pyramid $= k^3 \times$ volume of smaller pyramid $32 = (2.5)^3 \times$ volume of smaller pyramid $\text{Volume of smaller pyramid} = \dfrac{32}{(2.5)^3} = 2.048$
2 Write your answer in the required units and round to the required level of accuracy.	The volume of the smaller pyramid is $2.05\,\text{cm}^3$.

EXERCISE 4.2 Area and volume of similar figures ANSWERS p. 414

Recap

1 A triangle has sides of length 15 cm, 32 cm and 40 cm. A second triangle which is similar to the first triangle has a longest side of 60 cm. What is the perimeter of the second triangle?

2 The equilateral triangles below are similar. If the scale factor is 1.5, what are the dimensions of the second triangle?

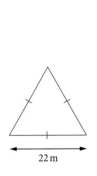
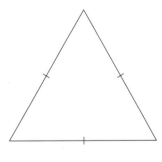

22 m

Mastery

3 WORKED EXAMPLE 4 For each of the following pairs of similar three-dimensional objects, find the
  i scale factor, k.
  ii surface area of the smaller object if the surface area of the larger object is $48\,\text{cm}^2$, to one decimal place.
  iii volume of the larger object if the volume of the smaller object is $937.5\,\text{cm}^3$.

  a Two similar cylinders with heights 24 cm and 36 cm respectively.

  b Two similar cones with bases of radius 24 cm and 20 cm respectively.

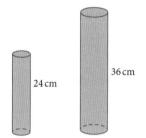
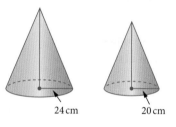

Calculator-free

4 (2 marks) The length and width of a rectangular photograph were increased by a scale factor of $k = 4$. What factor is the area of the photograph increased by?

5 (2 marks) An art piece has been designed in the shape of a triangular prism. The model has a surface area of $100\,\text{cm}^2$. If the lengths are increased by a scale factor of $k = 2$ for the actual piece, what is the surface area of the art piece?

Calculator-assumed

6 (2 marks) Cube A and cube B are shown. The side length of cube A is 1.5 times larger than the side length of cube B.

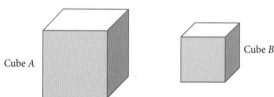

  a The surface area of cube B is $256\,\text{cm}^2$. What is the surface area of cube A? (1 mark)
  b The volume of cube A is $942\,\text{cm}^3$. What is the volume of cube B, correct to the nearest whole number? (1 mark)

7 (2 marks) A model car is a scaled down version of a real car. The real car is 20 times the length of the model car.
  a If it takes 15 mL of paint to cover the model, how much paint is needed to paint the real car? Give your answer in litres. (1 mark)
  b The fuel tank of the real car holds 60 L. What would be the capacity of the fuel tank on the model? Give your answer in millilitres. (1 mark)

8 (4 marks) A rectangular prism has a volume of $300\,\text{cm}^3$. A second rectangular prism has a volume of $2400\,\text{cm}^3$.
  a Determine the scale factor. (2 marks)
  b The surface area of the second rectangular prism is $1085\,\text{cm}^2$. What is the surface area of the smaller rectangular prism? (2 marks)

4.3 Scale drawings

Scales on house plans or maps are usually written in a ratio, for example, 1 : 100, or given by a statement such as '1 cm represents 2 km'. The first term refers to the scaled length on the plan or map and the second term refers to the actual length.

To convert between the scaled length and real length, we need to multiply the measurement by the scale factor and convert the answer to the required units.

WORKED EXAMPLE 5 Reading a scale diagram

The bedroom below is drawn to a scale of 1:120.

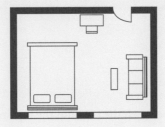

a Determine the internal dimensions of the room in metres.
b If new carpet is installed at a cost of $55/ m², determine the cost of the carpet.

Steps	Working

a 1 Measure the length of the room on the diagram and find the real length by multiplying by the scale.	scaled length = 36 mm
	real length = 36 × 120
	       = 4320 mm
	       = 4.32 m

2 Measure the width of the room on the diagram and find the real width by multiplying by the scale.	scaled width = 26 mm
	real width = 26 × 120
	      = 3120 mm
	      = 3.12 m

3 Write the answer	The dimensions of the room are 4.32 m by 3.12 m.

b 1 Calculate the area of the room using the real dimensions.	area = $l \times w$
	   = 4.32 × 3.12
	   = 13.4784 m²

2 Calculate the cost of the carpet.	cost = 13.4784 × 55 = $741.31
3 Write the answer.	The cost to install the carpet is $741.31.

Video playlist
Scale drawings

Video
Examination question analysis: Similar figures and similarity

EXAMINATION QUESTION ANALYSIS

Calculator-assumed (6 marks)

A flying fox is constructed between the top of a tree and a pile of stones. A post is put into the ground 1.5 m from the pile of stones. The length of cable from the top of the tree to the top of the post is 36 m. The length of cable from the pile of stones to the top of the post is 2 m.

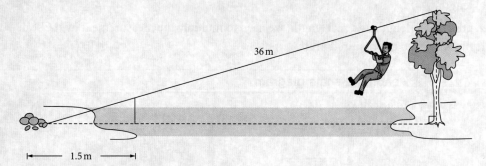

a Show the two triangles are similar. (2 marks)

b Determine the scale factor. (2 marks)

c What is the (horizontal) distance between the tree and the base of the post? (2 marks)

Reading the question

- Identify the triangles and determine which test proves they are similar.
- Determine the corresponding sides for the scale factor.

Thinking about the question

- Use SAS, SSS or AA to prove the triangles are similar.
- For part **b**, think about the length of the hypotenuse for both triangles.

Worked solution (✓ = 1 mark)

a Similar due to AA. ✓

The angle at the stones is the same. **Both have a corresponding 90° angle.** ✓ The third angle is corresponding and is therefore also equal.

b Find the length of the hypotenuse for both triangles. For the larger triangle, the length of the flying fox is 36 m plus the 2 m for the length of cable from the stones to the post. The length of the hypotenuse of the larger triangle = **38 m**. ✓

The hypotenuse of the smaller triangle has a length of 2 m.

Scale factor = $\dfrac{38}{2}$ = **19** ✓

c $1.5 \times 19 = 28.5$ m ✓

The distance from the tree to the base of the post is $28.5 - 1.5 = $ **27 m**. ✓

EXERCISE 4.3 Scale drawings

ANSWERS p. 414

Recap

1 The length, width and height of a rectangular box was increased by a scale factor of $k = 3$. What factor is the volume of the box increased by?

2 A triangle with an area of 40 cm² has its lengths increased by a scale factor of 1.5. What is the area of the new triangle?

Mastery

3 WORKED EXAMPLE 5 The plan of a one bedroom apartment is drawn to a scale of 1 : 110.

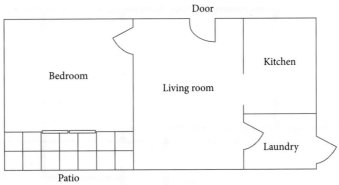

Scale 1 : 110

  a Determine the internal dimensions of the bedroom in metres.
  b If new carpet is to be installed at a cost of $52/ m², determine the cost to install the carpet.

Calculator-free

4 (4 marks) This house plan has a scale of 1 : 100.

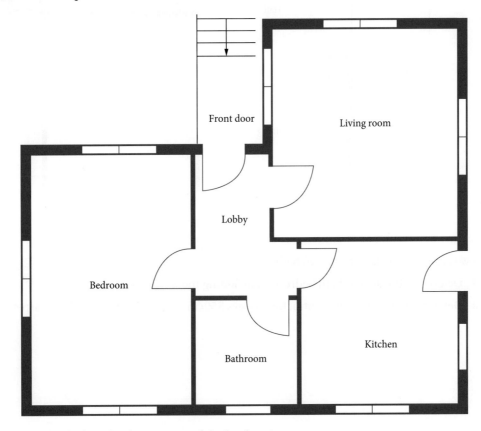

  a Measure and calculate the dimensions of the kitchen in metres. (2 marks)
  b New tiles are to be laid in the kitchen.
   i Determine the area of the kitchen. (1 mark)
   ii Determine the cost if the contractor charges $60/ m² to supply and install the tiles. (1 mark)

Calculator-assumed

5 (5 marks) This house plan has a scale of 1 : 120.

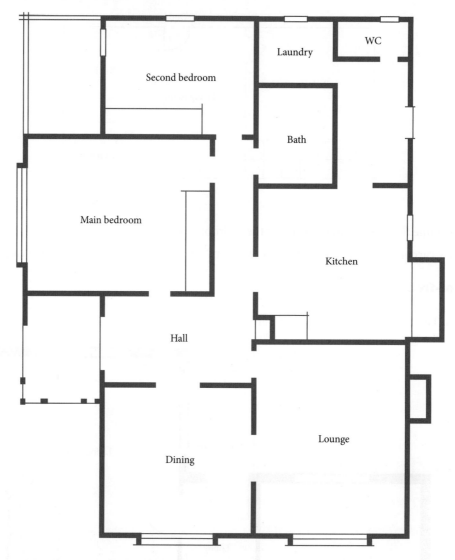

a Measure and calculate the dimensions of both bedrooms. (2 marks)

b New carpet is to be laid in the two bedrooms.

  i Determine the area of both bedrooms, including the wardrobes. (2 marks)

  ii Determine the cost if the contractor charges $50/ m^2 to supply and install the carpet. (1 mark)

6 (4 marks) The following house plans are shown with the scale included in the diagram.

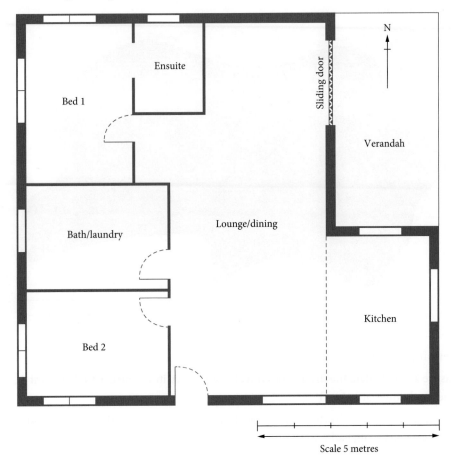

a If the fridge, stove, and kitchen cupboards take up $\frac{1}{3}$ of the area in the kitchen, determine the remaining floor space, to one decimal place. (2 marks)

b If the tiles to be used in the kitchen measure 30 cm by 30 cm. How many tiles would need to be purchased to cover the total floor space? Express your answer as a whole number. (2 marks)

7 (6 marks) The following diagram, drawn to scale, shows the plan for Farmer Ben's field. The width of the field is 600 m.

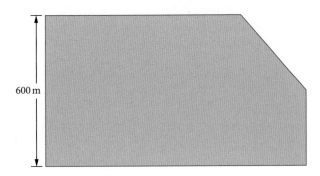

a Express the scale factor of the drawing in the form $1:x$. (2 marks)
b Determine the remaining dimensions of the field to the nearest metre. (2 marks)
c If 1 hectare = $10\,000\,m^2$, calculate the size of the field in hectares. (2 marks)

8 (5 marks) The following shows a map of Mackerel Island, which lies off the coast of Western Australia.

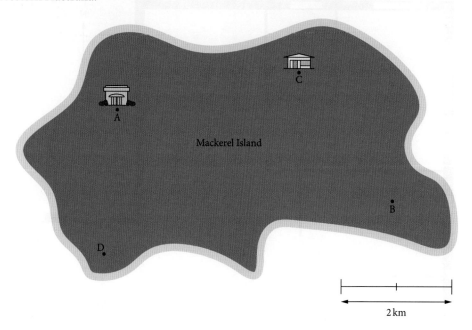

a Determine the scale as a ratio. (1 mark)

b Measure and calculate the distance to travel from the visitor centre, labelled point A, to the sand dunes at point B. (2 marks)

c Measure and calculate the distance to travel from the accommodation, labelled point C, to the lighthouse, labelled point D. (2 marks)

Chapter summary

Similarity and scale

- Two figures are **similar** if they have exactly the same shape but are different sizes.
- k = **scale factor** for similar figures = $\dfrac{\text{any length of second shape}}{\text{matching length of first shape}}$

  $k > 1$ gives a larger second shape

  $k < 1$ gives a smaller second shape

Circles	Squares	Rectangles

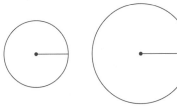   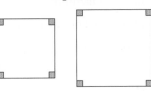   

All circles are similar. The radius is the only length measurement needed. | All squares are similar. The scale factors for each matching pair of sides will be the same because all sides in each shape are of equal length. | Rectangles are similar if the two pairs of matching sides have the same scale factor.

All figures with straight sides (polygons)

 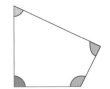 

Polygons are similar if every pair of matching sides have the same scale factor *and* all pairs of matching angles are equal.

Similar triangles

Triangles

   or   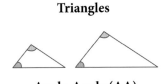   or

Side-Side-Side (SSS)   **Angle-Angle (AA)**   **Side-Angle-Side (SAS)**

Triangles are similar if the three pairs of matching sides have the same scale factor. | Triangles are similar if two pairs of matching angles are equal. | Triangles are similar if two pairs of matching sides have the same scale factor and the angles between them are equal.

The symbol for similar figures is ~. When two triangles are similar, we write a similarity statement such as $\triangle ABC \sim \triangle DFE$. Take note of the order of the letters so that the corresponding angles match.

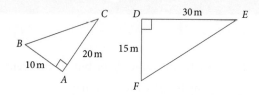

Area and volume of similar figures

- When a shape is scaled by a factor of k:

    area of second shape = k^2 × area of first shape

- When a three-dimensional object is scaled by a factor of k:

    surface area of second object = k^2 × surface area of first object

    volume of second object = k^3 × volume of first object

    where

    $$k = \frac{\text{any length of second shape/object}}{\text{matching length of first shape/object}}.$$

Scale drawings

Scales on house plans or maps are usually written in a ratio, for example, 1 : 200. To convert between the scaled length and real length, we need to multiply the measurement by the scale factor and convert the answer to the required units.

Cumulative examination: Calculator-free

Total number of marks: 24   Reading time: 3 minutes   Working time: 24 minutes

1 (5 marks) Let $P = \begin{bmatrix} 3 \\ 5 \\ -1 \end{bmatrix}$ and $Q = \begin{bmatrix} 2 & -3 & -4 \end{bmatrix}$.

    a Explain why $P + Q$ is not defined. (1 mark)

    b Which of the two matrix products, PQ or QP, are defined? Justify your answer. (2 marks)

    c Show how the matrices P and Q can be multiplied to obtain a resulting matrix with dimensions 3×3. (2 marks)

2 (3 marks) Convert the following to cm^3.

    a $6000 \, mm^3$ (1 mark)

    b $8.7 \, m^3$ (1 mark)

    c $0.000\,005\,6 \, km^3$ (1 mark)

3 (3 marks) The two triangles, ABC and FGH, are similar.

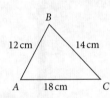 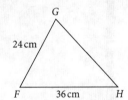

    a What is the scale factor? (1 mark)

    b What is the length of GH? (1 mark)

    c Calculate how many times greater the area of the larger triangle is compared to the area of the smaller triangle. (1 mark)

4 (3 marks) The following are similar objects.

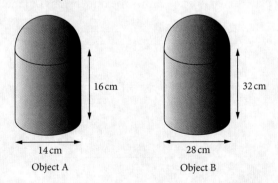

    a Determine the scale factor. (1 mark)

    b If the volume of A is $8210 \, cm^3$, what is the volume of object B? (2 marks)

5 (3 marks) The two triangles below are similar.

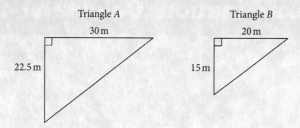

a Determine the missing length of Triangle B. (1 mark)

Triangle B is enlarged by a scale factor of 3.

b Determine the dimensions of the new triangle. (1 mark)

c Calculate how many times greater the area of enlarged Triangle B is compared to the area of Triangle A. (1 mark)

6 (4 marks) This plan of a bedroom has a scale of 1 : 60.

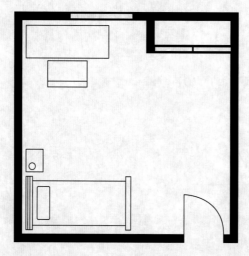

a Determine the internal dimensions of the bedroom, including the wardrobe, in metres. (2 marks)

b If new carpet is to be installed at a cost of $50/ m², determine the cost to install the carpet. (2 marks)

7 (3 marks) The following plan drawn to scale shows the dimensions of Jane's living room.

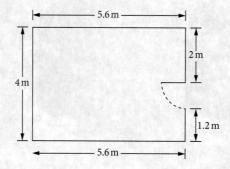

a Determine the scale in the form 1 : m, where m is a whole number. (1 mark)

b Determine the area of the living room. (1 mark)

c If new carpet is to be laid and costs $50/m², determine the cost to replace the carpet in the living room. (1 mark)

Cumulative examination: Calculator-assumed

Total number of marks: 30   Reading time: 4 minutes   Working time: 30 minutes

1 (6 marks) Calculate each percentage amount.

   a 9% of $25 000   **b** 5% of $800   **c** 2% of $300 000

   d 2.5% of $500 000   **e** 3.75% of $175 200   **f** 0.95% of $60 000

2 (2 marks) A right-angled triangle, XYZ, has side lengths $XY = 38.5$ cm and $YZ = 24.0$ cm, as shown in the diagram.

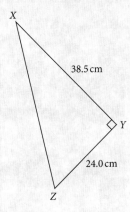

Calculate the length of XZ, in centimetres, correct to one decimal place.

3 (5 marks) A tent with semi-circular ends is in the shape of a prism. The diameter of the ends is 1.5 m. The tent is 2.5 m long.

   a What is the surface area (in m^2 correct to one decimal place) of the tent, including the base?   (2 marks)

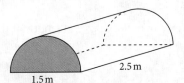

A second tent has the diameter and length increased by a scale factor of 1.5.

   b **i** What factor is the surface area of the tent increased by?   (1 mark)

       ii What is the surface area (in m^2 correct to one decimal place) of the second tent?   (2 marks)

4 (4 marks)

   a Determine the missing side length for the triangle shown, correct to one decimal place.   (2 marks)

   b A similar triangle to the one pictured has a scale factor of 2.5. Determine the perimeter of the second triangle?   (2 marks)

5 (2 marks) During the COVID-19 pandemic, people were restricted to travelling a limit of 5 km from their house. Calculate the area (in square kilometres) people were allowed to travel, to one decimal place.

6 (6 marks) John is standing next to a flagpole. John is 180 cm tall and casts a shadow 2.2 m long. The flagpole casts a shadow that is 8.9 m long.

    a Draw a diagram to represent this information. (2 marks)

    b Are the triangles formed similar? Justify your answer. (2 marks)

    c Determine the scale factor. (1 mark)

    d What is the height of the flagpole, correct to one decimal place? (1 mark)

7 (5 marks) An old wooden gate is 80 cm wide and needs a diagonal brace for support. The diagram is drawn to scale.

80 cm

    a Determine the scale in the form $1:a$. where a is a whole number. (1 mark)

    b Determine the height of the gate. (2 marks)

    c Use Pythagoras' theorem to calculate the diagonal length of the gate. Answer correct to one decimal place. (2 marks)

MAKING SENSE OF DATA

CHAPTER 5

Syllabus coverage
Nelson MindTap chapter resources

5.1 Introduction to data distributions
What is data?
Types of data
Types of categorical data
Types of numerical data
Measures of centre and spread

5.2 Grouped frequency tables and histograms
Frequency tables
Grouped frequency tables
Centre and spread of histograms
Shapes of histograms
Outliers
Using CAS 1: Constructing histograms for ungrouped data

5.3 Dot plots, stem plots and bar charts
Dot plots
Stem plots
Bar charts
Which display do we use?

5.4 The mean and standard deviation
The mean
Comparing the mean and median
The standard deviation
Using CAS 2: Finding the mean and standard deviation for ungrouped data
Using CAS 3: Finding the mean and standard deviation for grouped data

5.5 Bell-shaped distributions
The normal or bell-shaped distribution
Standard deviations from the mean
The 68–95–99.7% rule
Using CAS 4: Normal probability calculations
Using CAS 5: Inverse normal probability calculations

5.6 Standardised values
z-scores
Using z-scores to compare

Examination question analysis
Chapter summary
Cumulative examination: Calculator-assumed
Cumulative examination: Calculator-free

Syllabus coverage

TOPIC 2.1: UNIVARIATE DATA ANALYSIS AND THE STATISTICAL INVESTIGATION PROCESS

Making sense of data relating to a single statistical variable

2.1.2 classify a categorical variable as ordinal, such as income level (high, medium, low) or nominal, such as place of birth (Australia, overseas) and use tables and bar charts to organise and display data

2.1.3 classify a numerical variable as discrete, such as the number of rooms in a house, or continuous, such as the temperature in degrees Celsius

2.1.4 with the aid of an appropriate graphical display (chosen from dot plot, stem plot, bar chart or histogram), describe the distribution of a numerical data set in terms of modality (uni or multimodal), shape (symmetric versus positively or negatively skewed), location and spread and outliers, and interpret this information in the context of the data

2.1.5 determine the mean and standard deviation of a data set using technology and use these statistics as measures of location and spread of a data distribution, being aware of their limitations

2.1.6 use the number of deviations from the mean (standard scores) to describe deviations from the mean in normally distributed data sets

2.1.7 calculate quantiles for normally distributed data with known mean and standard deviation in practical situations

2.1.8 use the 68%, 95%, 99.7% rule for data one, two and three standard deviations from the mean in practical situations

2.1.9 calculate probabilities for normal distributions with known mean μ and standard deviation σ in practical situations

Mathematics Applications ATAR Course Year 11 syllabus p. 12 © SCSA

Video playlists (7):
5.1 Introduction to data distributions
5.2 Grouped frequency tables and histograms
5.3 Dot plots, stem plots and bar charts
5.4 The mean and standard deviation
5.5 Bell-shaped distributions
5.6 Standardised values

Examination question analysis Making sense of data

Skillsheet (1):
5.4 Statistical measures

Worksheets (13):
5.1 Statistical data match-up • Mean, median, mode and range • Mode, median and mean • Measures of central tendency
5.2 Histograms
5.3 Stem-and-leaf plots
5.4 • Standard deviation • Statistical calculations • Statistics review • Calculating and interpreting summary statistics • Statistics crossword
5.5 The normal curve
5.6 z-scores

Puzzle (1):
5.4 Data and statistics crossword

Nelson MindTap

To access resources above, visit
cengage.com.au/nelsonmindtap

5.1 Introduction to data distributions

What is data?

We live in a world where we can access **data** to make informed decisions. How much of climate change is caused by human activity? Let's look at the data. What illnesses does vaping cause? Let's look at the data. How can we reduce crime? Let's look at the data.

Data is information collected through observation that can then be used to make decisions. Here is some data based on the responses to survey questions by a group of ten students:

Survey question	Data
What is the colour of your best friend's eyes?	brown, brown, brown, blue, brown, green, blue, blue, hazel, brown
How many pets do you have at home?	0, 1, 1, 5, 2, 0, 3, 1, 3, 2
How would you rate the current most popular song on Spotify from 1 to 4, where 1 = great, 2 = okay, 3 = awful and 4 = haven't heard it?	2, 4, 2, 2, 1, 3, 1, 2, 4, 1
What is the length of your handspan?	8.0 cm, 7.3 cm, 6.8 cm, 9.2 cm, 8.0 cm, 8.9 cm, 9.2 cm, 7.1 cm, 9.3 cm, 6.5 cm
What is your house number?	23, 41, 6, 118, 51, 33, 2, 19, 12, 4
What was the coldest temperature (°C) you remember experiencing?	−1°C, 3°C, 5°C, 0°C, −8°C, 12°C, −4°C, 2°C, 7°C, −3°C

The information you are looking for such as *colour of eyes*, *number of pets* and *temperature* are called **variables**.

Types of data

We will be looking at six data types. We need to know the data type before we can decide on the right way to work with data.

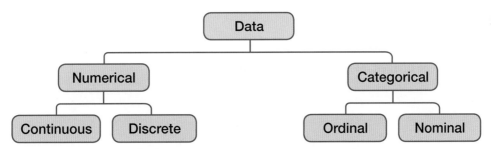

There are two main types of data:

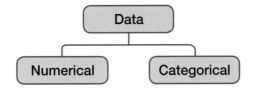

Video playlist
Introduction to data distributions

Worksheet
Statistical data match-up

Chapter 5 | Making sense of data   177

Numerical data involves numbers that can be measured or counted. One way to test this is to ask, 'Does it makes sense to add the numbers together?' For example:

- *number of pets* in your home

  We are counting the number of pets, and adding the numbers makes sense: 2 pets + 4 pets = 6 pets

- *length* of your handspan

  We are measuring length, and adding the numbers makes sense: 8.1 cm + 7.3 cm = 15.4 cm

Categorical data involves either numbers where adding makes no sense or categories that don't involve any numbers. For example:

- *colour* of your best friend's eyes

  The categories brown, blue etc. are not numbers.

- the number of your house

  Adding these numbers makes no sense:
  house number 23 + house number 41 ≠ house number 64

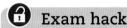 **Exam hack**

Don't fall into the trap of thinking that because numbers are involved, the data must be numerical!

Types of categorical data

There are two types of categorical data:

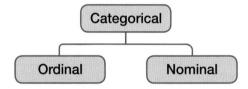

Ordinal data is categorical data that has a natural order. For example:

- *rating* of the current most popular song on Spotify from 1 to 4, where 1 = great, 2 = okay, 3 = awful and 4 = haven't heard it
- the *number* of your house

Nominal data is categorical data with *no* natural order. For example:

- *colour* of your best friend's eyes

Types of numerical data

Numerical data can be divided into these two types of data:

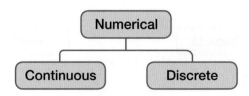

Continuous data is numerical data that can be measured to ever-increasing levels of accuracy. For example:

- *length* of your handspan

  This can be measured to ever-increasing levels of accuracy (8 cm, 7 cm, 7 cm … or 8.1 cm, 7.3 cm, 6.8 cm … or 8.14 cm, 7.31 cm, 6.79 cm … etc.).

Discrete data is numerical data that can only take specific values and *can't* be measured to ever-increasing levels of accuracy. For example:

- *number of pets* in your home

  This can only take whole number values 0, 1, 2, 3 … It is impossible, for example, to have 1.5 or 2.8 of a pet.

Classifying data

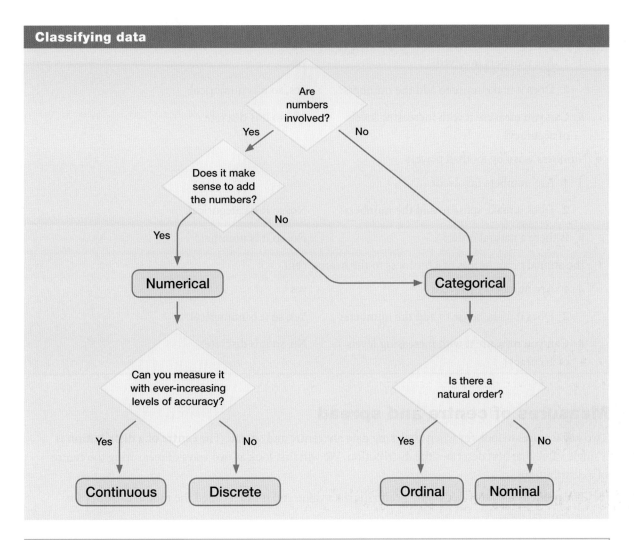

WORKED EXAMPLE 1	Deciding on the type of data
State whether the following data is	
i categorical or numerical	
ii nominal, ordinal, discrete or continuous.	
Steps	**Working**
a Height	
i 1 Are numbers involved?	yes
2 Does it make sense to add the numbers?	Yes, so it is numerical.
ii Can you measure it with increasing levels of accuracy?	Yes, so it is continuous.
b Hair colour	
i Are numbers involved?	No, so it is categorical.
ii Is there a natural order?	No, so it is nominal.
c Finishing position in a 100-metre race	
i 1 Are numbers involved?	yes
2 Does it make sense to add the numbers?	No, so it is categorical.
ii Is there a natural order?	Yes, so it is ordinal.

d Number of friends on a social media platform

i	**1** Are numbers involved?	yes
	2 Does it make sense to add the numbers?	Yes, so it is numerical.
ii	Can you measure it with increasing levels of accuracy?	No, so it is discrete.

e Numbers worn by football players

i	**1** Are numbers involved?	yes
	2 Does it make sense to add the numbers?	No, so it is categorical.
ii	Is there a natural order?	No, so it is nominal.

f The annual profit or loss of a business to the nearest cent

i	**1** Are numbers involved?	yes
	2 Does it make sense to add the numbers?	Yes, so it is numerical.
ii	Can you measure it with increasing levels of accuracy?	No, so it is discrete.

Worksheets
Mean, median, mode and range

Mode, median and mean

Measures of central tendency

Measures of centre and spread

Two key features to look for when analysing data are centre and spread. The **centre of a distribution** is a single value that best describes the distribution. We will first look at two ways of measuring the centre of a distribution:

- The **mode** is the most frequently occurring data value and is often called the **modal category** for categorical data.
  - There can be more than one mode.
  - Data with more than one mode is called **multimodal**.
  - If *every* data value appears exactly once, there is no mode.
- The **median** is the middle value. If there are two middle values, add them and divide the sum by 2.

Odd number of ordered data values	Even number of data values
Median is the 6th data value	Median is between the 6th and 7th data value
3, 6, 7, 7, 7, 8, 13, 19, 20, 20, 22 5 data values ↑ 5 data values Median	3, 6, 7, 7, 7, 8, 13, 17, 19, 20, 22, 26 6 data values ↑ 6 data values Median = $\frac{8 + 13}{2}$ = 10.5

When analysing data, we also look at the **spread of a distribution**. One measure of the spread of a distribution is the **range**:

range = largest value − smallest value

The more spread out the data is, the larger the range.

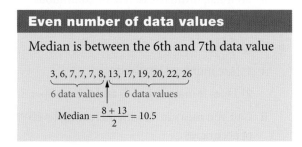

🔓 **Exam hack**

Not every measure of centre and spread can be used with every data type.

Centre, spread and data types

Categorical: Nominal data	Categorical: Ordinal data	Numerical data
Measures of centre		
mode	mode median	mode median
Measures of spread		
–	range	range

WORKED EXAMPLE 2 Finding the median, mode and range

For each of the following, find the

i median ii mode iii range.

Steps	Working
a Number of pets ten students have at home: 0, 1, 1, 5, 2, 0, 3, 1, 3, 2	
i 1 Order the values from smallest to largest.	0, 0, 1, 1, 1, 2, 2, 3, 3, 5
2 Find the middle value (or two middle values). If there are two middle values, then add them and divide by 2.	middle values are 1 and 2 median = $\dfrac{1+2}{2}$ = 1.5 pets
ii Find the most commonly occurring value.	mode = 1 pet
iii range = largest value − smallest value	range = 5 − 0 = 5 pets
b Marks in a Mathematics Applications test for nine students: 15, 19, 17, 6, 12, 18, 18, 17, 15	
i 1 Order the values from smallest to largest.	6, 12, 15, 15, 17, 17, 18, 18, 19
2 Find the middle value (or two middle values). If there are two middle values, then add them and divide by 2.	The median is the 5th value in the ordered list. median = 17 marks
ii Find the most commonly occurring value.	mode = 15, 17 and 18 marks
iii range = largest value − smallest value	range = 19 − 6 = 13 marks

EXERCISE 5.1 Introduction to data distributions   ANSWERS p. 415

Mastery

1 Copy and complete the following diagram:

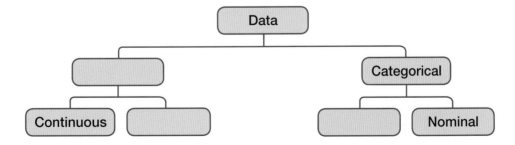

2 **WORKED EXAMPLE 1** State whether the following data is
   i categorical or numerical
   ii nominal, ordinal, discrete or continuous.
   a Number of books in a home
   b Distance to the nearest train station from your home
   c Years when a solar eclipse occurred
   d Amount of time spent sleeping last night
   e Speed of an aeroplane just after take-off
   f Sizes of pizzas from a takeaway (regular, large, family)
   g Numbers on the Matildas' soccer uniforms
   h Opinion of chocolate on a scale of 1 to 5, where 1 is don't like and 5 is love
   i Salary in dollars
   j Salary classified as high, medium or low
   k Number of people watching the Australian Tennis Open
   l Titles of surfing magazines
   m Foot length of swimmers in a squad
   n Classification ratings for movies (G, PG, M, MA, R)
   o Maximum temperature (Celsius) for a series of days
   p State of birth of each person born in Australia

3 **WORKED EXAMPLE 2** For each of the following, find the
   i median   ii mode   iii range.
   a Number of screens (phones, tablets and television) twelve students have in their home:
      5, 8, 3, 10, 4, 5, 7, 16, 9, 12, 4, 6
   b Marks in a Mathematics Applications test for nine students:
      12, 9, 14, 13, 20, 15, 18, 18, 15
   c Number of minutes it takes for eight students to get to school on a particular day:
      7, 3, 20, 9, 5, 3, 10, 12
   d Thumb length of ten students in centimetres:
      6, 6, 5, 5, 6, 7, 6, 7, 5, 5

Calculator-free

4 (2 marks) A survey was completed to collect the heights of students within a group of Year 11 students. Describe the type of data using the terms numerical, categorical, continuous, discrete, ordinal and/or nominal.

5 (2 marks) A survey was conducted about the colour of family cars, with the following results:
   grey, grey, white, grey, white, red, white, blue, red, white, red, silver
   a Is the data categorical (ordinal or nominal) or numerical (discrete or continuous)? (1 mark)
   b What is the median of this data? (1 mark)

6 (10 marks) Data relating to the following five variables was collected from insects that were caught overnight in a trap. Describe the type of data for each variable using the terms categorical, numerical, ordinal, nominal, discrete and/or continuous.

    a *colour* (2 marks)

    b *name of species* (2 marks)

    c *number of wings* (2 marks)

    d *body length* (in millimetres) (2 marks)

    e *body weight* (in milligrams) (2 marks)

7 (2 marks) The level of water usage of 250 houses was rated in a survey as low, medium or high, and the size of the houses as small, standard or large.

Classify the variables *level of water usage* and *size of house*, as recorded in this survey as either categorical or numerical.

Calculator-assumed

8 (2 marks) The marks for a spelling test of ten words were recorded as follows:

8, 6, 8, 4, 5, 6, 8, 5, 7, 4, 7, 8, 6, 8, 9

    a Calculate the median. (1 mark)

    b Calculate the mode. (1 mark)

9 (2 marks) The percentage investment returns of seven superannuation funds for the year 2022 were:

−4.6%, −4.7%, 2.9%, 0.3%, −5.5%, −4.4%, −1.1%

    a Determine the median investment return. (1 mark)

    b Determine the range of investment returns. (1 mark)

10 (10 marks) The maximum daily temperatures (°C to one decimal place) in Melbourne during the first week of the 2020 Australian Open are shown in the table.

Day	Date	Max temperature (°C)
Monday	20/01/2020	21.8
Tuesday	21/01/2020	23.5
Wednesday	22/01/2020	31.7
Thursday	23/01/2020	22.6
Friday	24/01/2020	24.1
Saturday	25/01/2020	27.4
Sunday	26/01/2020	23.6

    a Find, rounded to one decimal place, the

        i median   **ii** mode   **iii** range. (3 marks)

    b Round the original data to the nearest whole degree. Use these values to find, rounded to one decimal place, the

        i median   **ii** mode   **iii** range. (3 marks)

    c Explain why your answers for the median in parts **a** and **b** are different. Which of the two answers do you think is more accurate and why? (2 marks)

    d Which of your answers for the mode in parts **a** and **b** give more helpful information? Why would finding the mode of data rounded to a large number of decimal places usually not give helpful information? (2 marks)

Video playlist
Grouped frequency tables and histograms

Worksheet
Histograms

5.2 Grouped frequency tables and histograms

When we're dealing with a large number of data values to see patterns or draw conclusions, we need to organise and present the data in a manageable form. When choosing a display for the data, we have to decide which one best shows what we want to communicate.

Frequency tables

A **frequency table** can be used to display both categorical and numerical data. It involves counting the number of times each data value occurs.

Grouped frequency tables

For continuous data, or discrete data with a large number of values, it's not practical to list each individual value in a frequency table. For example, if we have whole number data values ranging between 0 and 99, we would need to include 100 rows in a frequency table. If our data included decimals, we would need even more rows.

To reduce the number of rows, we use intervals that describe a range of data values such as 20–<30, meaning any number from 20 to 30 but not including 30. A **grouped frequency table** is a frequency table that uses intervals.

The following grouped frequency table shows the times (in minutes) taken by the 122 participants in a fun run.

Time (min)	Number
0–<30	12
30–<60	52
60–<90	42
90–<120	16
Total	122

This means less than 30 minutes → 0–<30
This means 30 minutes or more but less than 60 minutes → 30–<60
This means 60 minutes or more but less than 90 minutes → 60–<90

The **modal interval** is the interval that occurs most frequently.

> **Grouped frequency tables**
> - When numerical data is continuous, or discrete with a large number of values, numbers can be grouped into intervals to form a grouped frequency table.
> - The intervals must all be the same size.
> - There should be a maximum of 15 intervals.

WORKED EXAMPLE 3 | Interpreting a grouped frequency table

The following grouped frequency table shows the results for an exam.

a Find the modal interval.
b Find the range.

Exam scores

Score	Frequency
0–<20	2
20–<40	1
40–<60	6
60–<80	3
80–100	2
Total	14

Steps	Working
a Find the interval that occurs most frequently.	The modal interval is 40–<60.
b range = largest value – smallest value	range = 100 – 0 = 100

184   Nelson WAmaths Mathematics Applications 11

A **histogram** is a graphical way of displaying data from a grouped frequency table. It is effective when dealing with large amounts of data.

Grouped frequency table	
Exam scores	
Score	Frequency
0–<20	23
20–<40	35
40–<60	71
60–<80	30
80–<100	19
Total	178

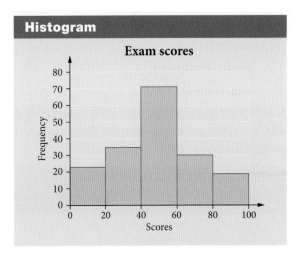

Following the pattern of the grouped frequency table, someone with a score of exactly 20 is recorded in the second interval, someone with a score of exactly 40 is recorded in the third interval, and so on.

> **Exam hack**
>
> Although at first glance a histogram looks similar to a bar chart, there are a number of differences:
>
> - Histograms display numerical data, whereas bar charts are best used to display categorical data.
> - Histograms do not have spaces between the columns.
> - Histograms are *always* vertical.

Centre and spread of histograms

Histograms make it easier to compare the centres or spreads of two distributions. We refer to the modal interval, or most frequently occurring interval, rather than the mode when looking at centres of histograms.

Here are two distributions with the same spread but different centres:

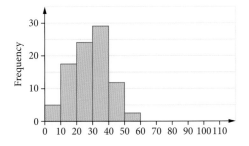 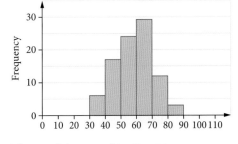

The modal interval is 30–<40.   The modal interval is 60–<70.

Here are two distributions with approximately the same centres but different spreads:

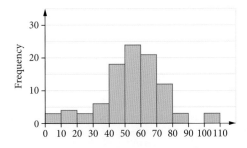 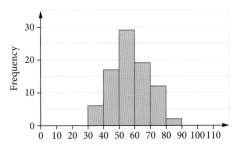

The modal interval for both of these histograms is 50–<60.

Shapes of histograms

A histogram can also be used to describe the **shape of a distribution**. The shape of a distribution can be
- approximately symmetric
- **asymmetric**
  - positively skewed
  - negatively skewed.

A histogram shows an approximately **symmetric distribution** when the left side closely mirrors the right side. For approximately symmetric distributions, the median is always near the line of symmetry but the modal interval can be somewhere else.

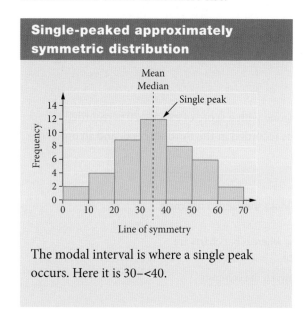

The modal interval is where a single peak occurs. Here it is 30–<40.

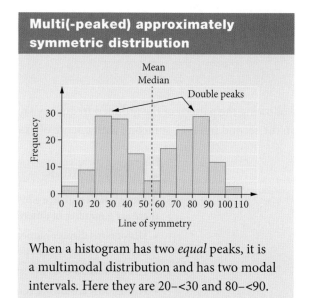

When a histogram has two *equal* peaks, it is a multimodal distribution and has two modal intervals. Here they are 20–<30 and 80–<90.

The following two histograms show asymmetric or **skewed distributions**.

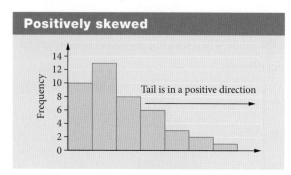

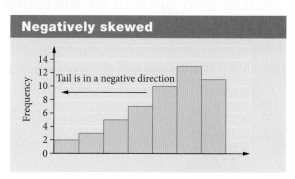

> 🔒 **Exam hack**
>
> To help you remember which skew is positive and which is negative, identify where the 'tail' of the histogram is. Think of the positive and negative directions of a number line. If the tail is in the positive direction, the distribution is positively skewed. If the tail is in the negative direction, the distribution is negatively skewed.

Outliers

We can also comment on a distribution by referring to its **outliers**. An outlier is an extreme high or low value in the data. Outliers can indicate an error made in dealing with the data and can sometimes contaminate calculations and conclusions drawn from data sets. However, they can also occur without an error being involved. Histograms often make it easier to identify possible outliers.

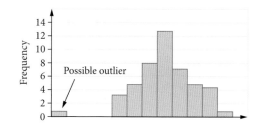

WORKED EXAMPLE 4 Reading and describing histograms

The following table shows the amount of sleep (in minutes) a group of 125 students had the night before an exam.

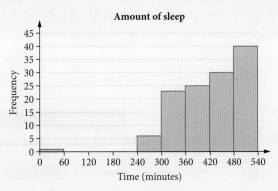

Steps	Working
a How many intervals are there?	
Count the number of intervals on the horizontal axis.	There are nine intervals.
b How many students slept for more than 7 hours?	
1 Convert the time to hours.	7 hours = 420 minutes
2 Read from the histogram.	30 + 40 = 70
	70 students slept for more than 7 hours.
c What percentage of students slept for less than 5 hours? Give your answer rounded to one decimal place.	
1 Convert the time to hours.	5 hours = 300 minutes
2 Read from the histogram.	1 + 6 = 7
	Seven students slept for less than 5 hours.
3 Convert to a percentage using $\text{percentage} = \frac{\text{frequency}}{\text{total}} \times 100\%$ rounding to one decimal place.	$\text{percentage} = \frac{7}{125} \times 100\%$ $= 0.056 \times 100\%$ $= 5.6\%$
d Is the histogram approximately symmetric, positively skewed or negatively skewed? Does it have any possible outliers?	
The histogram has a negative tail.	The histogram is negatively skewed with a possible outlier.
e What is the modal interval?	
Which interval has the highest frequency?	The modal interval is 480–<540 minutes.

 Exam hack

A ruler often helps when you are reading values from histograms and other statistical charts.

USING CAS 1 Constructing histograms for ungrouped data

Scores on an end-of-year mathematics examination, out of 100, achieved by 37 Year 11 students from a particular school were recorded as follows:

84, 79, 99, 52, 63, 70, 65, 78, 47, 72, 73, 60, 52, 76, 77, 65, 71, 61, 53,
62, 41, 88, 57, 71, 43, 89, 74, 50, 49, 61, 72, 70, 68, 95, 58, 67, 43

Use intervals of 40–<50, 50–<60 and so on to construct a histogram for this data.

ClassPad

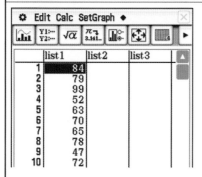

1. Tap **Menu** and open the **Statistics** application.
2. Clear all lists.
3. Enter the ungrouped scores into **list1** as shown above.

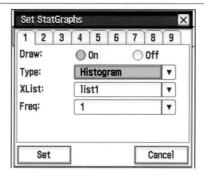

4. Tap **SetGraph** > **Setting**.
5. In the dialogue box, select the following:
   - Type: Histogram
   - XList: list1
   - Freq: 1
6. Tap **Set**.

7. Tap **Graph**.
8. In the dialogue box, set the following:
   - HStart: 40
   - HStep: 10
9. Tap **OK**.

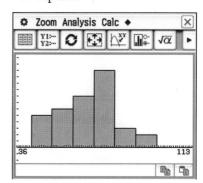

10. A histogram of the data will appear in the lower window.

TI-Nspire

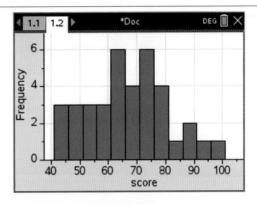

1. Start a new document and add a **Lists & Spreadsheet** page.
2. Label column **A** as **score**.
3. Enter the ungrouped scores as shown above.
4. Insert a **Data & Statistics** page.
5. For the horizontal axis, select **score**.
6. Press **menu > Plot Type > Histogram**.
7. A histogram of the data will be displayed.

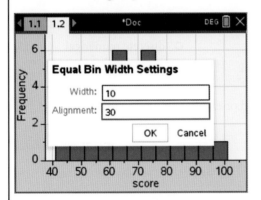

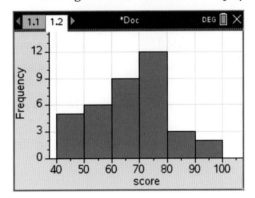

8. Press **menu > Plot Properties > Histogram Properties > Bin Settings > Equal Bin Width.**
9. Enter the following settings:
   - Width: 10
   - Alignment: 30
10. Select **OK**.
11. A histogram of the grouped data in intervals of 10 will be displayed.
12. Adjust the **Window/Zoom** settings or grab the vertical axis and drag it down to view the full histogram.

EXERCISE 5.2 Grouped frequency tables and histograms

ANSWERS p. 415

Recap

Use the following information to answer the next two questions.

The bar chart shows the results of a survey of opinions on a new online game called *Conquist*.

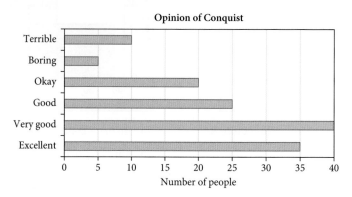

1 The type of data collected is best described as

   A ordinal numerical.
   B nominal categorical.
   C discrete numerical.
   D ordinal categorical.
   E continuous numerical.

2 The percentage of people that thought it was either very good or excellent, rounded to one decimal place, is

   A 13.5%   B 25.9%   C 29.6%   D 55.6%   E 75.0%

Mastery

3 For the histogram shown, state whether each of the statements are true or false.

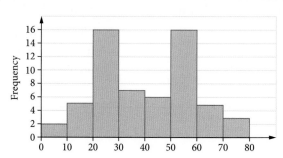

   a The histogram is approximately symmetric.
   b The histogram has a possible outlier.
   c The histogram is double-peaked.
   d The histogram is negatively skewed.
   e The modal interval is 70–<80.
   f The centre is around 40.
   g The histogram is multimodal.

4 **WORKED EXAMPLE 3**

i Find the modal interval for each of the grouped frequency tables below.

ii Find the range for each of the grouped frequency tables below.

a
Score	Frequency
0–<10	1
10–<20	3
20–<30	3
30–<40	4
40–50	5
Total	16

b
Score	Frequency
0–<5	5
5–<10	0
10–<15	4
15–<20	3
20–25	3
Total	15

c
Score	Frequency
0–<10	1
10–<20	3
20–<30	5
30–<40	2
40–<50	2
50–60	2
Total	14

5 **WORKED EXAMPLE 4** The following histogram shows the amount of time (in minutes) a group of students spent studying the night before an exam.

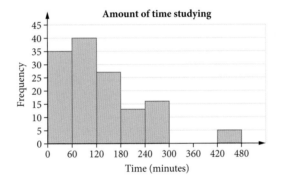

a How many intervals are there?

b How many students studied for less than 2 hours?

c What percentage of students studied for more than 4 hours? Give your answer rounded to one decimal place.

d Is the histogram approximately symmetric, positively skewed or negatively skewed? Does it have any possible outliers?

e What is the modal interval?

6 **Using CAS 1** Scores on an end-of-year science examination, out of 100, achieved by 25 Year 11 students from a particular school were recorded as follows:

52, 45, 40, 67, 60, 73, 88, 61, 64, 73, 71, 41, 57, 48, 76, 97, 79, 59, 83, 73, 66, 71, 76, 32, 67

Use intervals of 30–<40, 40–<50 and so on to construct a histogram for these data.

Calculator-free

7 (5 marks)

a Draw a histogram by hand from the grouped frequency table of test results shown. (4 marks)

b Use the histogram from part **a** to determine whether the data is approximately symmetric, positively skewed or negatively skewed. (1 mark)

Marks	Frequency
0–<10	0
10–<20	5
20–<30	16
30–<40	25
40–<50	14
50–<60	6
Total	66

8 (2 marks) The histogram below displays the distribution of the percentage of internet users in 160 countries in 2023.

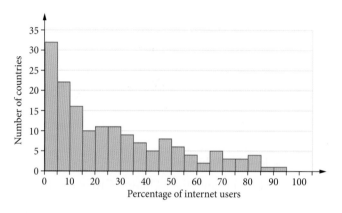

a Describe the shape of the histogram. (1 mark)

b Approximately what is the number of countries in which less than 10% of people are internet users? (1 mark)

9 (2 marks) The histogram below shows the number of hours of sleep Year 11 students had the previous night.

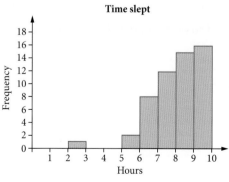

a What is the most appropriate description for the histogram showing how long students slept the previous night? (1 mark)

b What is the modal interval? (1 mark)

Calculator-assumed

10 (2 marks) The histogram shows the distribution of life expectancy of people for 183 countries.

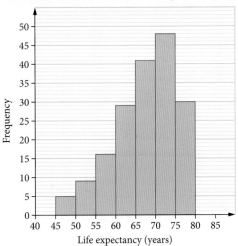

a In how many of these countries is life expectancy less than 55 years? (1 mark)

b In what percentage of these 183 countries is life expectancy between 75 and 80 years? Write your answer correct to one decimal place. (1 mark)

11 (8 marks) The age, in years, of employees at Burger Heaven were recorded in the following grouped frequency table.

Age (years)	Frequency
15–<20	16
20–<25	20
25–<30	4
30–<35	5
35–<40	1
40–<45	2
45–<50	2

a How many people work at Burger Heaven? (1 mark)

b Construct a histogram to display the above data by hand. (4 marks)

c How many employees are aged 40 or over? (1 mark)

d What percentage of employees are aged 40 or over? (1 mark)

e Describe what the histogram shows about Burger Heaven's employment practices. (1 mark)

Video playlist
Dot plots,
stem plots
and bar charts

5.3 Dot plots, stem plots and bar charts

Dot plots

Dot plots can be used for both categorical and numerical data. They are useful as long as there are not too many data values involved and the values are not too spread out.

WORKED EXAMPLE 5 | Using dot plots

The dot plot shows the ages of student council members at a school.

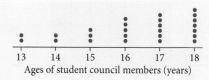
Ages of student council members (years)

a Find the

  i mode   **ii** range   **iii** median   **iv** mean.

b What could best describe the shape of the distribution: approximately symmetric, positively skewed or negatively skewed?

Steps	Working
a **i** Find the most common value.	mode = 18 years
ii range = largest value − smallest value	range = 18 − 13 = 5 years
iii **1** Count the number of dots n, note whether it's odd or even, and find the position of the median.	$n = 25$; odd $$\frac{n+1}{2} = \frac{25+1}{2} = \frac{26}{2} = 13$$ The median is the 13th ordered data value.
2 If n is odd, find the data value of the middle dot. If n is even, find the average of the data values for the two middle dots. Count each column of dots from the bottom up to reach the median.	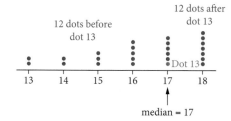
iv **1** Add all the test scores together.	Sum = 13 × 2 + 14 × 2 + 15 × 3 + 16 × 5 + 17 × 6 + 18 × 7 = 407
2 Count the number of data values n.	$n = 25$
3 Use the following to calculate the mean: $$\bar{x} = \frac{\text{sum of all values}}{\text{number of values}} = \frac{\Sigma x}{n}$$	$$\bar{x} = \frac{407}{25} = 16.28$$
b Picture the dot plot as a histogram.	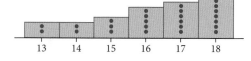
	The distribution is negatively skewed.

Stem plots

Stem plots, also known as **stem-and-leaf plots**, are an alternative to histograms where actual data values appear. The data values are ordered from smallest to largest, where the stem is made up of the leading digits and the leaf is the last digit. Stem plots are best used up to a maximum of 50 data values. When there are a small number of stems, we split the stem to see the distribution more clearly.

The following stem plots show the ordered data values:

80, 81, 84, 85, 89, 89, 91, 92, 92, 96, 105, 107, 108, 109, 109, 112, 114, 118

Stem plot

Stem	Leaf
8	0 1 4 5 9 9
9	1 2 2 6
10	5 7 8 9 9 ← Data values from 100 to 109
11	2 4 8

Key: 9|1 means 91

Split stem plot

Stem	Leaf
8	0 1 4
8	5 9 9
9	1 2 2
9	6
10	← Data values from 100 to 104
10	5 7 8 9 9 ← Data values from 105 to 109
11	2 4
11	8

Key: 9|1 means 91

Worksheet
Stem-and-leaf plots

To decide whether a stem plot is approximately symmetric, positively skewed or negatively skewed, visualise rotating it so that the stem forms the horizontal axis, and picture it as a histogram. For example, we can see after rotating the following stem plot that the distribution is negatively skewed.

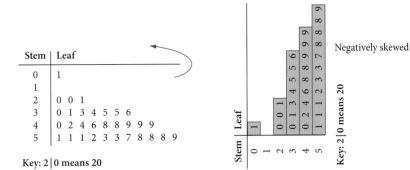

Stem	Leaf
0	1
1	
2	0 0 1
3	0 1 3 4 5 5 6
4	0 2 4 6 8 8 9 9 9
5	1 1 1 2 3 3 7 8 8 9

Key: 2|0 means 20

Stem plots

Stem plots
- can be used with both continuous and discrete data
- are best used with up to a maximum of 50 data values
- always require a key.

WORKED EXAMPLE 6 — Using stem plots

The stem plot shows the test scores out of 70 for a class of 37 students.

a Find the
  i mode   ii range
  iii median   iv mean.

b What could best describe the shape of the distribution: approximately symmetric, positively skewed or negatively skewed?

Stem	Leaf
0	5
1	
2	5 6 6 7 9 9
3	2 3 3 4 6 6 6 8 9
4	0 1 2 3 3 4 5 5 6 8
5	1 1 6 6 6 7 8
6	0 5 7 9

Key: 2 | 5 means 25

Steps	Working
a i Find the most common value.	Test scores of 36 and 56 each appear three times, so this data set is multimodal. Modes are scores of 36 and 56.
ii range = largest value − smallest value	range = 69 − 5 = 64
iii 1 Count the number of data values n, note whether it's odd or even, and find the position of the median.	$n = 37$; odd $$\frac{n+1}{2} = \frac{37+1}{2} = \frac{38}{2} = 19$$ The median is the 19th ordered data value.
2 If n is odd, find the middle data value. If n is even, find the average of the two middle data values.	Stem | Leaf 0 | 5 1 | 2 | 5 6 6 7 9 9 18 values before the median 3 | 2 3 3 4 6 6 6 8 9 4 | 0 1 3 3 4 5 5 6 8 median = 42 5 | 1 1 6 6 6 7 8 6 | 0 5 7 9 18 values after the median median = 42
iv 1 Add all the test scores together.	Sum = 5 + 25 + 2 × 26 + 27 + 2 × 29 + 32 + 2 × 33 + 34 + 3 × 36 + 38 + 39 + 40 + 41 + 42 + 2 × 43 + 44 + 2 × 45 + 46 + 48 + 2 × 51 + 3 × 56 + 57 + 58 + 60 + 65 + 67 + 69 = 1567
2 Count the number of data values n.	$n = 37$
3 Use the following to calculate the mean: $$\bar{x} = \frac{\text{sum of all values}}{\text{number of values}} = \frac{\Sigma x}{n}$$	$\bar{x} = \dfrac{1567}{37}$ = 42.35 (rounded to two decimal places)
b Picture the stem plot as a histogram.	The data is approximately symmetrical.

🔓 Exam hack

Always look to see if the total number of data values is given in the question, so you don't waste time counting them.

Bar charts

Bar charts help us to see patterns when dealing with categorical data. Categories can be represented on the horizontal or vertical axis, with their corresponding frequency on the other axis.

Bar chart with vertical bars

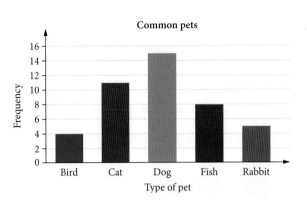

Bar chart with horizontal bars

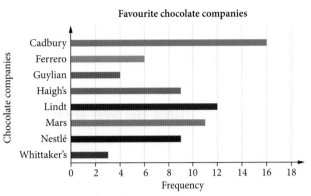

> **Bar charts**
>
> Bar charts
> - display categorical data
> - can have vertical or horizontal bars.

WORKED EXAMPLE 7 — Reading a bar chart

This bar chart shows the number of songs streamed over seven months by a student.

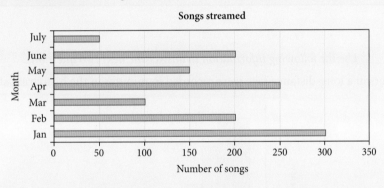

Use the graph to answer the following questions.

a In which month did the student stream the most songs?
b How many songs were streamed over the seven-month period?
c What percentage of songs did the student stream in March?

Steps	Working
a Find the longest bar.	The student streamed the most songs in January.
b Add the frequencies for each month.	300 + 200 + 100 + 250 + 150 + 200 + 50 = 1250 The student streamed 1250 songs over the seven-month period.
c Use percentage = $\frac{\text{frequency}}{\text{total}} \times 100\%$.	$\frac{100}{1250} \times 100\% = 8\%$, therefore 8% of the songs were streamed in March.

Chapter 5 | Making sense of data

Displays and data types

Categorical: Nominal data	Categorical: Ordinal data	Numerical data
Displays		
dot plot bar chart	dot plot bar chart	dot plot histogram stem plot

Which display do we use?

Often there is more than one suitable display. Here are some guidelines to help us decide which statistical display to use.

Display	Type of data	Guidelines
Bar chart	Categorical data	Categories can be represented on the horizontal or vertical axis.
Histogram	Numerical data	Best if data has been grouped into between 5 and 15 intervals. Can be used for a large number of data values.
Dot plot	Numerical data	Best used with a maximum of 50 data values and when the data values are not too spread out.
Stem plot	Numerical data	Best used with a maximum of 50 data values and to see all the actual data values.

EXERCISE 5.3 Dot plots, stem plots and bar charts ANSWERS p. 416

Recap

Use the following information to answer the next two questions.

The times of runners in a long-distance race were recorded in minutes in the histogram shown.

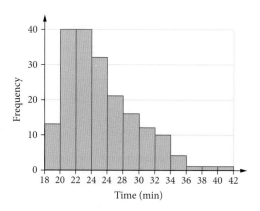

1 Which of the following best describes the histogram?
   A negatively skewed
   B positively skewed with three outliers
   C approximately symmetric
   D multimodal
   E single-peaked

2 Which of the following is **not** true?

   A The two modal intervals are 20–<22 and 22–<24.

   B The histogram is double-peaked.

   C There are no runners with times under 18 minutes.

   D The number of runners with times greater than 30 minutes is over 20.

   E Under 100 runners were in the race.

Mastery

3 **WORKED EXAMPLE 5** The dot plot shows the number of text messages sent by a group of students in one hour.

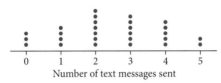

   a Find the

      i mode   ii range   iii median   iv mean.

   b What could best describe the shape of the distribution: approximately symmetric, positively skewed or negatively skewed?

4 **WORKED EXAMPLE 6** The stem plot shows the ages of 33 customers in a store recorded one day.

Stem	Leaf
2	3 5 7 7 7 7 8 9 9
3	0 1 1 1 1 3 4 4 5 6 7 8 8
4	0 2 3 4 4 8
5	7 9
6	3
7	7

Key: 2 | 3 means 23

   a Find the

      i mode   ii range   iii median   iv mean.

   b What could best describe the shape of the distribution: approximately symmetric, positively skewed or negatively skewed?

5 **WORKED EXAMPLE 7** This bar chart shows the results of a poll before an election.

   a How many people favour the Greens?

   b Which is the most popular party in the poll?

   c How many more people favour the Labor Party than the Liberal Party?

   d If there are 80 people who are in the 'Other' category, how many people were polled altogether?

   e What percentage of those polled favoured the Liberal Party?

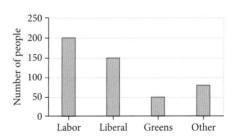

6 Describe the shape of the data for each of the following.

a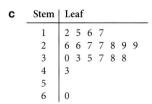

b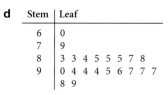

c
Stem	Leaf
1	2 5 6 7
2	6 6 7 7 8 9 9
3	0 3 5 7 8 8
4	3
5	
6	0

Key: 1|2 means 12

d
Stem	Leaf
6	0
7	9
8	3 3 4 5 5 5 7 8
9	0 4 4 4 5 6 7 7 7
	8 9

Key: 7|9 means 79

Calculator-free

7 (5 marks) This stem plot represents the number of goals scored per match by the Sapphires in a netball season. The leaves in this stem plot are not in order.

Stem	Leaf
3	7 5 8 4
4	9 0 4 1
5	3 6 3 7 4 3 1
6	2 8 2 0
7	9 3 5

Key: 6|8 means 68

a Rewrite the stem plot so that the leaves are in ascending order. (1 mark)
b How many matches were played in the season? (1 mark)
c What was the Sapphires' highest score for a match? (1 mark)
d In what percentage of matches did the Sapphires score below 55 goals? Round your answer to the nearest percentage. (1 mark)
e What could best describe the shape of the distribution: approximately symmetric, positively skewed or negatively skewed? (1 mark)

8 (4 marks) The stem plot shows the percentage of homes connected to broadband internet for 24 countries in 2023.

Stem	Leaf
1	
1	6 7
2	0 1 1 3 4 4
2	5 7 8 9
3	0 0 1 1 1 2 2 3
3	5 7 8 8
4	

Key: 1|6 means 16%

a Determine the number of these countries with more than 22% of homes connected to broadband internet in 2023. (1 mark)
b Determine the minimum. (1 mark)
c Determine the maximum. (1 mark)
d Determine the median. (1 mark)

9 (2 marks) The following bar chart shows the distribution of wind directions recorded at a weather station at 9:00 am on each of 214 days in 2024.

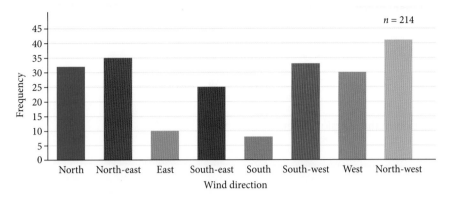

  a According to the bar chart, what was the most frequently observed wind direction? (1 mark)

  b According to the bar chart, what was the percentage of the 214 days on which the wind direction was observed to be east or south-east? (1 mark)

Calculator-assumed

10 (2 marks) The dot plot shows the distribution of the number of bedrooms in each of the 21 apartments advertised for sale in a new high-rise apartment block.

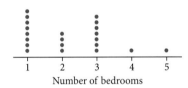

  a What is the mode of this distribution? (1 mark)

  b Find the median of this distribution. (1 mark)

11 (2 marks) The marks obtained by students who sat for a test are displayed as a stem plot, shown below.

Stem	Leaf
0	0
1	
2	0 1 2 5 6
3	0 1 1 1 3 5 5 7 8 9 9
4	1 2 3 4 4 6 7 7
5	0

Key: 2 | 1 means 21

  a How many students sat the test? (1 mark)

  b Determine the median test marks. (1 mark)

12 (2 marks) The development index for each country is a whole number between 0 and 100. The dot plot displays the values of the development index for each of the 28 countries that has a high development index.

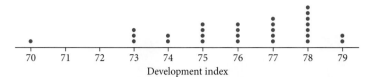

Using the information in the dot plot, determine the

a mode (1 mark)

b range. (1 mark)

13 (3 marks) Table 1 shows the number of rainy days recorded in a high rainfall area for each month during 2023.

Table 1

Month	Number of rainy days
January	12
February	8
March	12
April	14
May	18
June	18
July	20
August	19
September	17
October	16
November	15
December	13

The dot plot below displays the distribution of the number of rainy days for the 12 months of 2023.

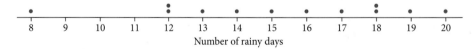

a Copy the dot plot and **circle** the dot that represents the number of rainy days in April 2023. (1 mark)

b For the year 2023, determine the

   i median number of rainy days per month. (1 mark)

   ii percentage of months that have more than 10 rainy days. Write your answer correct to the nearest percent. (1 mark)

14 (4 marks) The stem plot in Figure 1 shows the distribution of the average age, in years, at which women first marry in 17 countries.

Figure 1
Average age, in years, of women at first marriage

Stem	Leaf
24	
25	0
26	6
27	1 1 3 4 7
28	2 2 2 3 3 6
29	1 1
30	1 4
31	

Key: 27 | 3 means 27.3 years

  a For these countries, determine the
    i lowest average age of women at first marriage (1 mark)
    ii median average age of women at first marriage. (1 mark)

The stem plot in Figure 2 shows the distribution of the average age, in years, at which men first marry in 17 countries.

Figure 2
Average age, in years, of men at first marriage

Stem	Leaf
25	
26	0
27	
28	9
29	0 9 9
30	0 0 3 5 6 7 9
31	0 0 2
32	5 9
33	

Key: 32 | 5 means 32.5 years

  b For these countries, determine the
    i lowest average age of men at first marriage (1 mark)
    ii median average age of men at first marriage. (1 mark)

15 (3 marks) A number of teenagers were surveyed on how they rated a particular movie. The results are shown in the following bar chart.

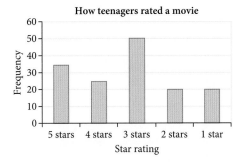

  a What is the most common star rating for the movie? (1 mark)
  b How many teenagers rated the movie? (1 mark)
  c Calculate the percentage of teenagers who rated the movie as 2 stars to the nearest whole percent. (1 mark)

5.4 The mean and standard deviation

The mean

The **mean** is another measure of the centre of a distribution. It is often referred to as the 'average'. The symbol for the mean of a data set is $\bar{x}$ (called 'x bar'). There are shortcuts for calculating the mean depending on how the data are displayed.

> **The mean**
>
> Mean for ungrouped data:
>
> $$\bar{x} = \frac{\text{sum of all values}}{\text{number of values}} = \frac{\Sigma x}{n}$$
>
> where Σ means 'sum of'.
>
> Mean for data in a grouped frequency table:
>
> $$\bar{x} = \frac{\text{sum of (each value} \times \text{its corresponding frequency)}}{\text{sum of frequencies}} = \frac{\Sigma xf}{\Sigma f}$$

Comparing the mean and median

We often need to choose between the mean and the median for the best measure of the centre.

- For symmetric distributions, the mean = the median.
- For distributions that are approximately symmetric, the mean and the median will be very close in value.
- The mean is greater than the median for positively skewed distributions.
- The mean is less than the median for negatively skewed distributions.
- Outliers usually don't affect the median, but they can often significantly affect the mean.

> **Mean vs Median**
>
> Choosing between the mean and the median as the measure of the centre of a distribution:
>
> | Shape of distribution | Choose |
> | --- | --- |
> | Approximately symmetric distributions with no outliers | mean or median |
> | Approximately symmetric distributions with outliers | median |
> | Skewed distributions | median |

WORKED EXAMPLE 8 | Calculating the mean

The scores for the players in two nine-hole golf tournaments were recorded. For each one

i find how many players were in each tournament

ii calculate the mean score, correct to one decimal place.

a Tournament 1 scores: 36, 44, 35, 47, 42, 37, 43, 39, 40, 38

b Tournament 2 scores:

Score (x)	Frequency
37	2
38	4
39	7
40	4
41	1

Steps	Working
a i Count the number of data values. ii Use the formula $\bar{x} = \dfrac{\Sigma x}{n}$, giving your answer correct to one decimal place.	There are 10 players in tournament 1. $\bar{x} = \dfrac{36 + 44 + 35 + 47 + 42 + 37 + 43 + 39 + 40 + 38}{10}$ $= \dfrac{401}{10} = 40.1$ The mean score for tournament 1 is 40.1.
b i Find the sum of the frequencies. ii 1 Add an extra column to the table and an extra row for totals. Fill in the $x \times f$ column and totals. 2 Use the formula $\bar{x} = \dfrac{\Sigma xf}{\Sigma f}$, giving your answer correct to one decimal place.	$\Sigma f = 2 + 4 + 7 + 4 + 1 = 18$ There were 18 players in tournament 2. | Score (x)	Frequency (f)	$x \times f$
---	---	---
37	2	74
38	4	152
39	7	273
40	4	160
41	1	41
Total	18	700	 $\bar{x} = \dfrac{74 + 152 + 273 + 160 + 41}{2 + 4 + 7 + 4 + 1} = \dfrac{700}{18} = 38.9$ The mean score for tournament 2 is 38.9.

The standard deviation

The **standard deviation**, like the range, is a measure of the spread of the data. The standard deviation measures the spread of the data around the mean. The symbol for the standard deviation of a data set is σ.

Standard deviation

The formula for standard deviation of a data set is

$$\sigma = \sqrt{\dfrac{\Sigma(x - \bar{x})^2}{n}}$$

where

σ = standard deviation

Σ = means the 'sum of'

x = each value from the data set

$\bar{x}$ = mean of all the values in the data set

n = total number of values in the data set

although calculations are done using CAS.

Worksheets
Standard deviation

Statistical calculations

Statistics review

Calculating and interpreting summary statistics

Statistics crossword

Puzzle
Data and statistics crossword

Chapter 5 | Making sense of data

USING CAS 2 — Finding the mean and standard deviation for ungrouped data

Find the mean $\bar{x}$ and the standard deviation σ, rounded to two decimal places, for the ungrouped data shown.

3, 4, 4, 5, 5, 7, 7, 7, 8, 9, 9, 9, 12, 12, 13, 15, 15, 16, 18, 20

ClassPad

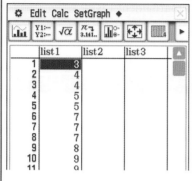  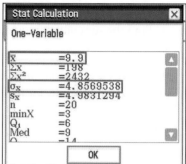

1. Tap **Menu** and open the **Statistics** application.
2. Clear all lists and enter the data as shown.
3. Tap **Calc** > **One-Variable**.
4. Leave the default settings of **XList** as **list1** and **Freq:** as **1**.
5. Tap **OK**.
6. The mean and standard deviation values will be displayed.

TI-Nspire

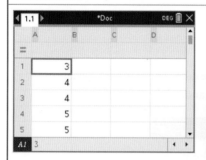 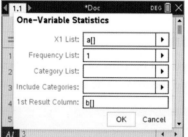 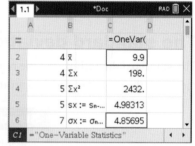

1. Start a new document and add a **Lists & Spreadsheet** page.
2. Enter values into column **A**.
3. Press **menu** > **Statistics** > **Stat Calculations** > **One-Variable Statistics**.
4. On the next screen, keep the number of lists default setting of **1** and select **OK**.
5. In the **X1 List:** field, keep the default setting of **a[ ]***.
6. Select **OK**.

* Alternatively, label the column and use the variable name.

7. The one-variable labels and values will be displayed in columns **B** and **C**.
8. Scroll down to view the mean and standard deviation values.

$\bar{x} = 9.90$, $\sigma = 4.86$

USING CAS 3 — Finding the mean and standard deviation for grouped data

Find the mean $\bar{x}$ and the standard deviation σ, rounded to two decimal places, for the grouped data shown.

Score	Frequency
1	5
2	11
3	4
4	3
5	7
6	3

ClassPad

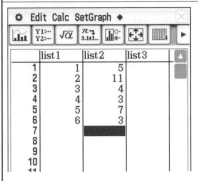

1. Tap **Menu** and open the **Statistics** application.
2. Clear all lists.
3. Enter the scores into **list1** and the frequencies into **list2** as shown above.
4. Tap **Calc > One-Variable**.
5. In the dialogue box, select the following:
   XList: list1
   Freq: list2
6. Tap **OK**.
7. The mean ($\bar{x}$) value and standard deviation (σ_x) will be displayed.

$\bar{x} = 3.15$, $\sigma = 1.64$

TI-Nspire

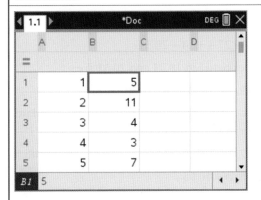
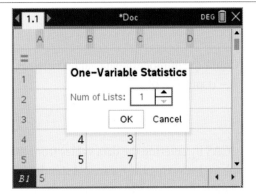

1. Start a new document and add a **Lists & Spreadsheet** page.
2. Enter the scores into column **A** and the frequencies into column **B**, as shown.
3. Press **menu > Statistics > Stat Calcuations > One–Variable Statistics**.
4. In the dialogue box, keep the default setting of the **Num of Lists:** as **1**.
5. Select **OK**.

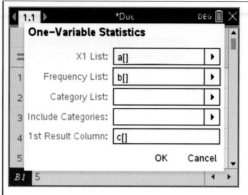

6 In the dialogue box, enter the following:

X1 List: a[ ]

Frequency List: b[ ]

7 Select OK.

8 The labels and values will be displayed in columns **C** and **D**.

9 Scroll down to view the mean ($\bar{x}$) and standard deviation (σx).

$\bar{x} = 3.15$, $\sigma = 1.64$

WORKED EXAMPLE 9 Working with the mean and standard deviation from a display

For each of the following displays, find the

  i mean and standard deviation, correct to two decimal places

  ii number of data values that are within one standard deviation from the mean.

a

(Dot plot on Age (years) axis from 17 to 22)

b

Stem	Leaf
1	5 6 7 8 9
2	0 1 7
3	2 4 6 6
4	4 8

Key: 1 | 5 means 15 years

Steps	Working
a i Use CAS by entering the data values from the graph and selecting the mean and standard deviation. Round to two decimal places.	17, 18, 18, 18, 19, 19, 19, 19, 19, 20, 20, 20, 20, 21, 22 $\bar{x} = 19.27$, $\sigma = 1.24$
ii 1 Find $\bar{x} - \sigma$ and $\bar{x} + \sigma$.	$\bar{x} - \sigma = 19.27 - 1.24 = 18.03$ $\bar{x} + \sigma = 19.27 + 1.24 = 20.51$
2 Count the number of data values between $\bar{x} - \sigma$ and $\bar{x} + \sigma$.	The values between 18.03 and 20.51 are: 19, 19, 19, 19, 19, 20, 20, 20, 20 There are 9 data values that are within one standard deviation from the mean.
b i Use CAS by entering the data values from the graph and selecting the mean and standard deviation. Round to two decimal places.	15, 16, 17, 18, 19, 20, 21, 27, 32, 34, 36, 36, 44, 48 $\bar{x} = 27.36$, $\sigma = 10.95$
ii 1 Find $\bar{x} - \sigma$ and $\bar{x} + \sigma$.	$\bar{x} - \sigma = 27.36 - 10.55 = 16.81$ $\bar{x} + \sigma = 27.36 + 10.55 = 37.91$
2 Count the number of data values between $\bar{x} - \sigma$ and $\bar{x} + \sigma$.	The values between 16.41 and 38.91 are: 17, 18, 19, 20, 21, 27, 32, 34, 36, 36 There are 10 data values that are within one standard deviation from the mean.

EXERCISE 5.4 The mean and standard deviation

ANSWERS p. 416

Recap

1 What is the correct data classification for a person's favourite ice-cream flavour?

　A　discrete　　　B　numerical　　　C　ordinal　　　D　continuous　　　E　nominal

2 The marks of 15 students for a maths test marked out of 10 were recorded as follows:

9, 7, 7, 7, 5, 10, 6, 8, 3, 10, 9, 6, 8, 10, 4

The median and mode respectively are

　A　median = 7, mode = 7

　B　median = 10, mode = 7 and 10

　C　median = 7, mode = 7 and 10

　D　7 and 8

　E　median = 7, mode = 10.

Mastery

3 ⚙ WORKED EXAMPLE 8　Two groups of students were surveyed about the number of movies they had streamed in the last month. For each group

　　i find how many students were surveyed

　　ii calculate the mean number of movies streamed, correct to one decimal place.

　a Group 1 movies streamed: 4, 7, 2, 0, 8, 3, 6, 2, 1, 0, 4, 5, 3

　b Group 2 movies streamed:

Movies (x)	Frequency
0	6
1	7
2	8
3	10
4	9
5	5
6	5

4 ⚙ Using CAS 2　Find the mean $\bar{x}$ and the standard deviation σ, rounded to two decimal places, for the ungrouped data shown.

32, 43, 35, 45, 31, 43, 34, 34, 35, 47, 37, 39, 42, 38, 36

5 ⚙ Using CAS 3　Find the mean $\bar{x}$ and the standard deviation σ, rounded to two decimal places, for the grouped data shown.

Score	Frequency
5	3
6	5
7	1
8	0
9	5
10	2

6 Find the mean, median, mode, range and standard deviation of the following data. Where necessary, round answers to one decimal place.

23, 28, 29, 25, 26, 25, 29, 28, 22, 24, 21, 31, 32, 24, 27, 24, 26

7 **WORKED EXAMPLE 9** For each of the following displays, find the
  i mean and standard deviation, correct to two decimal places
  ii number of data values that are within one standard deviation of the mean.

a

b
Stem	Leaf
1	2 4 6 9
2	1 1 4 6 7
3	0 3 5 8 9 9
4	3

Key: 1 | 2 means 12 years

Calculator-free

8 (2 marks) The total mass of nine oranges is 1.53 kg. Using this information,
  a calculate the mean mass of an orange. (1 mark)
  b If a tenth orange was added weighing 200 g, what is the mean of the ten oranges? (1 mark)

Calculator-assumed

9 (2 marks) The number of DVD players in each of 20 households is recorded in the frequency table.

Number of DVD players	Frequency
0	6
1	9
2	3
3	1
4	0
5	1
Total	20

  a For this sample of households, what is the percentage of households with **at least** one DVD player? (1 mark)
  b For this sample of households, what is the mean number of DVD players in these 20 households? (1 mark)

10 (2 marks) A sample of 14 people were asked to indicate the time (in hours) they had spent watching television on the previous night. The results are displayed in the dot plot.

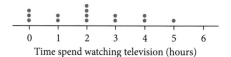

  a Calculate the mean of these times, correct to two decimal places. (1 mark)
  b Calculate the standard deviation of these times, correct to two decimal places. (1 mark)

5.5 Bell-shaped distributions

The normal or bell-shaped distribution

Many variables in real life have what is known as a **normal distribution** or **bell-shaped distribution**. This means they have an approximate bell shape as in this example.

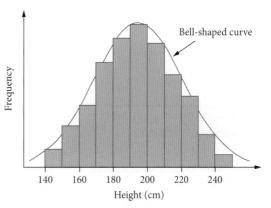

Standard deviations from the mean

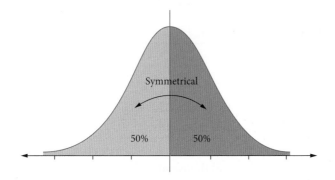

For normal distributions, we know that the following mean-standard deviation scale gives the percentages of data lying within 1, 2 or 3 standard deviations of the mean:

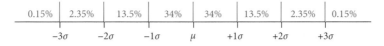

If $\mu = 10$ and $\sigma = 2$, then this becomes

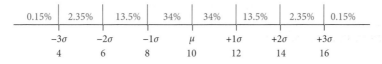

So we know, for example, that

- 34% of the data lies between 10 and 12
- 13.5% of the data lies between 12 and 14
- 2.35% of the data lies between 4 and 6
- 0.15% of the data is less than 4

Bell-shaped distributions

Bell-shaped distributions
- are symmetric about the mean
- have 50% of the data either side of the mean
- mode = mean = median
- have a peak in the centre and tail off towards zero on both sides.

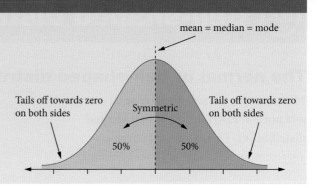

WORKED EXAMPLE 10 — Working with the mean-standard deviation scale

A study found that the number of orange hundreds and thousands on a freckle chocolate button has a normal distribution with a mean of 17 and a standard deviation of 4. Find the percentage of freckles that have

a between 9 and 13   b more than 29   c between 17 and 25   d less than 13

orange hundreds and thousands.

Steps	Working
a 1 Write a mean-standard deviation scale that includes the mean and standard deviation values given.	0.15% | 2.35% | 13.5% | 34% | 34% | 13.5% | 2.35% | 0.15% −3σ −2σ −1σ μ +1σ +2σ +3σ 5 9 13 17 21 25 29
2 Read from the mean-standard deviation scale.	13.5% of freckles have between 9 and 13 orange hundreds and thousands.
b Read from the mean-standard deviation scale.	0.15% of freckles have more than 29 orange hundreds and thousands.
c Add the required percentages from the mean-standard deviation scale.	Add the percentages between 17 and 25: 34% + 13.5% = 47.5% 47.5% of freckles have between 17 and 25 orange hundreds and thousands.
d Add the required percentages from the mean-standard deviation scale.	Add the percentages less than 13: 0.15% + 2.35% + 13.5% = 16% 16% of freckles have less than 13 orange hundreds and thousands.

The 68–95–99.7% rule

For a normal distribution, we can use the **68–95–99.7% rule**:

Around 68% of the data values lie within *one* standard deviation of the mean.

Around 95% of the data values lie within *two* standard deviations of the mean.

Around 99.7% of the data values lie within *three* standard deviations of the mean.

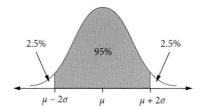
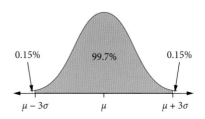

Representing the 68–95–99.7% rule

68–95–99.7% rule diagram

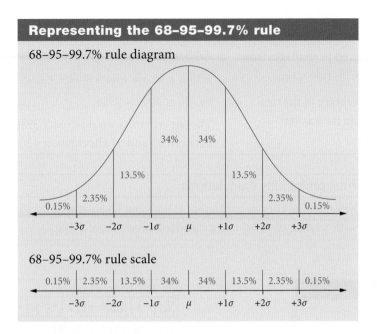

68–95–99.7% rule scale

WORKED EXAMPLE 11	Working with the 68–95–99.7% rule

After a lengthy study, it was found that the number of chockbits in packets approximated a bell-shaped distribution with a mean of 42 and a standard deviation of 3.

Steps	Working
a Write the 68–95–99.7% rule scale for this.	
Write a 68–95–99.7% rule scale that includes the mean and standard deviations given.	
b Find the percentage of packets that have between 36 and 48 chockbits.	
1 Add the required percentages from the 68–95–99.7% rule scale.	Adding the percentages between 36 and 48: 13.5% + 34% + 34% + 13.5% = 95%
2 Write the answer.	95% of packets have between 36 and 48 chockbits.
c Find the percentage of packets with more than 45 chockbits.	
1 Add the required percentages from the 68–95–99.7% rule scale.	Adding the percentages greater than 45: 13.5% + 2.35% + 0.15% = 16%
2 Write the answer.	16% of packets have more than 45 chockbits.
d Find the percentage of packets with more than 39 chockbits.	
1 Add the required percentages from the 68–95–99.7% rule scale.	Adding the percentages greater than 39: 34% + 34% + 13.5% + 2.35% + 0.15% = 84%
2 Write the answer.	84% of packets have more than 39 chockbits.
e A supermarket has bought 4000 chockbits packets.	
i How many of these would they expect to have fewer than 33 chockbits?	
1 Add the required percentages from the 68–95–99.7% rule scale.	Adding the percentages less than 33: 0.15%
2 Find this percentage of the total given and write the answer.	0.15% of 4000 = 6. The supermarket would expect 6 packets to have fewer than 33 chockbits.

ii How many of these would they expect to have between 33 and 51 chockbits?

1 Add the required percentages from the 68–95–99.7% rule scale.	Adding the percentages between 33 and 51: 2.35% + 13.5% + 34% + 34% + 13.5% + 2.35% = 99.7%
2 Find this percentage of the total given and write the answer.	99.7% of 4000 = 3988 The supermarket would expect 3988 packets to have between 33 and 51 chockbits.

USING CAS 4 — Normal probability calculations

Find the probability that a person chosen at random is between 65 kg and 72 kg, given the data is normally distributed with a mean of 70 kg and a standard deviation of 10 kg.

ClassPad

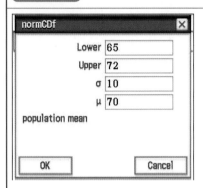

1 Tap **Interactive** > **Distribution/Inv. Dist** > **Continuous** > **normCDf**.

2 Enter the values in the dialogue box as shown above and tap **OK**.

3 Rounding to four decimal places, the probability is 0.2707.

TI-Nspire

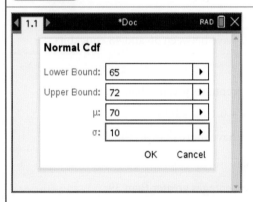
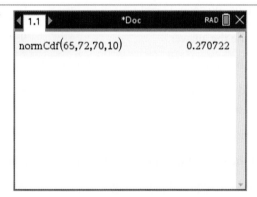

1 Press **menu** > **Probability** > **Distributions** > **Normal Cdf**.

2 Enter the values in the dialogue box as shown above and press **enter**.

3 Rounding to four decimal places, the probability is 0.2707.

USING CAS 5 | Inverse normal probability calculations

Find k where $P(x \geq k) = 0.3$, assuming we are working with a normal population of weights with a mean of 70 kg and standard deviation of 10 kg.

ClassPad

1. Tap **Interactive > Distribution/Inv. Dist > Inverse > InvNormCDf**.
2. Tap on the **Tail setting** drop down menu and select **Right**.
3. Enter the values in the dialogue box as shown above and tap **OK**.

4. Rounding to two decimal places, $k = 75.24$.

TI-Nspire

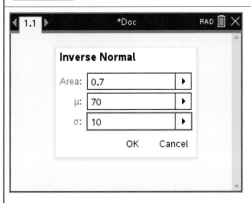
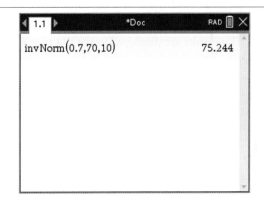

1. Press **menu > Probability > Distributions > Inverse Normal**.
2. Enter the values in the dialogue box as shown above and press **enter**.

3. Rounding to two decimal places, $k = 75.24$.

Quantiles are a statistical value that divide a data set into equal parts. For example, quartiles, deciles, and percentiles. If 0.6 (60%) of the distribution is below 80 then 80 is said to be the 0.6 quantile. This is also said to be the 60th percentile.

The 60th percentile or $P(X \leq 80) = 0.6$ quantile.

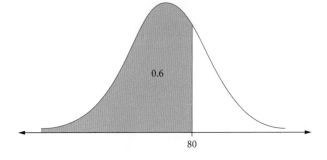

EXERCISE 5.5 Bell-shaped distributions

ANSWERS p. 416

Recap

1 The table displays the systolic blood pressure readings, in mmHg, that result from fifteen successive measurements of the same person's blood pressure.

Reading number	Systolic blood pressure	Reading number	Systolic blood pressure
1	121	9	125
2	126	10	121
3	141	11	118
4	125	12	134
5	122	13	125
6	126	14	127
7	129	15	119
8	130		

Correct to one decimal place, the mean and standard deviation of this person's systolic blood pressure measurements are respectively

A 124.9 and 4.4 **B** 125.0 and 5.8 **C** 125.0 and 6.0 **D** 125.9 and 5.8 **E** 125.9 and 6.0

2 A sample of 15 students were asked to indicate the time (in hours) they spent completing homework the previous night. The results are displayed in the dot plot. Correct to one decimal place, the mean and standard deviation of these times are respectively

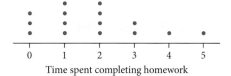

A $\bar{x} = 1.4$ $\sigma = 1.8$ **B** $\bar{x} = 1.8$ $\sigma = 1.4$ **C** $\bar{x} = 1.8$ $\sigma = 1.5$

D $\bar{x} = 1.5$ $\sigma = 1.8$ **E** $\bar{x} = 1.9$ $\sigma = 1.5$

Mastery

3 State whether each of these histograms represent an approximate normal distribution.

a
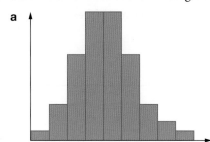

b
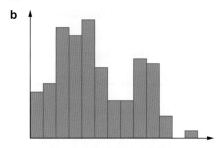

c
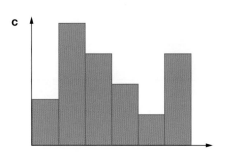

d
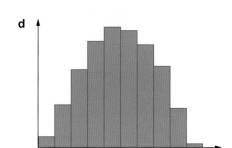

4 **WORKED EXAMPLE 10** A study found that the number of cheese puffles in a packet has a normal distribution with a mean of 55 and a standard deviation of 3. Find the percentage of packets that have
   a between 55 and 58
   b less than 46
   c between 55 and 64
   d more than 55 of cheese puffles.

5 **WORKED EXAMPLE 11** A restaurant chain records the total time customers spend in their restaurants. It was found that the times approximated a bell-shaped distribution with the mean amount of time spent in the restaurant being 73 minutes and the standard deviation 8 minutes.
   a Write the 68–95–99.7% rule scale for this.
   b Find the percentage of times spent in the restaurant that are less than 73 minutes.
   c Find the percentage of times spent in the restaurant that are between 49 and 97 minutes.
   d Find the percentage of times spent in the restaurant that are less than 57 minutes.
   e Find the percentage of times spent in the restaurant that are more than 49 minutes.
   f One of the restaurants in the chain had 426 customers in one day. How many of these customers would they expect to spend
      i more than 81 minutes in the restaurant?
      ii between 65 and 81 minutes in the restaurant?

6 In a particular school, the number of days that students are late to class in a year has an approximate normal distribution with a mean of 21 and a standard deviation of 3.
   a Write the 68–95–99.7% rule scale for this.

Use the scale to state whether the following are reasonable estimates.
   b 50% of students are late to class on more than 21 days in a year.
   c 25% of students are late to class on fewer than 21 days in a year.
   d 68% of students are late to class between 18 and 24 days in a year.
   e 99.7% of students are late to class between 12 and 30 days in a year.
   f 16% of students are late to class on fewer than 18 days in a year.
   g 2.5% of students are late to class on more than 27 days in a year.
   h 0.15% of students are late to class on more than 12 days in a year.
   i 84% of students are late to class on more than 18 days in a year.
   j 97.5% of students are late to class on fewer than 27 days in a year.
   k 99.85% of students are late to class on fewer than 30 days in a year.

7 **Using CAS 4** A normal distribution has a mean of 60 and standard deviation of 4.
   a Find the probability that a random sample selected is greater than 63. Round your answer correct to three decimal places.
   b Find the probability that a random sample selected is between 61 and 65. Round your answer correct to three decimal places.

8 **Using CAS 5** A normal distribution of weights has a mean of 70 kg and standard deviation of 5 kg.
   a Find k, where $P(x \geq k) = 0.25$.
   b Find a, such that 85% of the weights lie within a kg of the mean.

Calculator-free

9 (2 marks) The time taken to travel between two regional cities is approximately normally distributed with a mean of 70 minutes and a standard deviation of 2 minutes.

   a What is the percentage of travel times that are between 68 minutes and 72 minutes? (1 mark)

   b What is the percentage of travel times that are between 66 minutes and 72 minutes? (1 mark)

10 (4 marks) The *wing length* of a species of bird is approximately normally distributed with a mean of 61 mm and a standard deviation of 2 mm.

   a Using the 68–95–99.7% rule, for a random sample of 5000 of these birds, determine the number of these birds with a wing length between 59 mm and 63 mm. (2 marks)

   b Using the 68–95–99.7% rule, for a random sample of 10 000 of these birds, determine the number of these birds with a wing length of less than 57 mm. (2 marks)

11 (2 marks) The level of oil use in certain countries is approximately normally distributed with a mean of 42.2 units and a standard deviation of 10.2 units.

   a Determine the percentage of these countries in which the level of oil use is greater than 32 units. (1 mark)

   b Determine the percentage of these countries in which the level of oil use is between 32 and 52.4 units. (1 mark)

12 (2 marks) The pulse rates of a large group of 18-year-old students are approximately normally distributed with a mean of 75 beats/minute and a standard deviation of 11 beats/minute.

   a What is the percentage of 18-year-old students with a pulse rate less than 75 beats/minute? (1 mark)

   b What is the percentage of 18-year-old students with a pulse rate between 64 beats/minute and 86 beats/minute? (1 mark)

Calculator-assumed

13 (2 marks) The lifetime of a certain brand of light globe, in hours, is approximately normally distributed. It is known that 16% of the light globes have a lifetime of less than 655 hours and 50% of the light globes have a lifetime that is greater than 670 hours.

   a Determine the mean of this normal distribution. (1 mark)

   b Determine the standard deviation of this normal distribution. (1 mark)

14 (3 marks) The time, in hours, that each student spent sleeping on a school night was recorded for 1550 secondary school students. The distribution of these times was found to be approximately normally distributed with a mean of 7.4 hours and a standard deviation of 0.7 hours.

   a Determine the 0.95 quantile (95th percentile). (1 mark)

   b Determine the number of these students who spent more than 8.1 hours sleeping on a school night. (2 marks)

15 (4 marks) The volume of a cup of soup served by a machine is normally distributed with a mean of 240 mL and a standard deviation of 5 mL. A fast-food store used this machine to serve 160 cups of soup.

   a What is the number of these cups of soup that are expected to contain less than 230 mL of soup? (2 marks)

   b What is the number of these cups of soup that are expected to contain more than 245 mL of soup? (2 marks)

16 (2 marks) The length of 3-month-old baby boys is approximately normally distributed with a mean of 61.1 cm and a standard deviation of 1.6 cm.

    a Determine the percentage of 3-month-old baby boys with length greater than 59.5 cm. (1 mark)

    b Determine the percentage of 3-month-old baby boys with length less than 64.3 cm. (1 mark)

17 (3 marks) The dot plot below displays the maximum daily temperature (in °C) recorded at a weather station on each of the 30 days in November 2022.

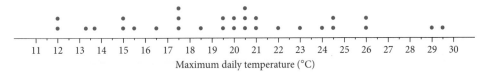

    a From this dot plot, determine the

        i median maximum daily temperature, correct to the nearest degree. (1 mark)

        ii percentage of days on which the maximum temperature was less than 16°C. Write your answer, correct to one decimal place. (1 mark)

Records show that the **minimum** daily temperature for November at this weather station is approximately normally distributed with a mean of 9.5°C and a standard deviation of 2.25°C.

    b Determine the percentage of days in November that are expected to have a minimum daily temperature less than 14°C at this weather station. Write your answer, correct to one decimal place. (1 mark)

18 (2 marks) The distribution of the weights of eggs produced by a chicken farm is approximately normal with a mean of 85 g and a standard deviation of 5 g.

    a Eggs weighing 95 g or more are classified as Extra Large. What is the percentage of eggs that would be classified as Extra Large? (1 mark)

    b Eggs weighing between 85 g to 95 g were classified as Large. If the chicken farm produced 5000 eggs a day, how many eggs would be classified as Large? (1 mark)

5.6 Standardised values

z-scores

Standardised values, also known as **z-scores**, allow us to compare values from different normal distributions. The standardised values

- always have mean = 0
- always have standard deviation = 1
- tell us the number of standard deviations each actual value lies from the mean.

> **Standardised values**
>
> To calculate standardised values from actual values, use the formula:
>
> $$\text{standardised value} = \frac{\text{actual value} - \text{mean}}{\text{standard deviation}} \quad \text{or} \quad z = \frac{x - \mu}{\sigma_x}$$
>
> To calculate actual values from standardised values, use the formula:
>
> $$\text{actual value} = (\text{standardised value} \times \text{standard deviation}) + \text{mean} \quad \text{or} \quad x = z \times \sigma_x + \mu$$

We can say the following about the actual value if we know its z-score.

z-score	What does it mean?
z-score is positive	The actual value is above the mean.
z-score is negative	The actual value is below the mean.
z-score = 0	The actual value equals the mean.
z-score = 1	The actual value is 1 standard deviation above the mean.
z-score = −2	The actual value is 2 standard deviations below the mean.

Exam hack

When dealing with z-scores rather than the original data, the three standardised 68–95–99.7% rule diagrams simplify to:

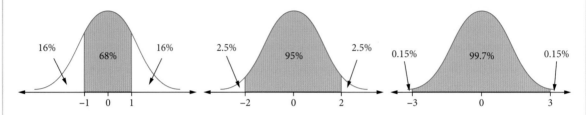

and the 68–95–99.7% rule scale simplifies to:

WORKED EXAMPLE 12 Working with z-scores

The lengths of the handspans of adults are known to be normally distributed with a mean of 21.5 cm and a standard deviation of 1.5 cm. A person has a handspan of 20 cm.

Steps | **Working**

a Calculate the standardised value for this actual value.

1 Write the z-score formula.

$$z = \frac{x - \mu}{\sigma_x}$$

2 Substitute in the actual value, mean and standard deviation.

$$z = \frac{20 - 21.5}{1.5} = \frac{-1.5}{1.5} = -1$$

b Show that this standardised value means that 84% of people have handspans wider than this person.

1 State how the z-score relates to the number of standard deviations.

$z = -1$ means the person's handspan is one standard deviation below the mean.

2 Sketch the relevant standardised 68–95–99.7% rule diagram and show the required region or use the 68-95-99.7% rule scale.

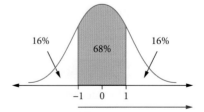

3 Read the percentage from the diagram.

From the diagram, 68% + 16% = 84%.

So, 84% of people have a handspan greater than this person.

c Another person has a standardised handspan of $z = 2.5$. What is this person's actual handspan?

Steps	Working
1 Write the formula for finding the actual value from the standardised value, and the values for z, σ_x and μ.	$x = z \times \sigma_x + \mu$ $z = 2.5$, $\sigma_x = 1.5$, $\mu = 21.5$
2 Substitute the values into the formula and solve for x.	$x = z \times \sigma_x + \mu$ $x = 2.5 \times 1.5 + 21.5$ $x = 25.25$
3 Write the answer.	The person's actual handspan is 25.25 cm.

Using z-scores to compare

Suppose you achieved a mark in the Mathematics Applications exam of 91% and a mark in the Mathematics Specialist exam of 49%. You couldn't necessarily say you performed better in Mathematics Applications because Mathematics Specialist is a more difficult subject. The way to compare the two results, assuming both distributions are bell-shaped, is to **standardise** them.

WORKED EXAMPLE 13 — Using z-scores to compare

The table below shows the marks out of 100 that a student has achieved on her exams in three mathematics subjects, plus the means and standard deviations for each of the subjects.

	Mark	Mean	Standard deviation
Mathematics Applications	91	73	6
Mathematics Methods	67	51	4
Mathematics Specialist	49	31	5

Assume the results for each of the three subjects approximates a bell-shaped distribution.

Steps	Working
a In which mathematics subject did the student perform the best?	
1 Write the z-score formula.	$z = \dfrac{x - \mu}{\sigma_x}$
2 Substitute the values for each subject.	Mathematics Applications $z = \dfrac{91 - 73}{6} = \dfrac{18}{6} = 3$ Mathematics Methods $z = \dfrac{67 - 51}{4} = \dfrac{16}{4} = 4$ Mathematics Specialist $z = \dfrac{49 - 31}{5} = \dfrac{18}{5} = 3.6$
3 State which z-score is the largest.	The student performed best in Mathematics Methods.

b In which of these subjects was she in the top 0.15% of students? Draw a diagram that shows how you obtained your answer.

1 Sketch the relevant standardised 68–95–99.7% rule diagram or use the 68–95–99.7% rule scale.

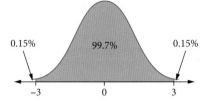

2 Read the answer from the diagram.

From the diagram, the top 0.15% of students had a z-score of 3 or more. The student had z-scores of 3, 4 and 3.6, so she was in the top 0.15% of students in all three mathematics subjects.

EXAMINATION QUESTION ANALYSIS

Video
Examination question analysis: Making sense of data

Calculator-assumed (7 marks)

The number of eggs counted in a sample of 12 clusters of moth eggs is recorded in the table.

Number of eggs	172	192	159	125	197	135	140	140	138	166	136	131

a From the information given, determine the
  i range (1 mark)
  ii percentage of clusters in this sample that contain more than 170 eggs. (1 mark)

In a large population of moths, the number of eggs per cluster is approximately normally distributed with a mean of 165 eggs and a standard deviation of 25 eggs.

b Using the 68–95–99.7% rule, determine the
  i percentage of clusters expected to contain more than 140 eggs (1 mark)
  ii number of clusters expected to have less than 215 eggs in a sample of 1000 clusters. (2 marks)

c The standardised number of eggs in one cluster is given by $z = -2.4$. Determine the actual number of eggs in this cluster. (2 marks)

Reading the question

- You will need to know both the range and standardised score formula.
- Make sure you are clear on what the 68–95–99.7% rule is.
- Identify in each question part whether you are being asked for a percentage or a number.

Thinking about the question

- Note that questions involving the 68–95–99.7% rule are best solved by drawing a scale or diagram.

Worked solution (✓ = 1 mark)

a i range = largest value − smallest value = 197 − 125 = **72** ✓

  ii There are 3 clusters with more than 170 eggs and 12 clusters in total, so $\frac{3}{12} \times 100\%$ = **25%** ✓ of clusters contain more than 170 eggs.

b i Write a 68–95–99.7% rule scale that includes the mean and standard deviations given.

| 0.15% | 2.35% | 13.5% | 34% | 34% | 13.5% | 2.35% | 0.15% |

−3σ −2σ −1σ   μ   +1σ +2σ +3σ
 90   115 140 165 190 215 240

From the scale, the percentage of clusters expected to contain more than 140 eggs is

34% + 34% + 13.5% + 2.35% + 0.15% = **84%** ✓

  ii From the scale, the percentage of clusters expected to have less than 215 eggs is

0.15% + 2.35% + 13.5% + 34% + 34% + 13.5% = 97.5%

Alternatively: 100% − (2.35% + 0.15%)
          = 100% − 2.5% = **97.5%** ✓

The number of clusters expected to have less than 215 eggs in a sample of 1000 clusters is

1000 × 97.5% = **975** ✓

> **🔒 Exam hack**
>
> Always note carefully if a question is asking for a number or a percentage. It's a common mistake in exams to calculate the wrong one.

c Use the formula for converting standardised values into actual values.

$x = z \times \sigma_x + \mu$

$z = -2.4, \sigma_x = 25, \mu = 165$

$x = -2.4 \times 25 + 165$ ✓

$x = 105$

The actual number in this cluster is **105** ✓ eggs.

EXERCISE 5.6 Standardised values ANSWERS p. 417

Recap

Use the following information to answer the next two questions.

The beak length of small birds in a large population is approximately normally distributed with a mean of 9.5 mm and a standard deviation of 0.50 mm.

1 Which one of the following statements relating to this population of birds is **not** true?

 A No bird will have a beak length that is less than 8.0 mm.

 B More than 99% of the birds will have a beak length that is less than 11 mm.

 C Approximately half of the birds will have a beak length that is less than 9.5 mm.

 D Approximately 2.5% of the birds will have a beak length that is greater than 10.5 mm.

 E Approximately 34% of the birds will have a beak length that is between 9.5 mm and 10.0 mm.

2 A random sample of 250 of these birds is captured and the beak length of each bird is measured. The expected number of these captured birds with beak lengths that are greater than 9 mm is closest to

 A 6   **B** 13   **C** 170   **D** 210   **E** 244

Mastery

3 **WORKED EXAMPLE 12** The heights of Year 12 teachers in Australia are known to be normally distributed with a mean of 175.6 cm and a standard deviation of 6.5 cm. A particular Year 12 teacher has a height of 162.6 cm.

   a Calculate the standardised value for this actual value.

   b Show that this standardised value means that 97.5% of Year 12 teachers are taller than this particular teacher.

   c Another Year 12 teacher has a standardised height of $z = -2.3$. What is this teacher's actual height?

4 **WORKED EXAMPLE 13** The table shows the marks out of 100 that a student has achieved on the final exams, plus the means and standard deviations for each of the subjects.

	Marks	Mean	Standard deviation
Mathematics Applications	78	72	5
German	47	48	2
Hospitality	77	71	2
Psychology	50	54	4
Systems Engineering	62	68	5

Assuming the results for each of the subjects approximate a bell-shaped distribution:

   a In which of the subjects did the student perform the best?

   b In which of these subjects was the student in the bottom 16%? Draw a diagram which shows how you obtained your answer.

Calculator-free

5 (4 marks) The lengths of the left foot of a large sample of Year 11 students were measured and recorded. These foot lengths were found to be approximately normally distributed with a mean of 24.2 cm and a standard deviation of 4.2 cm.

   a A Year 11 student has a foot length of 23 cm. Determine the student's standardised foot length (standard z-score), correct to one decimal place. (2 marks)

   b Determine the percentage of students with foot lengths between 20.0 and 24.2 cm. (2 marks)

6 (4 marks) A student obtains a mark of 56 on a test for which the mean mark is 67 and the standard deviation is 10.2.

   a Determine the student's standardised mark (standard z-score) to one decimal place. (2 marks)

   b A second student has a standard z-score of 1.50. Determine their mark for the same test. (2 marks)

7 (6 marks) The pulse rates of a population of Year 11 students are approximately normally distributed with a mean of 69 beats per minute and a standard deviation of 4 beats per minute.

   a A student selected at random from this population has a standardised pulse rate of $z = -2.5$. Determine the student's actual pulse rate. (2 marks)

   b A student selected at random from this population has a standardised pulse rate of $z = -1$. Determine the percentage of students in this population with a pulse rate greater than this student. (2 marks)

   c A sample of 200 students was selected at random from this population. Determine the number of these students with a pulse rate of less than 61 beats per minute or greater than 73 beats per minute. (2 marks)

8 (4 marks) A class of students sat for a Biology test and a Legal Studies test. Each test had a possible maximum score of 100 marks. The table shows the mean and standard deviation of the marks obtained in these tests. The class marks in each subject are approximately normally distributed.

	Subject	
	Biology	Legal Studies
Class mean	54	78
Class standard deviation	15	5

a Sashi obtained a mark of 81 in the Biology test. What is her standardised score for Biology? (2 marks)

b What mark does Sashi need to obtain on the Legal Studies test to achieve the same standard score for both Legal Studies and Biology? (2 marks)

9 (5 marks) The *weight* of a species of bird is approximately normally distributed with a mean of 71.5 g and a standard deviation of 4.5 g.

a What is the standardised weight (*z*-score) of a bird weighing 67.9 g? (2 marks)

b Use the 68–95–99.7% rule to estimate

  i the expected percentage of these birds that weigh less than 67 g (1 mark)

  ii the expected number of birds that weigh between 62.5 g and 76.0 g in a flock of 200 of these birds. (2 marks)

Chapter summary

Types of data

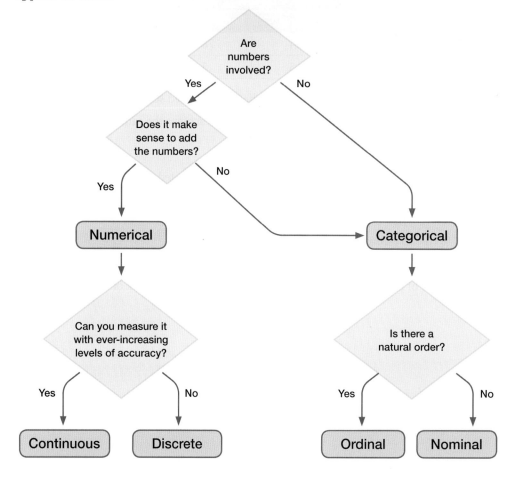

Describing data

Measure of centre

Measure of centre	Use for	Description
mode	numerical categorical	• most frequently occurring data value • is called the **modal category** for categorical data • data with more than one mode is called **multimodal** • if every data value appears exactly once, there's no mode
median	numerical ordinal	• the middle value • if there are two middle values, add them and divide by 2
mean	numerical	• the average of all the data values • $\bar{x} = \dfrac{\text{sum of all values}}{\text{number of values}}$ for a list of data values • $\bar{x} = \dfrac{\text{sum of (each value} \times \text{its corresponding frequency)}}{\text{sum of frequencies}}$ for data in a frequency table

Measures of spread

Measures of spread	Use for	Description
Range	numerical	• measures the spread of the entire data set • range = largest value − smallest value
Standard deviation	numerical	• measures the spread around the mean • $\sigma = \sqrt{\dfrac{\Sigma(x - \bar{x})^2}{n}}$ (use CAS)

Outliers

- An **outlier** is an extreme high or low data value.

Standard deviations of a data set from the mean

0.15%	2.35%	13.5%	34%	34%	13.5%	2.35%	0.15%
-3σ	-2σ	-1σ	μ	$+1\sigma$	$+2\sigma$	$+3\sigma$	

Centre, spread, display and data type summary

Categorical: Nominal data	Categorical: Ordinal data	Numerical data
Measures of centre		
mode	mode median	mode median mean
Measures of spread		
–	range	range standard deviation
Displays		
dot plot bar chart	dot plot bar chart	dot plot histogram stem plot

Data displays

Display	Description
Frequency table	- involves counting the number of times each data value occurs - frequencies often shown as percentages - percentage = $\dfrac{\text{frequency}}{\text{total}} \times 100\%$
Bar chart	- bars can be horizontal or vertical
Grouped frequency table	- involves numerical data that has been grouped into regular intervals - makes it easier to deal with large amounts of data
Histogram	- graphical display of data from a grouped frequency table - best if data is grouped into 5 to 15 intervals
Dot plot	- used for both categorical and numerical data - should not involve too many data values or a large data spread
Stem plot	- involves actual data values and requires a key - best used with a maximum of 50 data values and to see all the data values - back-to-back stem plots allow us to compare the distribution of numerical data for two groups

Describing distributions

Symmetric distributions

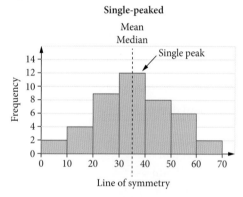

Modal interval: 30–<40   Multimodal: 20–<30 and 80–<90

The mean and median are both close to the line of symmetry.

Skewed distributions

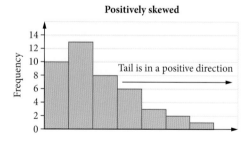 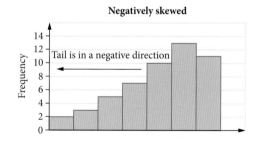

The mean is greater than the median.   The mean is less than the median.

Distributions with outliers

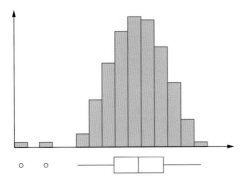

Mean vs median

Shape of distribution	Choose
Approximately symmetric distributions with no outliers	mean or median
Approximately symmetric distributions with outliers	median
Skewed distributions	median

Normal or bell-shaped distributions

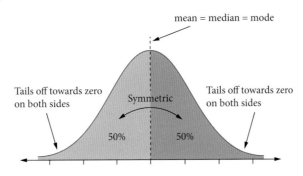

- The **68–95–99.7% rule** for approximate **bell-shaped distributions** says:

Around 68% of the data values lie within one standard deviation of the mean.

Around 95% of the data values lie within two standard deviations of the mean.

Around 99.7% of the data values lie within three standard deviations of the mean.

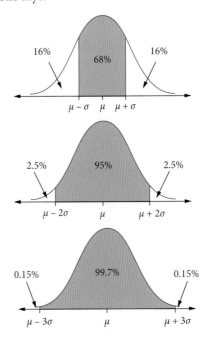

68–95–99.7% rule scale

```
 0.15% | 2.35% | 13.5% | 34% | 34% | 13.5% | 2.35% | 0.15%
   -3σ    -2σ    -1σ     μ    +1σ    +2σ    +3σ
```

Standardised values (z-scores)

- **standardised value** = $\dfrac{\text{data value} - \text{mean}}{\text{standard deviation}}$ or $z = \dfrac{x - \mu}{\sigma_x}$

- actual value = (standardised value × standard deviation) + mean or $x = z \times \sigma_x + \mu$

- 68–95–99.7% rule for *z*-scores:

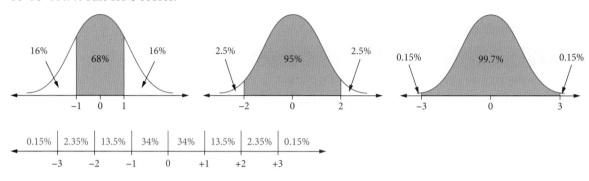

```
 0.15% | 2.35% | 13.5% | 34% | 34% | 13.5% | 2.35% | 0.15%
   -3     -2     -1     0    +1    +2     +3
```

Cumulative examination: Calculator-free

Total number of marks: 26   Reading time: 3 minutes   Working time: 26 minutes

1 (5 marks) Eric works as a farm-hand pruning fruit trees. He is paid $10 for each fruit tree he prunes. He can prune two fruit trees each hour.

  a How much can Eric earn per hour? (1 mark)

  b How many fruit trees does he need to prune to earn $200? (2 marks)

  c How much would Eric earn in a day if he works for eight hours? (2 marks)

2 (2 marks) A right-angled triangle has side lengths of 5 cm and 12 cm. Find the length of the longest side.

3 (3 marks) The following table shows the data collected from a random sample of seven drivers drawn from the population of all drivers who used a supermarket car park on one day. The variables in the table are:

- *distance* – the distance that each driver travelled to the supermarket from their home
- *sex* – the sex of the driver (female, male)
- *number of children* – the number of children in the car
- *type of car* – the type of car (sedan, wagon, other)
- *postcode* – the postcode of the driver's home.

Distance (km)	Sex (F = female, M = male)	Number of children	Type of car (1 = sedan, 2 = wagon, 3 = other)	Postcode
4.2	F	2	1	8148
0.8	M	3	2	8147
3.9	F	3	2	8146
5.6	F	1	3	8245
0.9	M	1	3	8148
1.7	F	2	2	8147
2.5	M	2	2	8145

  a List the discrete numerical variable/s in this data set. (1 mark)

  b List the categorical variables in this data set. (1 mark)

  c How many of the female drivers have three children in the car? (1 mark)

4 (2 marks) The following ordered stem plot shows the areas, in square kilometres, of 27 suburbs of a large city.

```
Stem | Leaf
  1  | 5 6 7 8
  2  | 1 2 4 5 6 8 9 9
  3  | 0 1 1 2 2 8 9
  4  | 0 4 7
  5  | 0 1
  6  | 1 9
  7  |
  8  | 4
```

Key: $1|6 = 1.6\,\text{km}^2$

  a What is the median area of these suburbs, in square kilometres? (1 mark)

  b What is the range area of these suburbs, in square kilometres? (1 mark)

5 (2 marks) The histogram shows the distribution of the number of billionaires per million people for 53 countries.

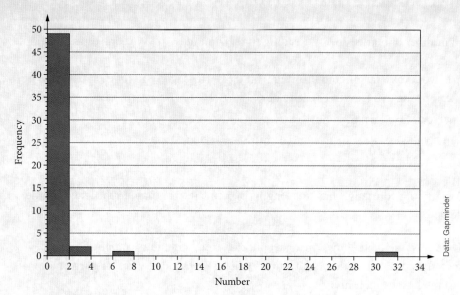

- **a** Using this histogram, what is the percentage of these 53 countries with less than two billionaires per million people? (1 mark)
- **b** What is the shape of the distribution? (1 mark)

6 (8 marks) The weights of male players in a basketball competition are approximately normally distributed with a mean of 78.6 kg and a standard deviation of 9.3 kg. There are 456 male players in the competition.
- **a** What is the percentage of males who would be expected to weigh less than 87.9 kg? (1 mark)
- **b** What is the percentage of males who would be expected to weigh between 60 and 87.9 kg? (1 mark)
- **c** What is the expected number of male players in the competition with weights above 60 kg? (2 marks)
- **d** Brett and Sanjeeva both play in the basketball competition. When the weights of all players in the competition are considered, Brett has a standardised weight of $z = -0.96$ and Sanjeeva has a standardised weight of $z = -0.26$. Explain these standardised weights. (4 marks)

 Exam hack

Always aim to have 60% of the exam done by halfway through the exam.

7 (4 marks) The birth weights of a large population of babies are approximately normally distributed with a mean of 3300 g and a standard deviation of 550 g.
- **a** A baby selected at random from this population has a standardised weight of $z = -0.75$. Show a calculation that will result in the *actual birth weight* of this baby. (1 mark)
- **b** Using the 68–95–99.7% rule, what is the percentage of babies with a birth weight of less than 1650 g? (1 mark)
- **c** A sample of 600 babies was drawn at random from this population. Using the 68–95–99.7% rule, what is the number of these babies with a birth weight between 2200 g and 3850 g? (2 marks)

Cumulative examination: Calculator-assumed

Total number of marks: 35   Reading time: 4 minutes   Working time: 35 minutes

1 (7 marks) Elena, a deli owner, imports three different brands of olive oil: Carmani (C), Linelli (L) and Ohana (O). The number of 1 litre bottles of these oils imported for sale in the month of January 2023 is shown in matrix J.

$$J = \begin{bmatrix} 2800 \\ 1700 \\ 2400 \end{bmatrix} \begin{matrix} C \\ L \\ O \end{matrix}$$

   a State the order of matrix J. (1 mark)

   b For sale in February 2023, Elena increases the number of 1 litre bottles she imported in January 2023 by 5%. Show use of scalar multiplication to state a matrix F, for the number of 1 litre bottles Elena imports for sale in February 2023. (2 marks)

   c Show the use of two appropriate matrix operations to determine the total number of 1 litre bottles of olive oil imported for sale in January and February 2023 combined. (2 marks)

   d If a 1 litre bottle of Carmani costs $5 to import, a 1 litre bottle of Linelli costs $6 to import and a 1 litre bottle of Ohana costs $8 to import, show the use of appropriate matrix operations to determine the cost of all imports over the two-month period. (2 marks)

2 (3 marks) A one-on-one basketball court is a composite shape made up of a rectangle and a semicircle, as shown.

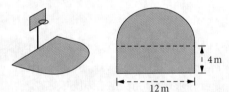

A boundary line is painted around the perimeter of the shape.

   a What is the total length of the boundary line? (2 marks)

   b If it costs $12/m to paint the boundary line, determine the total cost to paint the perimeter of the shape. (1 mark)

3 (2 marks) A chocolate bar is in the shape of a triangular prism. The manufacturer wants to increase profits by reducing the size of the bar.

   a If the dimensions of the new bar are reduced by half, what is the scale factor? (1 mark)

   b If the original bar was 600 g, and weight is directly proportional to volume, how much does the new bar weigh? (1 mark)

4 (5 marks) The table displays the *average sleep time*, in hours, for a sample of 19 types of mammals.

Type of mammal	Average sleep time (hours)
cat	14.5
squirrel	13.8
mouse	13.2
rat	13.2
grey wolf	13.0
arctic fox	12.5
raccoon	12.5
gorilla	12.0
jaguar	10.8
baboon	9.8
red fox	9.8
rabbit	8.4
guinea pig	8.2
grey seal	6.2
cow	3.9
sheep	3.8
donkey	3.1
horse	2.9
roedeer	2.6

Data: T Allison and DV Cicchetti, 'Sleep in Mammals: Ecological and Constitutional Correlates', in Science, American Association for the Advancement of Science, vol. 194, no. 4266, pp. 732–734, 12 November 1976; accessed from OzDASL, StatSci.org, www.statsci.org/data/general/sleep.html

- **a** Which of the two variables, *type of mammal* or *average sleep time*, is a nominal variable? (1 mark)
- **b** Determine the mean and standard deviation of the variable *average sleep time* for this sample of mammals. Round your answers to one decimal place. (2 marks)
- **c** The average sleep time for a human is eight hours. What percentage of this sample of mammals has an *average sleep time* that is less than the average sleep time for a human? Round your answer to one decimal place. (1 mark)
- **d** The sample is increased in size by adding in the average sleep time of the little brown bat. Its average sleep time is 19.9 hours. By how many hours will the range for *average sleep time* increase when the average sleep time for the little brown bat is added to the sample? (1 mark)

5 (6 marks) The dot plot shows the distribution of daily rainfall, in millimetres, at a weather station for 30 days in September.

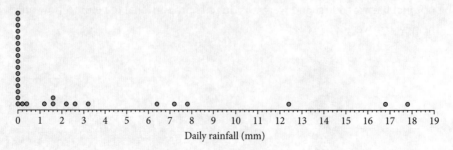

- **a** Find the
  - **i** range (1 mark)
  - **ii** median. (1 mark)

b Determine the
  i number of days on which no rainfall was recorded (1 mark)
  ii percentage of days on which the daily rainfall exceeded 12 mm. (1 mark)
c Construct a histogram on a grid like the one shown that displays the distribution of daily rainfall for the month of September. Use interval widths of two with the first interval starting at 0. (2 marks)

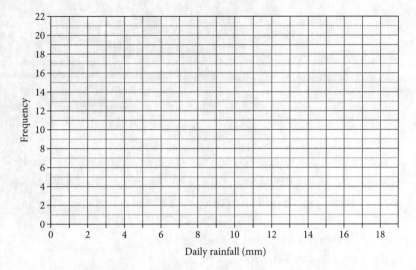

6 (4 marks) The *neck size*, in centimetres, of 250 men was recorded and displayed in the dot plot below.

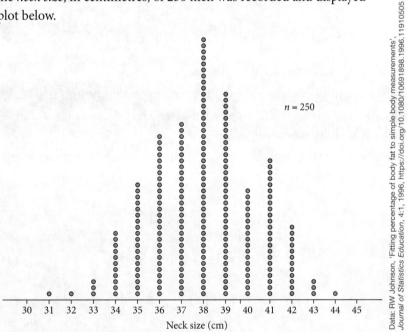

a Write down the modal *neck size*, in centimetres, for these 250 men. (1 mark)

b Assume that this sample of 250 men has been drawn at random from a population of men whose *neck size* is normally distributed with a mean of 38 cm and a standard deviation of 2.3 cm.

  i How many of these 250 men are **expected** to have a *neck size* that is more than three standard deviations above or below the mean? Round your answer to the nearest whole number. (2 marks)

  ii How many of these 250 men **actually** have a *neck size* that is more than three standard deviations above or below the mean? (1 mark)

7 (3 marks) The histogram shows the distribution of mean yearly rainfall (in mm) for Australia over 103 years.

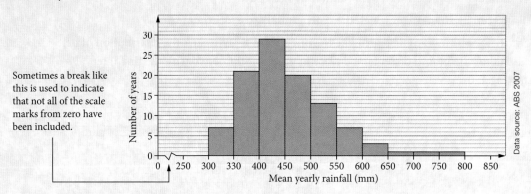

a   Describe the shape of the histogram. (1 mark)

b   Use the histogram to determine

   i   the number of years in which the mean yearly rainfall was 500 mm or more (1 mark)

   ii the percentage of years in which the mean yearly rainfall was between 500 mm and 600 mm. Write your answer correct to one decimal place. (1 mark)

8 (5 marks) The stem plot shows the number of customers in a store each day over a two-week period.

Stem	Leaf
2	3 7
3	1 1 4 6 7
4	0 3 5 8 9
5	2 6

Key: 1 | 2 means 12 years

a   Use the data to find the

   i   mean and standard deviation, correct to two decimal places (2 marks)

   ii number of data values that are one standard deviation from the mean. (1 mark)

b   Assuming the data has a normal distribution, what percentage of data values would lie one standard deviation from the mean? (1 mark)

c   What is the actual percentage of data values that lie within one standard deviation from the mean, correct to one decimal place? (1 mark)

COMPARING DATA

CHAPTER

Syllabus coverage
Nelson MindTap chapter resources
6.1 Five-number summary and outliers
   The five-number summary
   Using CAS 1: Finding the five-number summary
   IQR, outliers and fences
6.2 Box plots
   Box plots
   Using CAS 2: Constructing box plots
   Comparing box plots and histograms
6.3 Back-to-back stem plots and parallel box plots
   Back-to-back stem plots
   Parallel box plots
   Using CAS 3: Constructing parallel box plots
   Which display is best to compare data?
6.4 Comparing data using measures of centre and spread
Examination question analysis
6.5 Statistical investigation process
Chapter summary
Cumulative examination: Calculator-free
Cumulative examination: Calculator-assumed

Syllabus coverage

TOPIC 2.1: UNIVARIATE DATA ANALYSIS AND THE STATISTICAL INVESTIGATION PROCESS

The statistical investigation process

2.1.1 review the statistical investigation process; identifying a problem and posing a statistical question, collecting or obtaining data, analysing the data, interpreting and communicating the results

Comparing data for a numerical variable across two or more groups

2.1.10 construct and use parallel box plots (including the use of the 'Q1 − 1.5 × IQR' and 'Q3 + 1.5 × IQR' criteria for identifying possible outliers) to compare groups in terms of location (median), spread (IQR and range) and outliers, and interpret and communicate the differences observed in the context of the data

2.1.11 compare groups on a single numerical variable using medians, means, IQRs, ranges or standard deviations, and as appropriate; interpret the differences observed in the context of the data and report the findings in a systematic and concise manner

2.1.12 implement the statistical investigation process to answer questions that involve comparing the data for a numerical variable across two or more groups; for example, are Year 11 students the fittest in the school?

Mathematics Applications ATAR Course Year 11 syllabus pp. 11–12 © SCSA

Video playlists (6):

6.1 Five-number summary and outliers
6.2 Box plots
6.3 Back-to-back stem plots and parallel box plots
6.4 Comparing data using measures of centre and spread
6.5 Statistical investigation process

Examination question analysis Comparing data

Worksheets (13):

6.1 Five-number summaries 1 • Interquartile range
6.2 Box plots • Box plots 1 • Box plots 2 • Five-number summaries 2
6.3 Box-and-whisker plots • Back-to-back stem plots
6.4 Comparing group measures • Comparing city temperatures • Comparing word lengths • Comparing sports scores • Investigating young drivers

Puzzle (1):

6.1 Statistical measures puzzle

Nelson MindTap

To access resources above, visit
cengage.com.au/nelsonmindtap

6.1 Five-number summary and outliers

In Chapter 5 we looked at data relating to a single variable. In this chapter we will be comparing data for a numerical variable across two or more groups. One way of doing this is by using the **five-number summary**, which involves five numbers that give us information about the distribution.

The five-number summary

The five-number summary provides a good overview of a distribution. It is created using the maximum and minimum data values and the three quartiles.

The median is the value that divides an ordered data set in half. The **quartiles** are three values that divide an ordered data set into quarters.

- Q_1 or the **lower quartile** is the median of the lower half of the data.
- Q_2 is the **median** of the whole data set.
- Q_3 or the **upper quartile** is the median of the upper half of the data.

Example with an odd number of data values:

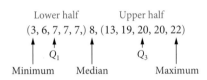

Example with an even number of data values:

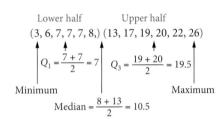

> ### The five-number summary
>
> The five-number summary divides the data into segments of 25%:
>
> - 25% of the data is less than Q_1.
> - 50% of the data is less than the median (Q_2).
> - 75% of the data is less than Q_3.
>
>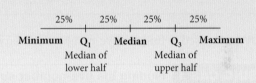
>
> These percentages are not exact for an odd-numbered data set because the median is one of the data values and we can't split a data value in half.

WORKED EXAMPLE 1 | Finding the five-number summary by hand

For the following data

23, 25, 35, 35, 22, 49, 7, 26, 24, 31, 22, 30

a find the five-number summary by hand

b use a diagram to show that

 i   25% of the data is less than the lower quartile (Q_1)

 ii 50% of the data is less than the median (Q_2)

 iii 75% of the data is less than the upper quartile (Q_3).

Steps	Working
a 1 Order the data from smallest to largest.	7, 22, 22, 23, 24, 25, 26, 30, 31, 35, 35, 49
2 Find the minimum and maximum value.	min = 7, max = 49

	3 Find the median. There is an even number of data values (12), so find the average of the two middle points.	The median is between the 6th and 7th data values in the ordered list. $Q_2 = \text{median} = \dfrac{25 + 26}{2} = 25.5$
	4 Find Q_1, the median of the lower half of the data.	The lower half of the data is 7, 22, 22, 23, 24, 25. $Q_1 = \text{lower quartile} = \dfrac{22 + 23}{2} = 22.5$
	5 Find Q_3, the median of the upper half of the data.	The upper half of the data is 26, 30, 31, 35, 35, 49. $Q_3 = \text{upper quartile} = \dfrac{31 + 35}{2} = 33$
	6 List the five-number summary.	min = 7, $Q_1 = 22.5$, median = 25.5, $Q_3 = 33$, max = 49
b	1 Draw a diagram showing the three quartiles.	7, 22, 22, | 23, 24, 25, | 26, 30, 31, | 35, 35, 49 Lower quartile = 22.5 Upper quartile = 33 Median = 25.5
	2 Use the diagram to calculate the percentage of data less than each quartile.	i 3 out of a total of 12 data values are less than the lower quartile: $\dfrac{3}{12} = 25\%$ ii 6 out of a total of 12 data values are less than the median: $\dfrac{6}{12} = 50\%$ iii 9 out of a total of 12 data values are less than the upper quartile: $\dfrac{9}{12} = 75\%$

CAS can be used to find the five-number summary of a set of data.

USING CAS 1 Finding the five-number summary

Calculate the five-number summary for the following data:

65, 47, 61, 44, 63, 56, 65, 52, 58

ClassPad

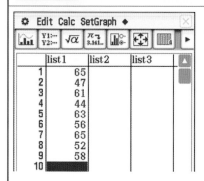

1 Tap **Menu** and open the **Statistics** application.

2 Clear all lists and enter the data as shown.

3 Tap **Calc > One-Variable**.

4 Leave the **XList** default setting of **list1**.

5 Tap **OK**.

6 The one-variable statistics will be displayed.

7 Scroll down to view the five-number summary values.

The five number summary is min = 44, $Q_1 = 49.5$, median = 58, $Q_3 = 64$, max = 65.

TI-Nspire

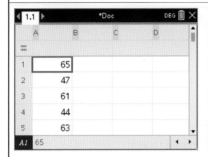
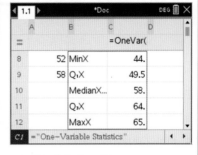

1. Start a new document and add a **Lists & Spreadsheet** page.
2. Enter the data into column **A** as shown.
3. Press **menu** > **Statistics** > **Stat Calculations** > **One-Variable Statistics**.
4. On the next screen, keep the number of lists default setting of **1** and select **OK**.
5. Leave the **X1 List** default setting of **a[ ]** then select **OK**.
6. The labels will appear in column **B** and the corresponding one-variable statistics will appear in column **C**.
7. Scroll down to view the five-number summary values.

The five number summary is min = 44, Q_1 = 49.5, median = 58, Q_3 = 64, max = 65.

> **Exam hack**
>
> If a data set is small and already ordered, it can sometimes be quicker to find the five-number summary by hand than by using CAS.

IQR, outliers and fences

The **interquartile range** (**IQR**) is the measure of the spread of the middle 50% of the data values.

$$IQR = Q_3 - Q_1$$

The IQR is usually a better measure of spread than the range because, by looking at only the middle 50% of data, we avoid taking outliers into account. The IQR is also used in a calculation to identify possible outliers, which allows us to do more than simply say something 'looks like an outlier'.

> **Interquartile range and fences**
>
> $$IQR = Q_3 - Q_1$$
>
> A data value is a possible outlier if it is
>
> less than the **lower fence**: $Q_1 - 1.5 \times IQR$
>
> or
>
> greater than the **upper fence**: $Q_3 + 1.5 \times IQR$

Worksheet Interquartile range

Puzzle Statistical measures puzzle

WORKED EXAMPLE 2 Finding outliers

For the ordered data set

3, **7**, 20, 22, 22, 22, 25, 25, 28, 31, 34, 34, **49**

perform a calculation to show whether the **blue** values are possible outliers.

Steps	Working
1 Find Q_1 and Q_3 by using CAS or by hand.	Q_1 = 21 and Q_3 = 32.5
2 Calculate the IQR.	IQR = 32.5 − 21 = 11.5
3 Calculate the lower and upper fences.	Lower fence: $Q_1 - 1.5 \times IQR = 21 - 1.5 \times 11.5 = 3.75$ Upper fence: $Q_3 + 1.5 \times IQR = 32.5 + 1.5 \times 11.5 = 49.75$
4 Check each of the **blue** values to see if they are less than the lower fence or greater than the upper fence.	**3** is less than 3.75 so it *is* a possible outlier. **7** isn't less than 3.75, so it's *not* an outlier. **49** isn't greater than 49.75, so it's *not* an outlier.

WORKED EXAMPLE 3 — IQR from a dot plot

The dot plot shows the ages of Student Council members at a school.

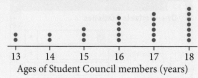

Find the

a median

b lower quartile (Q_1)

c upper quartile (Q_3)

d interquartile range (IQR).

Steps	Working
a 1 Count the number of dots, n, note whether it's odd or even, and find the position of the median.	$n = 25$; odd $$\frac{n+1}{2} = \frac{25+1}{2} = \frac{26}{2} = 13$$ The median is the 13th ordered data value.
2 If n is odd, find the data value of the middle dot. If n is even, find the average of the data values for the two middle dots. Count each column of dots from the bottom up to reach the median.	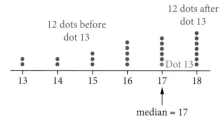 median = 17 years
b 1 To find the lower quartile, Q_1, find the median of the lower half. Count the number of dots, n, note whether it's odd or even, and find the position of Q_1.	$n = 12$; even $$\frac{n+1}{2} = \frac{12+1}{2} = \frac{13}{2} = 6.5$$ Q_1 is between the 6th and 7th ordered data values in the lower half.
2 If n is odd, find the data value of the middle dot. If n is even, find the average of the data values for the two middle dots. Count each column of dots from the bottom up to reach the lower quartile.	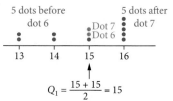 $$Q_1 = \frac{15+15}{2} = 15$$ $Q_1 = 15$ years
c 1 To find the upper quartile, Q_3, find the median of the upper half. Count the number of dots n, note whether it's odd or even, and find the position of Q_3.	$n = 12$; even $$\frac{n+1}{2} = \frac{12+1}{2} = \frac{13}{2} = 6.5$$ Q_3 is between the 6th and 7th ordered data values in the upper half.
2 If n is odd, find data value of the middle dot. If n is even, find the average of the data values for the two middle dots. Count each column of dots from the bottom up to reach the upper quartile.	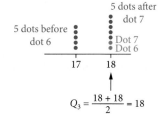 $$Q_3 = \frac{18+18}{2} = 18$$ $Q_3 = 18$ years
d Use Q_3 and Q_1 to calculate the interquartile range.	IQR = $Q_3 - Q_1 = 18 - 15 = 3$ years

WORKED EXAMPLE 4 — Outliers in a stem plot

The stem plot shows the test scores out of 70 for a class of 37 students.

a Find the
  i median
  ii lower quartile (Q_1)
  iii upper quartile (Q_3)
  iv interquartile range (IQR).

b Is there a possible outlier? Justify your answer.

Stem	Leaf
0	5
1	
2	5 6 6 7 9 9
3	2 3 3 4 6 6 6 8 9
4	0 1 2 3 3 4 5 5 6 8
5	1 1 6 6 6 7 8
6	0 5 7 9

Key: 2 | 5 means 25

Steps	**Working**

a i 1 Count the number of data values n, note whether it's odd or even, and find the position of the median.

$n = 37$; odd

$$\frac{n+1}{2} = \frac{37+1}{2} = \frac{38}{2} = 19$$

The median is the 19th ordered data value.

2 If n is odd, find the middle data value. If n is even, find the average of the two middle data values.

Stem	Leaf	
0	5	
1		18 values before the median
2	5 6 6 7 9 9	
3	2 3 3 4 6 6 6 8 9	
4	0 1 **2** 3 3 4 5 5 6 8	median = 42
5	1 1 6 6 6 7 8	
6	0 5 7 9	18 values after the median

median = 42

ii 1 To find the lower quartile, Q_1, find the median of the lower half.

Count the number of data values, n, note whether it's odd or even, and find the position of Q_1.

$n = 18$; even

$$\frac{n+1}{2} = \frac{18+1}{2} = \frac{19}{2} = 9.5$$

Q_1 is between the 9th and 10th ordered data values in the lower half.

2 If n is odd, find the middle data value. If n is even, find the average of the two middle data values.

Stem	Leaf	
0	5	8 values before the 9th value
1		
2	5 6 6 7 9 9	
3	2 **3** 3 4 6 6 6 8 9	$Q_1 = \frac{33+33}{2} = 33$
4	0 1	8 values after the 10th value

$Q_1 = 33$

iii 1 To find the upper quartile, Q_3, find the median of the upper half.

Count the number of data values, n, note whether it's odd or even, and find the position of Q_3.

$n = 18$; even

$$\frac{n+1}{2} = \frac{18+1}{2} = \frac{19}{2} = 9.5$$

Q_3 is between the 9th and 10th ordered data values in the upper half.

2 If n is odd, find the middle data value. If n is even, find the average of the two middle data values.

Stem	Leaf	
		8 values before the 9th value
4	3 3 4 5 5 6 8	
5	1 **1** 6 6 6 7 8	$Q_3 = \frac{51+56}{2} = 53.5$
6	0 5 7 9	8 values after the 10th value

$Q_3 = 53.5$

iv Use Q_3 and Q_1 to calculate the interquartile range.

IQR = $Q_3 - Q_1$ = 53.5 − 33 = 20.5

b Check any value that appears to be an outlier against the upper or lower fence.

The score of 5 may be an outlier. Check using the lower fence.

$Q_1 - 1.5 \times$ IQR $= 33 - 1.5 \times 20.5 = 2.25$

5 isn't less than 2.25, so it's not an outlier.

EXERCISE 6.1 Five-number summary and outliers

ANSWERS p. 418

Mastery

1 WORKED EXAMPLE 1 | Using CAS 1 | For the following data

39, 49, 76, 61, 42, 65, 62, 35, 78, 80, 59, 54

a find the five-number summary by hand and verify your answers by using CAS

b use a diagram to show that

  i 25% of the data is less than the lower quartile (Q_1)

  ii 50% of the data is less than the median (Q_2)

  iii 75% of the data is less than the upper quartile (Q_3).

2 WORKED EXAMPLE 2 | For each of the following data sets, do a calculation to show whether the **blue** values are possible outliers

a **52**, 73, 76, 81, 81, 90, 90, 92, 95, 96, 96, 105, **110**

b **9**, 16, 19, 20, 23, 23, 24, 24, 25, 25, 26, 26, 27, 27, 27, **33**, **35**

3 WORKED EXAMPLE 3 | The dot plot shows the number of text messages sent by a group of students during a one-hour period.

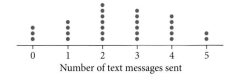

Find the

a median

b lower quartile (Q_1)

c upper quartile (Q_3)

d interquartile range (IQR).

4 WORKED EXAMPLE 4 | The stem plot shows the ages of 33 customers in a store recorded one day.

Stem	Leaf
2	3 5 7 7 7 7 8 9 9
3	0 1 1 1 1 1 3 4 4 5 6 7 8 8
4	0 2 3 4 4 8
5	7 9
6	3
7	7

Key: 2 | 3 means 23

a Find the

  i median

  ii lower quartile (Q_1)

  iii upper quartile (Q_3)

  iv interquartile range (IQR).

b Is there a possible outlier? Justify your answer.

Calculator-free

5 (3 marks) Complete a calculation to show whether the **blue** values are possible outliers.

12, **23**, 33, 36, 38, 40, 44, 45, 45, 48, 49, 49, 50, 52, 55, **71**

6 (7 marks)

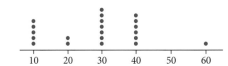

a For the dot plot, find the

  i median (1 mark)

  ii lower quartile (Q_1) (1 mark)

  iii upper quartile (Q_3) (1 mark)

  iv interquartile range (IQR). (2 marks)

b Is there a possible outlier? Justify your answer. (2 marks)

244 Nelson WAmaths Mathematics Applications 11

Calculator-assumed

7 (4 marks) Given the following dot plot

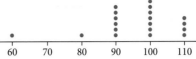

   a find the IQR   (2 marks)

   b do a calculation to show whether 60 is an outlier.   (2 marks)

8 (4 marks) Given the following stem plot

Stem	Leaf
6	0
7	9
8	3 3 4 5 5 5 7 8
9	0 4 4 4 5 6 7 7 7
	8 9

Key: 7 | 9 means 79

   a find the IQR   (2 marks)

   b do a calculation to show whether 60 is an outlier.   (2 marks)

6.2 Box plots

Video playlist
Box plots

Worksheets
Box plots
Box plots 1
Box plots 2
Five-number summaries 2

Box plots

Box plots, also known as **box-and-whisker plots**, display numerical data based on the five-number summary, IQR and outliers. If there are no outliers, the **whiskers** show the minimum and maximum values.

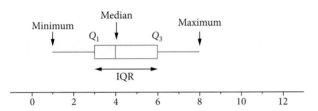

> 🔒 **Exam hack**
>
> The lengths of the boxes and whiskers depend on the spread of the data values in each quartile.

Outliers are shown as dots. If there are outliers, the lowest outlier is the minimum value and the highest outlier is the maximum value.

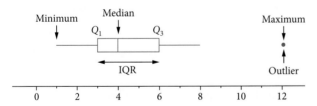

> 🔒 **Exam hack**
>
> You need to include outliers when finding the minimum and maximum values.

Box plots

Box plots provide the following information:

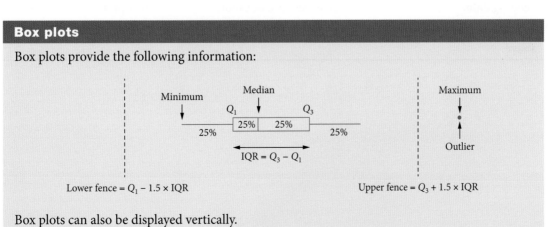

Box plots can also be displayed vertically.

USING CAS 2 Constructing box plots

Construct a box plot for the data set:

6, 21, 21, 22, 23, 24, 25, 29, 30, 34, 34, 48

ClassPad

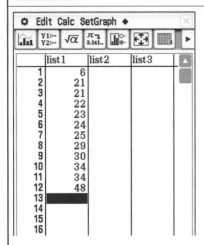  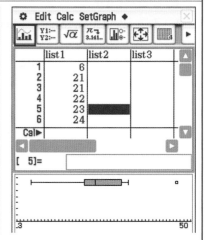

1. Tap **Menu** and open the **Statistics** application.
2. Clear all lists.
3. Enter the data into **list1** as shown above.
4. Tap **SetGraph** > **Setting**.
5. In the dialogue box, change the **Type**: field to **MedBox**.
6. Tap the box to select **Show Outliers**.
7. Tap **Set**.
8. Tap on the **Graph** tool to display the data as a box plot in the lower window.
9. With the graph window highlighted, tap **Analysis** > **Trace** to display the key values.

TI-Nspire

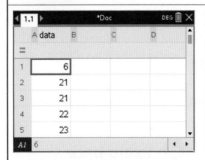 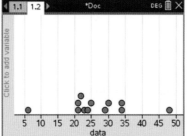 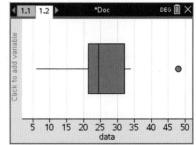

1. Start a new document and add a **Lists & Spreadsheet** page.
2. Label column **A** with the heading data.
3. Enter the data as shown above.
4. Insert a **Data & Statistics** page.
5. For the horizontal axis, select **data**. The data will be displayed as a dot plot.
6. Press **menu** > **Plot Type** > **Box Plot**. The data will be displayed as a box plot.
7. Move the cursor over the box plot to display the key values.

WORKED EXAMPLE 5 | Reading box plots

The box plot shows the distribution of 48 student test scores marked out of 100.

Find the

a five-number summary

b percentage of students who scored more than 65

c percentage of students who scored less than 55

d percentage of students who scored between 30 and 65

e number of students who scored less than 45

f scores at the lower end that would be considered outliers

g scores at the upper end that would be considered outliers.

Steps	Working
a Read directly from the box plot.	min = 30, Q_1 = 45, median = 55, Q_3 = 65, max = 90
Use the fact that quartiles divide data into four equal groups, so 25% of the data is in each group.	**b** Q_3 = 65, so 25% of students scored more than 65.
	c median = 55, so 50% of students scored less than 55.
	d Q_3 = 65, so 75% of students scored between 30 and 65.
e Find the percentage first and then multiply by the total number.	Q_1 = 45, so 25% of students scored less than 45. Total number of students = 48 Number of students who scored less than 45 = 48 × 25% = 12
f Use the IQR to calculate the lower fence. lower fence = Q_1 − 1.5 × IQR	IQR = Q_3 − Q_1 = 65 − 45 = 20 lower fence = 45 − 1.5 × 20 = 45 − 30 = 15 Scores less than 15 would be considered outliers.
g Use the IQR to calculate the upper fence. upper fence = Q_3 + 1.5 × IQR	upper fence = 65 + 1.5 × 20 = 65 + 30 = 95 Scores greater than 95 would be considered outliers.

Comparing box plots and histograms

If we know what the histogram of a distribution looks like, we can often have some idea of what the box plot will look like.

Approximately symmetric distributions

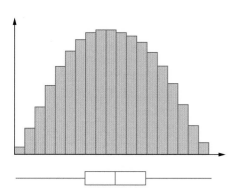

The median is approximately in the middle of the box and the whiskers are about the same length.

Distributions with outliers

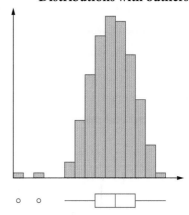

The box plot matches the histogram with the outliers shown by dots.

Positively skewed distributions

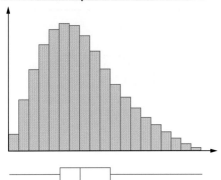

The box and whisker in the positive direction are longer than the box and whisker in the negative direction.

Negatively skewed distributions

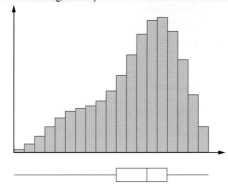

The box and whisker in the negative direction are longer than the box and whisker in the positive direction.

EXERCISE 6.2 Box plots

ANSWERS p. 418

Recap

Use the following information to answer the next two questions.

The stem plot shows the times (in seconds) of skiers who finished a slalom ski race.

Stem	Leaf
9	1 5 7 9
10	2 4 5 6 6 8
11	0 2 2 3 4 4 5
12	1 2 3 3 3 7 9
13	2 3 4 5 7 7
14	3 6 9
15	0 1 2

Key: 14 | 3 means 143 seconds

1 Determine the median for the stem plot.

2 Determine the IQR for the stem plot.

Mastery

3 Using CAS 2   For each of the following data sets, use CAS to construct a box plot.

  a 12, 23, 33, 36, 38, 40, 44, 45, 45, 48, 49, 49, 50, 52, 55, 71
  b 52, 73, 76, 81, 81, 90, 90, 92, 95, 96, 96, 105, 110
  c 9, 16, 19, 20, 23, 23, 24, 24, 25, 25, 26, 26, 27, 27, 27, 33, 35

4 WORKED EXAMPLE 5   For each of the box plots below showing the distribution of 60 student test scores marked out of 15, find the

  i five-number summary
  ii percentage of students who scored more than 7
  iii percentage of students who scored less than 10
  iv percentage of students who scored between 7 and 9
  v number of students who scored less than 5
  vi scores at the lower end that would be considered outliers
  vii scores at the upper end that would be considered outliers.

a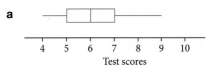

b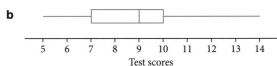

c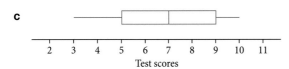

Calculator-free

5 (3 marks) For each of the following box plots, state whether the distribution is approximately symmetric, positively skewed or negatively skewed, and whether it has outliers, giving a reason for your answer.

a (1 mark)

b (1 mark)

c (1 mark)

6 (3 marks) The box plot shows the distribution of the time, in seconds, that 79 customers spent moving along a particular aisle in a large supermarket.

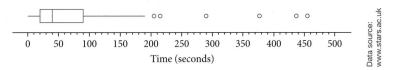

a Describe the shape of the distribution. (1 mark)
b What is the longest time, in seconds, spent moving along this aisle? (1 mark)
c Determine the number of customers who spent more than 90 seconds moving along this aisle. (1 mark)

> 🔒 **Exam hack**
>
> With box plot questions, always check whether the value you are being asked about is one of the quartiles.

Calculator-assumed

7 (3 marks) The following data was recorded from measurements made on 12 men.

For these men, determine the

a median age (1 mark)
b interquartile range (IQR), in years. (2 marks)

Age (years)	Mass (kg)	Waist (cm)
26	84	84
29	72	74
32	67	89
32	59	75
34	97	106
37	112	114
39	67	80
40	91	101
41	98	101
43	89	94
45	117	126
51	62	82

8 (6 marks) The box plot shows the number of cigarettes smoked per day by a sample of 120 smokers who were trying to quit.

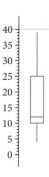

- **a** What is the median number of cigarettes smoked per day? (1 mark)
- **b** What is the interquartile range? (2 marks)
- **c** Is the distribution approximately symmetric, positively skewed or negatively skewed? (1 mark)
- **d** What percentage of people smoked less than 12 cigarettes per day? (1 mark)
- **e** Estimate how many people smoked more than 10 cigarettes per day. (1 mark)

9 (6 marks) For the following box plot

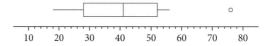

- **a** find Q_3 (1 mark)
- **b** calculate the IQR (2 marks)
- **c** state the value of the outlier shown (1 mark)
- **d** explain why the outlier is incorrect. (2 marks)

6.3 Back-to-back stem plots and parallel box plots

Video playlist
Back-to-back stem plots and parallel box plots

Worksheets
Box-and-whisker plots

Back-to-back stem plots

Back-to-back stem plots

Sometimes we want to compare the distribution of numerical data for two groups. For example, we may want to compare the test scores for two classes or the heights of players in two football teams. One way of displaying and comparing two sets of data is to construct **back-to-back stem plots**.

A back-to-back stem plot has two sets of leaves, one on the left of the stem and one on the right. This allows us to display all the data values for the two groups being compared, as in the example below.

Test results for two classes

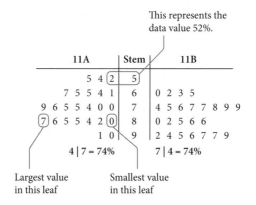

WORKED EXAMPLE 6 — Working with back-to-back stem plots

Two speed cameras on different roads recorded the following car speeds (in km/h).

Camera 1: 136, 140, 123, 135, 112, 120, 116, 131, 127, 125, 130, 116, 131, 120, 130, 117, 130, 134, 123, 148

Camera 2: 67, 72, 73, 78, 90, 84, 63, 69, 71, 89, 102, 86, 69, 71, 93, 83, 80, 65, 73, 69

a Display the data using a back-to-back stem plot.

b Comment on the shape of the data for Camera 2.

c Calculate the median, range and IQR for the car speeds recorded by each of the two speed cameras.

d If the speed limit on the first road is 100 km/h and on the second road is 80 km/h, which road has the greater speeding problem? Provide statistical evidence for your answer.

Steps	Working
a Order the data and display it using a back-to-back stem plot.	Camera 1 | | Camera 2 ⠀⠀⠀⠀⠀⠀⠀⠀⠀⠀⠀⠀6 | 3 5 7 9 9 9 ⠀⠀⠀⠀⠀⠀⠀⠀⠀⠀⠀⠀7 | 1 1 2 3 3 8 ⠀⠀⠀⠀⠀⠀⠀⠀⠀⠀⠀⠀8 | 0 3 4 6 9 ⠀⠀⠀⠀⠀⠀⠀⠀⠀⠀⠀⠀9 | 0 3 ⠀⠀⠀⠀⠀⠀⠀⠀⠀⠀10 | 2 ⠀⠀⠀⠀⠀⠀7 6 6 2 | 11 | ⠀⠀⠀7 5 3 3 0 0 | 12 | 6 5 4 1 1 0 0 0 | 13 | ⠀⠀⠀⠀⠀⠀⠀⠀⠀8 0 | 14 | ⠀⠀⠀⠀2 | 11 = 112 km/h⠀⠀⠀10 | 2 = 102 km/h
b To see the shape of the data in the right leaf, rotate the page 90° anticlockwise so that the stem forms the horizontal axis and picture it as a histogram.	Camera 2 ⠀6 | 3 5 7 9 9 9 ⠀7 | 1 1 2 3 3 8 ⠀8 | 0 3 4 6 9 ⠀9 | 0 3 10 | 2 11 | 12 | 13 | 14 | The data for Camera 2 is positively skewed.
c Calculate the median, range and IQR for each set of data in the back-to-back stem plot.	Camera 1: median = 128.5 km/h, range = 36 km/h, IQR = 12.5 km/h Camera 2: median = 73 km/h, range = 39 km/h, IQR = 16 km/h
d Compare the medians in relation to the speed limits to see if there are any noticeable differences.	The first road has the greater speeding problem. The median speed for the first road is 128.5 km/h, which is 28.5 km/h *above* the 100 km/h speed limit. The median speed for the second road is 73 km/h, which is 7 km/h *under* the 80 km/h speed limit.

WORKED EXAMPLE 7 Interpreting back-to-back stem plots

Arthur and Stella deliver flyers. The number of flyers delivered per hour over 12 hours is shown below.

Arthur: 20, 38, 23, 31, 14, 38, 28, 30, 37, 30, 24, 37

Stella: 35, 27, 25, 31, 27, 30, 35, 31, 24, 31, 23, 26

a Display the data using a back-to-back stem plot with split stems.
b Comment on the shape of Stella's data.
c Calculate the median, range and IQR for the number of flyers delivered by each person.
d Who would you say is the better delivery person? Justify your answer by quoting appropriate data statistics.

Steps	Working
a Order the data and display it with a back-to-back stem plot, splitting the stems. List leaves in the range 0–4 in the first half of the split stem, and leaves in the range 5–9 in the second half.	Arthur | | Stella 　　　　4 | 1 | 　　　　　| 1 | 　　　3 0 | 2 | 3 4 　　　8 4 | 2 | 5 6 7 7 　　1 0 0 | 3 | 0 1 1 1 　8 8 7 7 | 3 | 5 5 3 | 2 = 23 flyers　　2 | 3 = 23 flyers
b To see the shape of the data in the right leaf, rotate the page 90° anticlockwise so that the stem forms the horizontal axis and picture it as a histogram.	Stella 1 | 1 | 2 | 3 4 2 | 5 6 7 7 3 | 0 1 1 1 3 | 5 5 Stella's data is symmetric.
c Calculate the medians, ranges and IQRs from the back-to-back stem plot.	Arthur: median = 30, range = 24, IQR = 14.5 Stella: median = 28.5, range = 12, IQR = 5.5
d Use the results to decide who is the better delivery person.	Although the medians (30 and 28.5) are similar, Arthur's range (24) and IQR (14.5) are considerably higher than Stella's range (12) and IQR (5.5). This means Stella's deliveries have less variability and are more consistent than Arthur's, which indicates that Stella is the better delivery person.

Parallel box plots

Parallel box plots are a good choice if we want to compare the distribution of numerical data for two or more groups, particularly when the data set is large. It is also easier to compare medians and quartiles from parallel box plots than from back-to-back stem plots.

When comparing parallel box plots, we can comment on **location**, which includes the median, minimum, maximum, Q_1 and Q_3. We can also compare the spread of the data by looking at the range and IQR and comment on the shape of the distribution. Take note of any other relevant observations including if there are any outliers.

Here's an example based on the test results of four classes. It's relatively easy to find which class has the highest median, lowest Q_1 or highest maximum value etc.

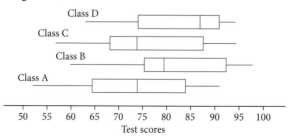

USING CAS 3 | Constructing parallel box plots

The test results of two Year 11 classes in Mathematics Applications are shown below.

Class A: 58, 46, 53, 52, 67, 36, 61, 49, 47, 59, 66, 53, 94, 69, 46, 44, 57

Class B: 60, 50, 70, 69, 86, 43, 60, 60, 44, 56, 49, 50, 56, 56, 42, 65, 47, 67, 25, 46

Construct parallel box plots for the data.

ClassPad

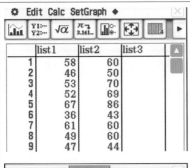

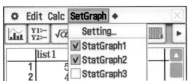

1. Tap **Menu** and open the **Statistics** application.
2. Clear all lists.
3. Enter the data into **list1** and **list2**.
4. Tap **SetGraph** to ensure both **StatGraph1** and **StatGraph2** are selected.

5. Tap **SetGraph > Setting**.
6. In the dialogue box, change the **Type:** field to **MedBox**.
7. Leave **XList:** as **list1**.
8. Tap the box to select **Show Outliers**.

9. Tap the **2** tab at the top of the window.
10. Ensure **Draw:** is set to **On**.
11. Change the **Type:** field to **MedBox**.
12. Change **XList:** to **list2**.
13. Tap the box to select **Show Outliers**.
14. Tap **Set**.

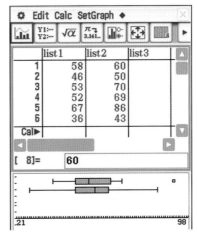

15. Tap the **Graph** tool. The parallel box plots will be displayed in the lower window.

TI-Nspire

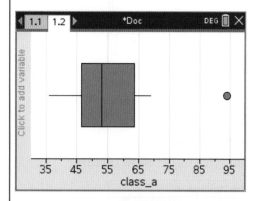

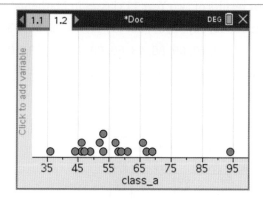

1. Start a new document and add a **Lists & Spreadsheet** page.
2. Label column **A** as **class_a** and column **B** as **class_b**.
3. Enter the data as shown above.
4. Insert a **Data & Statistics** page.
5. For the horizontal axis, select **class_a**. A dot plot of the data will be displayed.

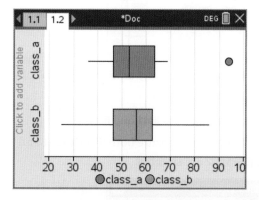

6. Press **menu > Plot Type > Box Plot**. A box plot of the data will be displayed.
7. Press **menu > Plot Properties > Add X Variable**.
8. Select **class_b**. Parallel box plots of the data will be displayed.

WORKED EXAMPLE 8 Working with parallel box plots

The parallel box plots below represent the average temperature (°C) in June for Victoria, New South Wales and Western Australia over a number of years.

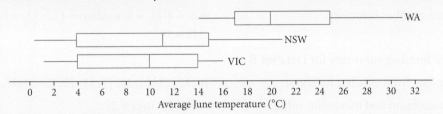

Average June temperature (°C)

a Which state has the highest median average June temperature?
b Which state has the largest range of average June temperatures?
c Which state's data is best described as positively skewed?
d Which state had the lowest average June temperature?
e Which of the three states has noticeably higher average June temperatures than the other two? Refer to medians as evidence in your answer.

Steps	Working
a Look for the state whose median line is furthest along the scale.	Western Australia
b Look for the longest box plot, including whiskers.	New South Wales
c Look for the box plot with its median to the left of the box and the right whisker being longer than the left one.	Western Australia
d Look for the lowest left endpoint.	New South Wales
e Compare the medians shown on the states' box plots.	Western Australia has noticeably higher average June temperatures than Victoria and New South Wales. Western Australia's median (20°C) is much higher than Victoria's (10°C) and NSW's (11°C).

WORKED EXAMPLE 9 Constructing parallel box plots by hand

Construct parallel box plots for the data sets below and compare the distributions.

Data set A: 4, 9, 7, 14, 12, 11, 15, 18, 19, 5, 11, 14

Data set B: 15, 13, 10, 17, 18, 25, 23, 23, 21, 27, 12, 13

Steps	Working
Find the five-number summary for Data set A	
1 Order the data from smallest to largest.	4, 5, 7, 9, 11, 11, 12, 14, 14, 15, 18, 19
2 Find the minimum and maximum value.	min = 4, max = 19
3 Find the median. There is an even number of data values, so find the average of the two middle points.	The median is between the 6th and 7th data values in the ordered list. $Q_2 = \text{median} = \dfrac{11 + 12}{2} = 11.5$
4 Find Q_1, the median of the lower half of the data.	The lower half of the data is 4, 5, 7, 9, 11, 11. $Q_1 = \text{lower quartile} = \dfrac{7 + 9}{2} = 8$

	5 Find Q_3, the median of the upper half of the data.	The upper half of the data is 12, 14, 14, 15, 18, 19. Q_3 = upper quartile = $\frac{14 + 15}{2}$ = 14.5
	6 List the five-number summary.	min = 4, Q_1 = 8, median = 11.5, Q_3 = 14.5, max = 19
Find the five-number summary for Data set B		
	1 Order the data from smallest to largest.	10, 12, 13, 13, 15, 17, 18, 21, 23, 23, 25, 27
	2 Find the minimum and maximum value.	min = 10, max = 27
	3 Find the median. There is an even number of data values, so find the average of the two middle points.	The median is between the 6th and 7th data values in the ordered list. Q_2 = median = $\frac{17 + 18}{2}$ = 17.5
	4 Find Q_1, the median of the lower half of the data.	The lower half of the data is 10, 12, 13, 13, 15, 17. Q_1 = lower quartile = $\frac{13 + 13}{2}$ = 13
	5 Find Q_3, the median of the upper half of the data.	The upper half of the data is 18, 21, 23, 23, 25, 27. Q_3 = upper quartile = $\frac{23 + 23}{2}$ = 23
	6 List the five-number summary.	min = 10, Q_1 = 13, median = 17.5, Q_3 = 23, max = 27
Draw the parallel box plots		Construct a parallel box plot with an appropriate scale. 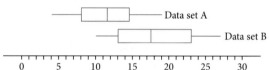
Comment on the distributions		
	1 Compare the location of the median, minimum, maximum, and significant difference with the Q_1 and Q_3.	Data set A has a lower median of 8 compared to a median of 17.5 for Data set B. Data set A also has a much lower minimum of 4 compared to 10 for Data set B. While the maximum is greater for Data set B at 27 compared to 19 for Data set A. The median value for Data set B is 17.5 compared to the Q_3 value of Data set A at 14.5. This indicates that more than 50% of the data in Data set B is greater than 25% of the data in Data set A.
	2 Compare the spread of data by looking at the range and IQR.	The range for Data set A is 15 and the range for Data set B is 17. The interquartile range is also similar with Data set A having an IQR of 6.5 and Data set B having an IQR of 5.5. This indicates that the spread of data is similar between the two sets.
	3 Compare the shape of the distribution.	Both box plots are approximately symmetrical.
	4 Comment on any other relevant observations including if there are any outliers.	Neither set has outliers.

Which display is best to compare data?

Often there is more than one suitable display. Here are some guidelines to help us decide which statistical display to use.

Display	Type of data	Guidelines
Back-to-back stem plots	Categorical and numerical data	Used to compare the distribution of numerical data for two groups, where the two data sets are small.
Parallel box plots	Categorical and numerical data	Used to compare the distribution of numerical data for two or more groups, where the data sets may be large and we want to compare the five-number summaries.

EXERCISE 6.3 Back-to-back stem plots and parallel box plots   ANSWERS p. 419

Recap

Use the following information to answer the next two questions.

To test the temperature control on an oven, the control is set to 180°C and the oven is heated for 15 minutes. The temperature of the oven is then measured. Three hundred ovens were tested in this way. Their temperatures were recorded and the results are displayed using a box plot.

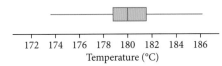

1 The interquartile range for temperature is closest to

   A 1.3°C   **B** 1.5°C   **C** 2.0°C   **D** 2.7°C   **E** 4.0°C

2 The range for temperature is closest to

   A 2.7°C   **B** 11.5°C   **C** 12.7°C   **D** 14.0°C   **E** 18.0°C

Mastery

3 **WORKED EXAMPLE 6** Two speed cameras on different roads recorded the following car speeds (in km/h).

   Camera 1: 75, 83, 75, 84, 91, 82, 64, 69, 73, 89, 105, 88, 76, 72, 95, 84, 82, 68, 74, 68

   Camera 2: 92, 100, 96, 119, 109, 109, 116, 84, 109, 100, 110, 97, 110, 114, 115, 118, 96, 111, 103, 88

   a Display the data using a back-to-back stem plot.
   b Comment on the shape of the data for Camera 2.
   c Calculate the median, range and IQR for the car speeds recorded by each of the two speed cameras.
   d If the speed limit on the first road is 80 km/h and 100 km/h on the second road, which road has the greater speeding problem? Provide statistical evidence for your answer.

4 **WORKED EXAMPLE 7** Irina and Steven deliver pamphlets. The number of pamphlets delivered per hour over 12 hours is shown below.

Irina: 32, 29, 32, 37, 33, 36, 29, 27, 24, 33, 22, 28

Steven: 25, 33, 16, 38, 30, 32, 22, 38, 39, 32, 26, 39

   a Display the data using a back-to-back stem plot with split stems.

   b Comment on the shape of Steven's data.

   c Calculate the median, range and IQR for the number of pamphlets delivered by each person.

   d Who would you say is the better delivery person? Justify your answer by quoting appropriate data statistics.

5 **Using CAS 3** Two speed cameras on different roads recorded the following car speeds (in km/h).

Camera 1: 83, 80, 69, 94, 92, 98, 63, 95, 69, 91, 90, 83, 65, 98, 69, 91, 93, 69, 89, 96, 132, 83

Camera 2: 91, 120, 116, 98, 55, 116, 96, 106, 118, 112, 98, 112, 100, 60, 120, 116, 95, 125, 90, 90, 94, 123

Construct parallel box plots for the data.

6 **WORKED EXAMPLE 8** The parallel box plots below represent the maximum daily temperatures (°C) recorded in July for three towns, Ashville, Ballinga and Colebrook, over a number of years.

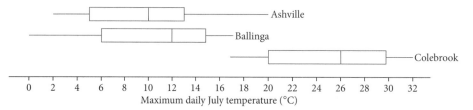

   a Which town has the highest median average July temperature?

   b Which town has the largest range of average July temperatures?

   c Which town's data is best described as negatively skewed?

   d Which town had the lowest average July temperature?

   e Which of the three towns has noticeably higher average July temperatures than the other two? Refer to medians as evidence in your answer.

7 **WORKED EXAMPLE 9** Construct parallel box plots for the data sets below and compare the distributions.

Data set A: 5, 10, 8, 13, 11, 15, 17, 18, 19, 4, 9, 12

Data set B: 14, 12, 11, 16, 19, 25, 22, 22, 20, 28, 12, 13

Calculator-free

8 (5 marks) The back-to-back stem plot shows the female and male smoking rates, expressed as a percentage, in 18 countries.

```
         Female       |   | Male
     9 9 9 7 7 6 5    | 1 | 7 9
8 6 5 5 5 5 5 3 2 1 0 | 2 | 2 4 4 4 5 6 7 7 7
                     | 3 | 0 0 1 1 6 9
                     | 4 | 7
```
Key: 2 | 7 means 27%

a For these 18 countries, what is the lowest female smoking rate? (1 mark)

b For these 18 countries, what is the interquartile range (IQR) of the female smoking rates? (2 marks)

c For these 18 countries, comment on the smoking rates of females compared to males. (2 marks)

9 (4 marks) The parallel box plots summarise the distribution of population density, in people per square kilometre, for 27 inner suburbs and 23 outer suburbs of a large city.

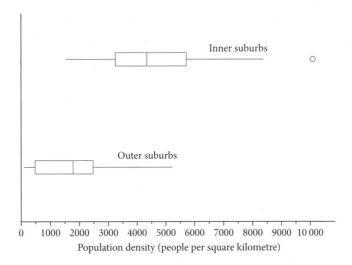

Compare the population density for the inner and outer suburbs using the information displayed in the parallel box plots.

Calculator-assumed

10 (2 marks) As part of an experiment, three samples of pine trees were planted. Each sample contained 50 trees. One sample was grown under hot conditions, one sample was grown under mild conditions and one sample was grown under cool conditions. The parallel box plots show the rate of growth (in centimetres per year) of these three samples.

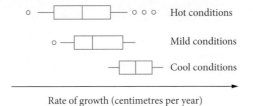

From the parallel box plots, what can be concluded, as conditions change from hot to mild to cool, about the rate of growth for these trees.

11 (4 marks) A weather station records the wind speed and the wind direction each day at 9:00 am. The wind speed is recorded, correct to the nearest whole number. The parallel box plots have been constructed from data that was collected on the 214 days from June to December in 2022.

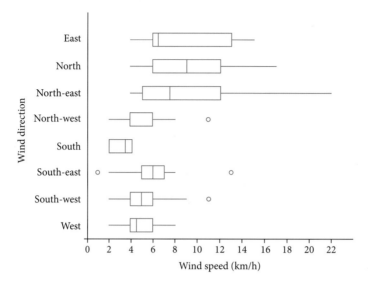

  a Copy and complete the following statements.

  The wind direction with the lowest recorded wind speed was _____ .

  The wind direction with the largest range of recorded wind speeds was _____ . (1 mark)

  b Which wind direction had the greatest variability? (1 mark)

  c Which wind directions had 50% of their speeds between 4 and 6 km/h? (1 mark)

  d Which wind directions had outliers of 11 km/h? (1 mark)

Video playlist
Comparing data using measures of centre and spread

Worksheets
Comparing group measures

Comparing city temperatures

Comparing word lengths

Comparing sports scores

Investigating young drivers

6.4 Comparing data using measures of centre and spread

When comparing data sets, we can comment on:
- location, which includes measures of centre such as mean, median and mode
- spread of the data, including range, IQR and standard deviation
- shape of the distribution
- any other relevant observations such as outliers, gaps or clusters in the data.

Standard deviation verses IQR

Choosing between the standard deviation and IQR as the measure of the spread of a distribution:

Shape of distribution	Choose
Approximately symmetric distributions with no outliers	standard deviation or IQR
Approximately symmetric distributions with outliers	IQR
Skewed distributions	IQR

WORKED EXAMPLE 10 Comparing distributions

Compare the two distributions below.

Data set A

Stem	Leaf
12	3
13	
14	
15	4 7 9 9 9
16	0 2 3 7 8 8
17	0 0 2 4 4 5 5 7 7 7 9
18	3 9
19	0 1 2

Key: 12 | 3 means 123

Data set B

Stem	Leaf
7	1 5 6 8 9 9
8	0 0 2 3 4 6 8 9
9	0 2 5 7 9
10	2 3 5 6 7 7 8

Key: 7 | 9 means 79

ClassPad

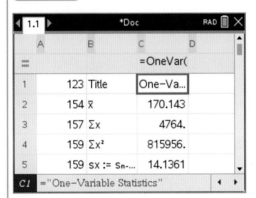

TI-Nspire

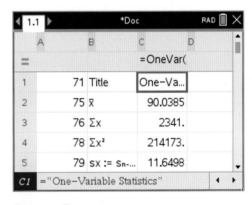

Data set A
Mean = 170.14
σ = 13.88
Mode = 159
Median = 171
Min = 123
Max = 192
Range = 192 − 123 = 69
IQR = $Q_3 - Q_1$ = 177 − 161 = 16

Data set B
Mean = 90.04
σ = 11.42
Mode = 80
Median = 88.5
Min = 71
Max = 108
Range = 108 − 71 = 37
IQR = $Q_3 - Q_1$ = 102 − 80 = 22

Steps	Working
Compare the measures of centre and spread in the context of the question	
1 Compare the location of the median, minimum, maximum, and significant difference with the Q_1 and Q_3.	Set A has a larger mean, 170.14, and median, 171, compared to set B which has a mean of 90.04 and a median of 88.5. The minimum value for set A, at 123, is larger than the maximum value for set B, at 108.

2 Compare the spread of data by looking at the range and IQR.	Set A has a larger range of 69 compared to 37, which indicates the data is more spread out in set A. The interquartile range is a more relevant measurement to compare than the standard deviation for this data because of the potential outlier in set A. The IQR for set A is slightly lower at 16 compared to 22 for set B.
3 Compare the shape of the distribution.	Both distributions are symmetrical with set A having an outlier.
4 Comment on any other relevant observations including if there are any outliers.	Set A has an outlier of 123.

Video
Examination question analysis: Comparing data

EXAMINATION QUESTION ANALYSIS

Calculator-assumed (7 marks)

The five-number summary for the distribution of minimum daily temperature for the months of February, May and July in 2017 are shown in the table. The associated box plots are shown below the table.

Five-number summary for minimum daily temperature

Month	Minimum	Q_1	Median	Q_3	Maximum
February	5.9	9.5	10.9	13.9	22.2
May	3.3	6.0	7.5	9.8	12.7
July	1.6	3.7	5.0	5.9	7.7

Data: Australian Government, Bureau of Meteorology, www.bom.gov.au

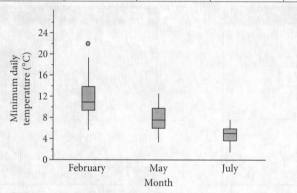

a Copy and complete the following sentence.

The middle 50% of the July minimum daily temperatures are between _____°C and _____°C. (1 mark)

b Describe the shape of the distribution of the minimum daily temperatures (including outliers) for February. (1 mark)

c What is the range of the May minimum daily temperatures? (1 mark)

d Is there a noticeable difference in minimum daily temperatures between the three months? Refer to medians as evidence in your answer. (2 marks)

> 🔓 **Exam hack**
>
> Don't give more information than asked for. If the extra information is wrong, you will lose marks.

e Determine the value of the upper fence for the February box plot and hence show that the outlier is correct. (2 marks)

Reading the question

- The data is presented in two ways: as five-number summaries and as box plots.
- Make sure you are clear on what the 'middle 50%' is referring to.
- Note that you are asked to 'refer' to something in part **d** and to 'show' something in part **e**.

Thinking about the question

- Decide whether it's best to use the five-number summaries or the box plots for each question part.
- A 'show that' answer must have whatever needs to be shown at the very end.
- Think about why part **e** is worth two marks.

Worked solution (✓ = 1 mark)

a The middle 50% lies between Q_1 and Q_3.

From the five-number summary table for July, $Q_1 = 3.7$ and $Q_3 = 5.9$. This means that the middle 50% of the July minimum daily temperatures are between **3.7°C** and **5.9°C**. ✓

b The box and whisker in the positive direction are longer than the box and whisker in the negative direction, so the shape is: **positively skewed with an outlier**. ✓

c range = largest value − smallest value = 12.7 − 3.3 = **9.4°C** ✓

d Yes, there is a noticeable difference. ✓ The February median (10.9°C) is noticeably higher than the May median (7.5°C), which is noticeably higher than the July median (5.0°C). ✓

e upper fence = $Q_3 + 1.5 \times IQR = 13.9 + 1.5 \times 6.6 = 20.5$. ✓ The outlier is 22.2, the maximum given in the five-number summary table. **This outlier is correct because 22.2 > 20.5**. ✓

EXERCISE 6.4 Comparing data using measures of centre and spread ANSWERS p. 420

Recap

1 The parallel box plots shown compare the distribution of life expectancy for 183 countries for the years 1983, 2003 and 2023.

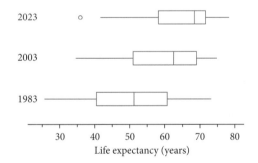

The shape of the distribution of life expectancy for 2003 can best be described as

A positively skewed with no outliers.

B negatively skewed with no outliers.

C approximately symmetric.

D positively skewed with outliers.

E negatively skewed with outliers.

2 The parallel box plots display the distribution of height for three groups of athletes: rowers, netballers and basketballers.

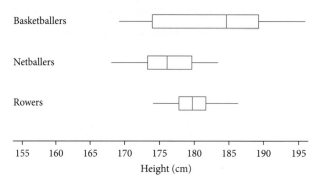

Which one of the following statements is **not** true?

A The shortest athlete is a netballer.
B The rowers have the least variable height.
C More than 25% of the netballers are shorter than all rowers.
D The basketballers are the tallest athletes in terms of median height.
E More than 50% of the basketballers are taller than any of the rowers or netballers.

Mastery

3 WORKED EXAMPLE 10 Compare the distributions below.

Data set A

Data set B

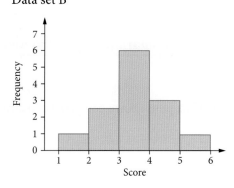

Calculator-free

4 (4 marks) Compare the distributions below.

Data set A

Data set B

Stem	Leaf
1	2 4 6 9
2	1 1 4 6 7
3	0 3 5 8 9 9
4	3

Key: 1 | 2 means 12 years

▶ **Calculator-assumed**

5 (4 marks) Two groups of students were surveyed about the number of movies they had streamed in the last month.

Compare the two data sets.

Group 1: 4, 7, 2, 0, 8, 3, 6, 2, 1, 0, 4, 5, 3

Group 2:

Movies (x)	Frequency
0	6
1	7
2	8
3	10
4	9
5	5
6	5

6 (4 marks) Compare the exam results of two Year 11 classes for mathematics and comment on which class performed better, using statistics to justify your answer.

Class 1:   61%, 72%, 63%, 65%, 63%, 70%, 52%, 56%, 71%, 72%, 66%, 63%

Class 2:   62%, 81%, 71%, 81%, 74%, 83%, 65%, 85%, 74%, 76%, 74%, 77%

7 (4 marks) The following shows the reaction times for two groups of students. Compare the data and comment on the reaction times for the two groups.

Reaction times (milliseconds)

Group 1	152	155	168	173	185	200	210	220	230
Group 2	158	183	187	191	205	218	232	241	256

6.5 Statistical investigation process

The **statistical investigation process** is used to investigate and solve real-world problems using course-related mathematics. It is a cyclical process that involves the collection and analysis of primary or secondary data. During the statistical investigation process the focus should be on the collection, analysis, interpretation and communication of the results to solve the problem in the context of the question. You will be expected to evaluate and interpret the data to formulate an evidence-based conclusion.

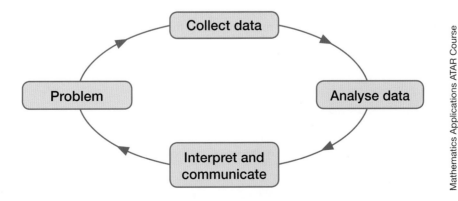

Mathematics Applications ATAR Course Year 11 syllabus p. 23 © SCSA

The below steps are an example of the statistical investigation process:
- Clarify the problem by identifying the key information, stating relevant assumptions to the task, and formulating at least one question that can be answered with evidence from the data.
- Outline the strategies used to collect, display and analysis the data.
- Show evidence of your mathematical knowledge and strategies used to analyse the data. This includes the use of appropriate calculations and graphical representations (graphs, tables, etc.)
- Communicate and interpret the findings in a systematic and concise manner, using appropriate mathematical language. You will need to ensure the analysis of the results relate to the original context of the question and that you consider the reasonableness and limitations of these conclusions.

When comparing data, we can comment on:
- location, such as measures of centre (including mean, median and mode)
- spread of the data such as minimum, maximum, standard deviation, IQR and range
- shape of the distribution
- other relevant observations such as outliers, gaps or clusters in the data.

EXERCISE 6.5 Statistical investigation process

Statistical investigation task (39 marks)

A local horticulturist wants to compare two different fertilisers on the growth of their tomato plants. They have asked you to analyse the data below and, using the statistical investigation process, determine which fertiliser produces the best growth.

Growth after 30 days using fertiliser (cm)

Fertiliser A	20.1	18.5	17.4	21.4	22.8	16.5	19.8	18.6	19.1	18.9	19.2	20.1	17.9	19.5
Fertiliser B	20.2	17.5	17.1	18.1	18.4	16.4	19.5	18.4	19.3	18.1	19.1	18.4	19.1	18.2

Your investigation report should include the following:
- An introduction to the investigation (6 marks)
  - A brief overview in your own words
  - Two questions that can be answered as part of the investigation
  - Two assumptions made about the task or problem
- An outline of the mathematical strategies you are planning to use in your investigation (5 marks)
- Numerical and graphical analysis (15 marks)
- Interpretation of the results, relating your answer to the original question (9 marks)
  - Discuss any trends in the data
  - Make comparisons using the numerical and graphical analysis
  - Link the interpretation to the analysis
- Conclusion (4 marks)
  - Summarise your findings in a concise manner, linking the conclusion back to the original question.

Chapter summary

The five-number summary

Measures of spread

Measures of spread	Use for	Description
Range	numerical	• measures the spread of the entire data set • range = largest value − smallest value
Interquartile range	numerical	• measures the spread of the middle 50% of the data values • IQR = $Q_3 - Q_1$
Standard deviation	numerical	• measures the spread around the mean • $\sigma = \sqrt{\dfrac{\Sigma(x-\bar{x})^2}{n}}$ (use CAS) where • σ = standard deviation • Σ = means the 'sum of' • x = each value from the data set • $\bar{x}$ = mean of all the values in the data set • n = total number of values in the data set

Outliers

- An **outlier** is an extreme high or low data value.
- A data value is a possible outlier if it is less than the **lower fence**: $Q_1 - 1.5 \times \text{IQR}$ or greater than the **upper fence**: $Q_3 + 1.5 \times \text{IQR}$.

Standard deviations from the mean

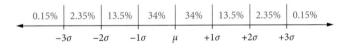

Centre, spread, display and data type summary

Display	Description
Box plot	• best if we want to read the five-number summary easily • helps to find IQR and outliers • parallel box plots allow us to compare the distribution of numerical data for several groups

Chapter 6 | Comparing data

Comparing box plots and histograms

Symmetric distributions

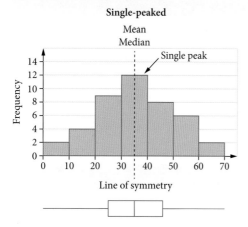

Modal interval: 30–<40

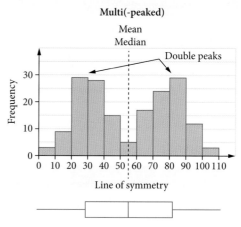

Multimodal: 20–<30 and 80–<90

The mean and median are both close to the line of symmetry.

Skewed distributions

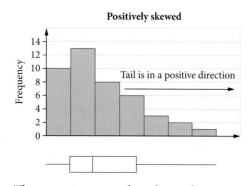

The mean is greater than the median.

The mean is less than the median.

Distributions with outliers

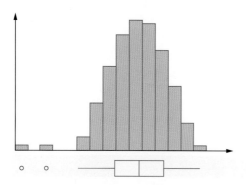

The box plot matches the histogram with the outliers shown as dots.

Mean vs median / Standard deviation vs IQR

Shape of distribution	Choose	
Approximately symmetric distributions with no outliers	mean or median	standard deviation or IQR
Approximately symmetric distributions with outliers	median	IQR
Skewed distributions	median	IQR

Comparing data

When comparing data, we can comment on:
- location, such as measures of centre (including mean, median and mode)
- spread of the data such as minimum, maximum, standard deviation, IQR and range.
- shape of the distribution
- any other relevant observations such as outliers, gaps or clusters in the data.

Statistical investigation process

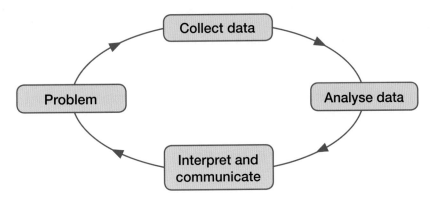

Mathematics Applications ATAR Course
Year 11 syllabus p. 23 © SCSA

Cumulative examination: Calculator-free

Total number of marks: 36   Reading time: 4 minutes   Working time: 36 minutes

1 (2 marks) Jenny's normal junior rate of pay is $15.00 per hour. How much will she earn when she works 4 hours at double-time?

2 (9 marks) Consider the four matrices shown.

$$A = \begin{bmatrix} 2 \\ 4 \end{bmatrix} \quad B = \begin{bmatrix} -1 & 1 & -1 \end{bmatrix} \quad C = \begin{bmatrix} 6 & 3 \\ 9 & 3 \end{bmatrix} \quad D = \begin{bmatrix} 5 \\ 10 \end{bmatrix}$$

  a State the dimensions of matrix B. (1 mark)

  b Evaluate:

    i $-2B$ (1 mark)

    ii $\frac{1}{3}C$ (1 mark)

  c It can be shown that $A + D = m \begin{bmatrix} 1 \\ 2 \end{bmatrix}$ and $A - D = n \begin{bmatrix} 1 \\ 2 \end{bmatrix}$, where m and n are scalar values. Determine the value of m and the value of n. (4 marks)

  d Consider the matrix product AB. If the product is defined, determine the resulting matrix and its size. If the product is not defined, explain why. (2 marks)

3 (2 marks) A triangle has side lengths of 9 cm, 12 cm and 15 cm. Is this a right-angle triangle? Justify your answer.

4 (2 marks) A triangular prism has a surface area of 400 cm². If the length of all sides of the prism is increased by a scale factor of 2, what is the surface area of the new prism?

5 (4 marks) 1526 students sat for an examination. The histogram shows the distribution of marks.

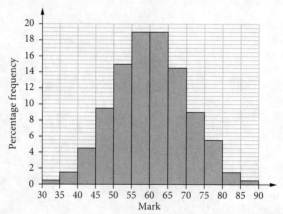

  a Determine the approximate percentage of students with marks 80 or higher. (1 mark)

  b Describe the shape of the histogram. (1 mark)

  c Determine the modal class for the histogram. (1 mark)

  d What is the centre of this distribution? (1 mark)

6 (3 marks) The back-to-back stem plot shows the distribution of maximum temperatures (in °Celsius) of two towns, Beachside and Flattown, over 21 days in January.

```
        Beachside    |   | Flattown
            9 8 7 5  | 1 | 8 9
    4 3 2 2 1 1 0 0  | 2 |
        9 9 8 7 6 5  | 2 | 8 9
                3 2  | 3 | 3 3 4
                  8  | 3 | 5 5 6 7 7 7 8 8
                     | 4 | 0 0 1 2
                     | 4 | 5 6
```

a Are the variables temperature (°Celsius) and town (Beachside or Flattown) categorical or numerical? (1 mark)

b For **Beachside**, what is the range of maximum temperatures? (1 mark)

c Describe the shape of the distribution of maximum temperatures for **Flattown**. (1 mark)

7 (3 marks) The box plots display the distribution of average pay rates, in dollars per hour, earned by workers in 35 countries for the years 1980, 1990 and 2000.

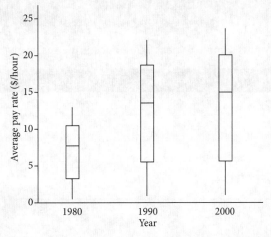

Based on the information contained in the box plots, determine the following.

a What year had an average pay rate less than $8.00 per hour? (1 mark)

b In which year/s did over 75% of the countries have an average pay rate greater than $5.00 per hour? (1 mark)

c What was the median pay rate in 2000? (1 mark)

8 (2 marks) Joel is training for a triathlon. He swam the following times, in minutes, in his last ten races.

28, 34, 22, 24, 25, 24, 26, 26, 24, 27

a What is Joel's mean swim time? (1 mark)

b If the outlier score of 34 was removed from the set of data, what would happen to the mean? (1 mark)

9 (4 marks) The *neck size*, in centimetres, of 250 men was recorded and displayed in the dot plot below.

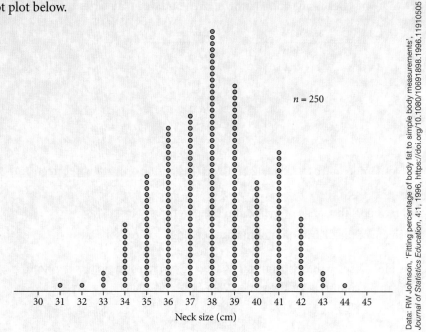

The five-number summary for this sample of neck sizes, in centimetres, is given below.

Minimum	First quartile (Q_1)	Median	Third quartile (Q_3)	Maximum
31	36	38	39	44

a Determine if there are any outliers. (2 marks)

b Copy the grid below and use the five-number summary to construct a box plot, showing any outliers if appropriate. (2 marks)

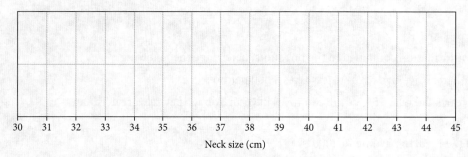

10 (5 marks) Construct a box plot for the following data:

22, 30, 55, 60, 60, 70, 73, 73, 75, 80, 80, 85

Cumulative examination: Calculator-assumed

Total number of marks: 41   Reading time: 4 minutes   Working time: 41 minutes

1 (2 marks) In her job as a real estate agent, Pauline is paid a retainer of $600 per month plus a commission of 2% of her sales over $800 000. How much did Pauline earn for a month when her sales totalled $1 300 000?

2 (10 marks) Five work colleagues, Alison (A), Benji (B), Cynthia (C), Daithi (D) and Elina (E), are working on a project together. Each person has the ability to delegate responsibilities to different members of the team for different reasons. The delegation of responsibilities between the colleagues is shown in the matrix M.

$$M = \text{From} \begin{array}{c} \\ A \\ B \\ C \\ D \\ E \end{array} \overset{\displaystyle \text{To}}{\begin{bmatrix} A & B & C & D & E \\ 0 & 1 & 0 & 1 & 0 \\ 0 & 0 & 1 & 0 & 0 \\ 0 & 0 & 0 & 0 & 1 \\ 0 & 0 & 1 & 0 & 1 \\ 0 & 0 & 1 & 0 & 0 \end{bmatrix}}$$

A '1' in this matrix represents a person can delegate to someone and a '0' means they cannot.

  a Interpret the significance of the row labelled C. (2 marks)

  b Draw a communication network for the work project. (2 marks)

  c Justify who is most likely to be the director of this project. (2 marks)

  d Evaluate M^2. (2 marks)

  e Hence, explain how it may be possible for Cynthia to be delegated a role from Alison. (2 marks)

3 (2 marks) A cake is in the shape of a rectangular prism, as shown in the diagram.

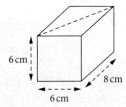

The cake is cut in half to create two equal portions. The cut is made along the diagonal, as represented by the dotted line. Determine the surface area, in square centimetres, of one portion of the cake.

4 (2 marks) A triangle has sides of length 17 cm, 35 cm and 40 cm. A second triangle which is similar has the longest side of 55 cm. Determine the dimensions of the second triangle.

5 (2 marks) A development index is used as a measure of the standard of living in a country. The bar chart displays the development index for 153 countries in four categories: low, medium, high and very high.

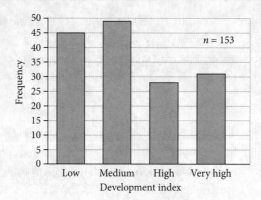

- **a** How many of these countries have a very high development index? (1 mark)
- **b** What percentage of the 153 countries has either a low or medium development index? Write your answer, correct to the nearest percentage. (1 mark)

6 (3 marks) The dot plot and box plot display the distribution of *skull length*, in millimetres, for a sample of the same species of bird.

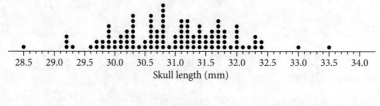

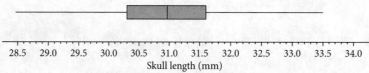

- **a** What is the modal skull length? (1 mark)
- **b** Use information from the plots above to show why the bird with a skull length of 33.5 mm is **not** plotted as an outlier in the box plot. (2 marks)

7 (5 marks) Edgar has a theory that everyone always underestimates the length of a piece of string. He asked his friends to estimate the lengths of several pieces of string and then measured their actual lengths. He then showed the results as the following parallel box plots.

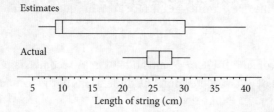

Find the following for both the estimated lengths and the actual lengths.

- **a** median (1 mark)
- **b** range (1 mark)
- **c** IQR (1 mark)
- **d** Describe the shape of each data set. (1 mark)
- **e** Do Edgar's results support his theory? Justify your answer by referring to medians. (1 mark)

8 (6 marks) The stem plot in Figure 1 shows the distribution of the average age, in years, at which women first marry in 17 countries.

Figure 1: Average age, in years, of women at first marriage

Stem	Leaf
24	
25	0
26	6
27	1 1 3 4 7
28	2 2 2 3 3 6
29	1 1
30	1 4
31	

Key: 27 | 3 = 27.3 years

a For these countries, determine the

   i lowest average age of women at first marriage (1 mark)

   ii median average age of women at first marriage. (1 mark)

The stem plot in Figure 2 shows the distribution of the average age, in years, at which men first marry in 17 countries.

Figure 2: Average age, in years, of men at first marriage

Stem	Leaf
25	
26	0
27	
28	9
29	0 9 9
30	0 0 3 5 6 7 9
31	0 0 2
32	5 9
33	

Key: 32 | 5 = 32.5 years

b For these countries, determine the interquartile range (IQR) for the average age of men at first marriage. (2 marks)

c If the data values displayed in Figure 2 were used to construct a box plot with outliers, then the country for which the average age of men at first marriage is 26.0 years would be shown as an outlier.

Explain why this is so. Show an appropriate calculation to support your explanation. (2 marks)

9 (5 marks) In the sport of heptathlon, athletes compete in seven events. These events are the 100 m hurdles, high jump, shot-put, javelin, 200 m sprint, 800 m run and long jump. Fifteen female athletes competed to qualify for the heptathlon at the Olympic Games. Their results for three of the heptathlon events – high jump, shot-put and javelin – are shown in the table.

Athlete number	High jump (metres)	Shot-put (metres)	Javelin (metres)
1	1.76	15.34	41.22
2	1.79	16.96	42.41
3	1.83	13.87	46.53
4	1.82	14.23	40.62
5	1.87	13.78	45.64
6	1.73	14.50	42.33
7	1.68	15.08	40.88
8	1.82	13.13	39.22
9	1.83	14.22	42.51
10	1.87	13.62	42.75
11	1.87	12.01	38.12
12	1.80	12.88	42.65
13	1.83	12.68	45.68
14	1.87	12.45	41.32
15	1.78	11.31	42.88

a State the number of numerical variables in the table. (1 mark)

b In the qualifying competition, the heights jumped in the high jump are expected to be approximately normally distributed. Chara's jump in this competition would give her a standardised score of $z = -1.0$. Use the 68–95–99.7% rule to calculate the percentage of athletes who would be expected to jump higher than Chara in the qualifying competition. (1 mark)

c The box plot on the right shows the distribution of high jump heights for all 15 athletes in the qualifying competition.

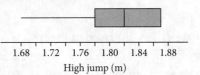

High jump (m)

Explain why the box plot has no whisker at its upper end. (1 mark)

d For the javelin qualifying competition, another box plot is used to display the distribution of athletes' results. An athlete whose result is displayed as an outlier at the upper end of the plot is considered to be a potential medal winner in the event. What is the minimum distance that an athlete needs to throw the javelin to be considered a potential medal winner? (2 marks)

10 (4 marks) The stem plot shows the *height*, in centimetres, of 20 players in a junior football team.

```
Stem | Leaf
  14 | 2 2 4 7 8 8 9
  15 | 0 0 1 2 5 5 6 8
  16 | 0 1 1 2
  17 | 9
```

Key: 14|2 = 142 cm $n = 20$

a Determine the IQR. (2 marks)

b Determine if there are any outliers. (2 marks)

CHAPTER 7
APPLICATIONS OF TRIGONOMETRY

Syllabus coverage
Nelson MindTap chapter resources

7.1 The trigonometric ratios
Reviewing the features of a right-angled triangle
Calculating an unknown side length of a right-angled triangle
Using CAS 1: Determining side lengths
Calculating an unknown angle of a right-angled triangle

7.2 General applications of right-angled trigonometry
Practical problems involving unknown lengths and angles

7.3 Angles of elevation and depression
Angles of elevation
Angles of depression

7.4 Bearings and navigation
True bearings
Compass bearings

7.5 Area of a non-right-angled triangle
Using $A = \frac{1}{2}ab\sin C$
Using Heron's rule
Using CAS 2: Choosing and using area formulas

7.6 The sine and cosine rules for non-right-angled triangles
The sine rule
Using CAS 3: Solving for a side length or an angle using the sine rule
The cosine rule
Using CAS 4: Solving for a side length or an angle using the cosine rule
Practical problems involving non-right-angled triangles

Examination question analysis
Chapter summary
Cumulative examination: Calculator-free
Cumulative examination: Calculator-assumed

Syllabus coverage

TOPIC 2.2: APPLICATIONS OF TRIGONOMETRY

2.2.1 use trigonometric ratios to determine the length of an unknown side, or the size of an unknown angle in a right-angled triangle

2.2.2 determine the area of a triangle, given two sides and an included angle by using the rule area = $\frac{1}{2}ab\sin C$, or given three sides by using Heron's rule, and solve related practical problems

2.2.3 solve problems involving non-right-angled triangles using the sine rule (acute triangles only when determining the size of an angle) and the cosine rule

2.2.4 solve practical problems involving right-angled and non-right-angled triangles, including problems involving angles of elevation and depression and the use of bearings in navigation

Mathematics Applications ATAR Course Year 11 syllabus pp. 12–13 © SCSA

Video playlists (7):
7.1 The trigonometric ratios
7.2 General applications of right-angled trigonometry
7.3 Angles of elevation and depression
7.4 Bearings and navigation
7.5 Area of a non-right-angled triangle
7.6 The sine and cosine rules for non-right-angled triangles

Examination question analysis Applications of trigonometry

Worksheets (23):
7.1 Trigonometric ratios • Identifying the correct trigonometric ratio • Calculating lengths and angles • Trigonometry review • Mixed trig questions
7.3 Angles of elevation and depression
7.4 Bearings 1 • Bearings 2 • Identifying bearings • A page of bearings • Bearings match-up • 16 points of the compass • Elevations and bearings
7.6 Trigonometric calculations • Discovering the sine rule • The sine rule – Finding lengths of sides • The sine rule – Finding angles • The cosine rule – Angles and sides • Applying trigonometry • Finding an unknown side • Finding an unknown angle • Sine rule problems • Cosine rule problems

Puzzles (5):
7.1 Trigonometry crossword
7.3 Angles of elevation and depression • Trigonometric match-up
7.4 Every which way • Three-figure bearings

Nelson MindTap

To access resources above, visit
cengage.com.au/nelsonmindtap

7.1 The trigonometric ratios

Reviewing the features of a right-angled triangle

The term **trigonometry** is derived from Greek; *trigonon* meaning 'three-angled' or 'tri-angle' and *-metry* is the suffix used to describe 'the study of the measurement' of something. So together, trigonometry is the study of the measurement of triangles. The right-angled triangle is a particular type of triangle with exactly one angle of size 90°, shown using the box in the 90° corner. We should also remember that the sum of the interior angles is 180°, meaning that if one of the angles is already 90°, the other two angles add together to give 90°; that is, they are **complementary angles**.

For example, consider the right-angled triangle $\triangle ABC$ with $\angle ACB = 90°$. We can also say that $\angle BAC + \angle ABC = 90°$.

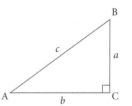

We also know that because $\angle ACB$ is the largest angle in the triangle, the side opposite the right-angle labelled c must be the longest side, the hypotenuse, which was introduced in Chapter 3.

In Chapter 3, we saw the use of Pythagoras' theorem for right-angled triangles, which says the square of the longest side (c) is equal to the sum of the square of the other two sides, or $c^2 = a^2 + b^2$.

We know that the sides and angles of right-angled triangles can be of all different sizes, but for any right-angled triangle, we can compare any two of the three sides to form a **trigonometric ratio**. These ratios between two side lengths tell us how an angle relates to those two side lengths, but first we need to identify this **reference angle**.

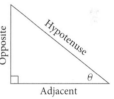

For a right-angled triangle, let the reference angle be represented by θ. We already know the hypotenuse is the longest side of the triangle, but from θ we can identify the

- **opposite side**: the side opposite θ
- **adjacent side**: the side next to θ (which isn't the hypotenuse).

Once θ has been labelled, there are three possible ratios that we can establish.

The trigonometric ratios

Trigonometric ratio	Notation	Meaning	Ratio
sine of θ	$\sin \theta$	Compares how long the opposite side is when compared to the hypotenuse.	$\sin \theta = \dfrac{\text{opposite}}{\text{hypotenuse}}$
cosine of θ	$\cos \theta$	Compares how long the adjacent side is when compared to the hypotenuse.	$\cos \theta = \dfrac{\text{adjacent}}{\text{hypotenuse}}$
tangent of θ	$\tan \theta$	Compares how long the opposite side is when compared to the adjacent side.	$\tan \theta = \dfrac{\text{opposite}}{\text{adjacent}}$

Consider the 3–4–5 right-angled triangle, with $\angle BAC = \theta$.

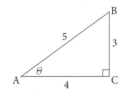

> **Exam hack**
>
> You can use the mnemonic 'SOH–CAH–TOA' to remember the three trigonometric ratios.

Trigonometric ratio	Value of the ratio	Interpretation
sine	$\sin \theta = \dfrac{\text{opposite}}{\text{hypotenuse}}$ $= \dfrac{3}{5}$ $= 0.6$	The length of the opposite side BC is 60% of the length of the hypotenuse AB.
cosine	$\cos \theta = \dfrac{\text{adjacent}}{\text{hypotenuse}}$ $= \dfrac{4}{5}$ $= 0.8$	The length of the adjacent side AC is 80% of the length of the hypotenuse AB.
tangent	$\tan \theta = \dfrac{\text{opposite}}{\text{adjacent}}$ $= \dfrac{3}{4}$ $= 0.75$	The length of the opposite side BC is 75% of the length of the adjacent side AC.

WORKED EXAMPLE 1   Establishing trigonometric ratios from a triangle

For the right-angled triangle with the reference angle α, state the value of each of the trigonometric ratios exactly and to four decimal places. Interpret their meaning geometrically.

a $\sin \alpha$ **b** $\cos \alpha$ **c** $\tan \alpha$

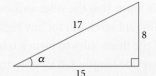

Steps	Working
Identify the opposite, adjacent and hypotenuse with respect to α. Label these O, A and H.	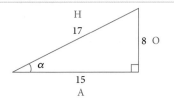
a 1 Use SOH to establish the exact value of $\sin \alpha$.	$\sin \alpha = \dfrac{8}{17}$
2 Evaluate the fraction to four decimal places.	$= 0.4706$
3 Interpret the decimal as a percentage of side lengths.	The length of the opposite side is 47.06% of the length of the hypotenuse.
b 1 Use CAH to establish the exact value of $\cos \alpha$.	$\cos \alpha = \dfrac{15}{17}$
2 Evaluate the fraction to four decimal places.	$= 0.8824$
3 Interpret the decimal as a percentage of side lengths.	The length of the adjacent side is 88.24% of the length of the hypotenuse.
c 1 Use TOA to establish the exact value of $\tan \alpha$.	$\tan \alpha = \dfrac{8}{15}$
2 Evaluate the fraction to four decimal places.	$= 0.5333$
3 Interpret the decimal as a percentage of side lengths.	The length of the opposite side is 53.33% of the length of the adjacent side.

For the 3–4–5 triangle, the trigonometric ratios would still be the same if the respective sides of the triangles were 6–8–10 or 30–40–50 or 3000–4000–5000. That is, if the triangles are similar (as seen in Chapter 4), the ratio between the sides will always be such that $\sin\theta = 0.6$, $\cos\theta = 0.8$ and $\tan\theta = 0.75$. This is called the **constancy of the trigonometric ratios**, which says that for similar triangles, triangles with equal angles and respective side lengths in proportion, the values of their trigonometric ratios won't change.

At this point, it is useful to check that your CAS is set to degrees mode, as all of our angles will be in degrees.

ClassPad	TI-Nspire

In the bottom right-hand corner of the screen, click on the angle setting to the left of the battery icon to toggle between degrees, gradians and radians. Make sure it says **Deg**.	In the top right-hand corner of the screen, click on the angle setting to the left of the battery icon to toggle between degrees and radians.
Note that Standard mode gives answers as fractions. Click this to change it to Decimal mode.	

It is useful to also ensure your calculator is set to degrees mode as well.

The trigonometric functions on CAS can be accessed as follows.

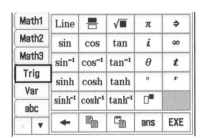

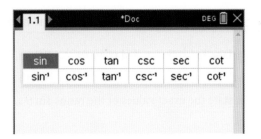

Open the **Keyboard** and select **Trig**.

Press the **trig** key.

Now, if we type in $\sin 30°$ into CAS, we will obtain the value of $\frac{1}{2}$ or 0.5. This is saying, for any right-angled triangle with a reference angle of 30°, the length of the opposite side will always be 50% of the length of the hypotenuse.

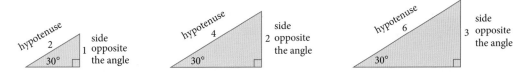

But now something interesting happens if we type $\cos 60°$ into CAS. We will also get the value of $\frac{1}{2}$ or 0.5.

Why is this the case? This is saying that because 60° is the complementary angle of 30°, if we changed the location of the reference angle, now the length of the adjacent side will always be 50% of the length of the hypotenuse.

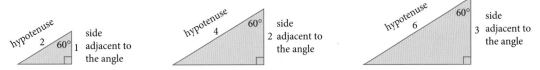

Cosine stands for the 'the complement of sine', meaning that $\sin 70° = \cos 20°$ because $70° + 20° = 90°$ and $\sin 45° = \cos 45°$ because $45° + 45° = 90°$.

The complementary relationship between sine and cosine
$\sin\alpha = \cos\beta$ if $\alpha + \beta = 90°$.

Chapter 7 | Applications of trigonometry 281

WORKED EXAMPLE 2 — Using the complementary relationship between sine and cosine

a State the value of β if $\sin 35° = \cos \beta$.
b State the value of α if $\sin \alpha = \cos 15°$.

Steps	Working
a 1 Write the relationship $\alpha + \beta = 90°$.	$\alpha + \beta = 90°$
2 Substitute α and solve for β.	$35° + \beta = 90°$
	$\beta = 55°$
b 1 Write the relationship $\alpha + \beta = 90°$.	$\alpha + \beta = 90°$
2 Substitute β and solve for α.	$\alpha + 15° = 90°$
	$\alpha = 75°$

However, if we were to type in something like $\tan 30°$, the answer is 0.5774. This is saying that for a reference angle of 30°, the length of the opposite side is 57.74% of the length of adjacent side. What about $\tan 45°$? The answer is 1, which makes sense because if the reference angle is 45° then it is a right-isosceles triangle whereby the length of the opposite side is 100% of the length of the adjacent side; that is, they are the same size.

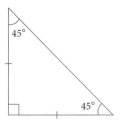

WORKED EXAMPLE 3 — Drawing a right-angled triangle from sufficient information

a Draw a labelled right-angled triangle, $\triangle ABC$, such that $\angle ACB = \theta$, AC is the hypotenuse and $\tan \theta = \dfrac{5}{12}$.
b Use Pythagoras' theorem to calculate the length of AC.
c Hence, state the exact values of the ratios $\sin \theta$ and $\cos \theta$.

Steps	Working
a 1 Draw a right-angled triangle with AC as the hypotenuse and B as the corner with the right-angle. 2 Identify $\angle ACB$ as θ. 3 Recognise $\tan \theta = \dfrac{5}{12}$ means that the opposite side is 5 and the adjacent side is 12 (or any similar triangle).	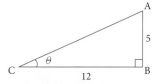 🔒 **Exam hack** Don't forget the right-angle symbol!
b 1 Establish $c^2 = a^2 + b^2$. 2 Substitute in the values of a and b and solve for c.	$c^2 = a^2 + b^2$ $AC^2 = 5^2 + 12^2$ $AC^2 = 25 + 144$ $AC^2 = 169$ $AC = \sqrt{169} = 13$
c 1 Modify the diagram to show the hypotenuse as 13. 2 Use SOH to establish the value of $\sin \theta$. 3 Use CAH to establish the value of $\cos \theta$.	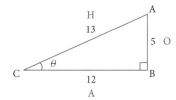 $\sin \theta = \dfrac{\text{opposite}}{\text{hypotenuse}} = \dfrac{5}{13}$ $\cos \theta = \dfrac{\text{adjacent}}{\text{hypotenuse}} = \dfrac{12}{13}$

As a result, the trigonometric ratios allow us to calculate lengths and angles that we can't measure directly.

Calculating an unknown side length of a right-angled triangle

A trigonometric ratio can be used to calculate the length of an unknown side in a right-angled triangle, as long as one angle (other than the right-angle) and one side length are known.

WORKED EXAMPLE 4 Calculating an unknown side length

Show the use of the trigonometric ratios to find the value of the pronumerals, correct to two decimal places, in the right-angled triangle.

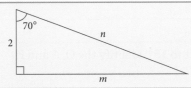

Steps	Working
1 With respect to 70°, identify the O, A and H.	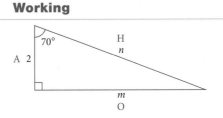
2 To calculate m, identify the trigonometric ratio needed using SOH–CAH–TOA and the known side length of 2.	m is the opposite side, 2 is adjacent side, so tan is needed.
3 Establish the tangent ratio.	$\tan 70° = \dfrac{\text{opposite}}{\text{adjacent}} = \dfrac{m}{2}$ $\tan 70° = \dfrac{m}{2}$
4 Rearrange to solve for m.	$m = 2 \times \tan 70°$ $= 5.49$ (to two decimal places)

Note: $\tan 70° = 2.7475$ means that when the reference angle is 70°, the opposite side is 274.75% of the adjacent side! Almost triple the length!

🔓 **Exam hack**

You can think of this calculation as 'Find 274.75% of 2.'

5 To calculate n, identify the trigonometric ratio needed using SOH–CAH–TOA and the known side length of 2.	n is the hypotenuse, 2 is adjacent, so cos is needed. $\cos 70° = \dfrac{\text{adjacent}}{\text{hypotenuse}} = \dfrac{2}{n}$

🔓 **Exam hack**

Avoid using the determined value of m as this could introduce a rounding error. Always go back to the initial values provided in the diagram where possible.

6 Establish the cosine ratio.	$\cos 70° = \dfrac{2}{n}$
7 Rearrange to solve for n.	$n \times \cos 70° = 2$ $n = \dfrac{2}{\cos 70°}$ $= 5.85$ (to two decimal places)

Note: Because the hypotenuse is unknown, we are dividing by $\cos 70°$ which has a value of 0.342.

🔓 **Exam hack**

Dividing 2 by 0.342 is like saying, 'what is the value of 100% if 2 represents 34.2%?'

USING CAS 1 — Determining side lengths

Find the value of the pronumerals, correct to two decimal places, in the right-angled triangle.

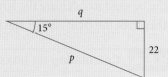

Steps	Working
1 With respect to 15°, identify the O, A and H.	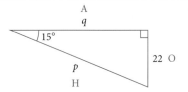
2 Establish the appropriate equations involving the correct trigonometric ratios.	$\sin 15° = \dfrac{22}{p}$ $\tan 15° = \dfrac{22}{q}$

ClassPad

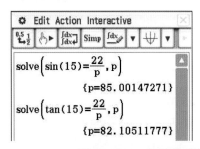

1. Open the **Keyboard** and tap **Math1** or **Trig** to access the trigonometric ratios.
2. Enter and highlight the equation.
3. Tap **Interactive** > **Equation/Inequality**.
4. Change the variable to **p** and tap **OK**.

TI-Nspire

solve$\left(\sin(15)=\dfrac{22}{p}, p\right)$ $p = 85.0015$

solve$\left(\tan(15)=\dfrac{22}{p}, p\right)$ $p = 82.1051$

1. Press **menu** > **Algebra** > **Solve**.
2. Press **trig** to access the mini-palette with the trigonometric ratios.
3. Enter the equation followed by **,p**.
4. Press **enter**.

Thus, $p = 85.00$ and $q = 82.11$ (to two decimal places).

Calculating an unknown angle of a right-angled triangle

Recall the 3–4–5 triangle whereby we had the trigonometric ratios of $\sin\theta = 0.6$, $\cos\theta = 0.8$ and $\tan\theta = 0.75$. Suppose we want to know what the value of θ needs to be to obtain these values. If we think of sin, cos and tan as operations, we can introduce an **inverse trigonometric ratio** to find the value of θ given a ratio of two sides. Much like if we had $x + 2 = 5$, we would apply the inverse of -2 from both sides such that $x = 3$.

To represent the inverse trigonometric ratios, we use the notation $\sin^{-1}$, $\cos^{-1}$ and $\tan^{-1}$.

For example:

- if $\sin\theta = 0.6$, then $\theta = \sin^{-1}(0.6)$.
- if $\cos\theta = 0.8$, then $\theta = \cos^{-1}(0.8)$.
- if $\tan\theta = 0.75$, then $\theta = \tan^{-1}(0.75)$.

Inverse trigonometric ratios

Inverse trigonometric ratios can be used to find unknown angles in right-angled triangles when we have the appropriate ratio of two sides.

If $\sin \theta = \dfrac{\text{opposite}}{\text{hypotenuse}}$, then $\theta = \sin^{-1}\left(\dfrac{\text{opposite}}{\text{hypotenuse}}\right)$.

If $\cos \theta = \dfrac{\text{adjacent}}{\text{hypotenuse}}$, then $\theta = \cos^{-1}\left(\dfrac{\text{adjacent}}{\text{hypotenuse}}\right)$.

If $\tan \theta = \dfrac{\text{opposite}}{\text{adjacent}}$, then $\theta = \tan^{-1}\left(\dfrac{\text{opposite}}{\text{adjacent}}\right)$.

WORKED EXAMPLE 5 Determining angles

Find the value of θ, correct to the nearest degree.

Steps	Working
1 With respect to θ, identify the O, A and H.	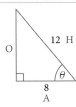
2 To calculate θ identify the trigonometric ratio needed using SOH–CAH–TOA.	8 is the adjacent, 12 is hypotenuse, so cos is needed.
3 Establish the cosine ratio.	$\cos \theta = \dfrac{8}{12}$
4 Apply the inverse trigonometric ratio using CAS to solve for θ.	$\theta = \cos^{-1}\left(\dfrac{8}{12}\right)$

> **Exam hack**
>
> Still show all working out when you are using CAS to enable follow-through marks in case the wrong ratio is used.

ClassPad

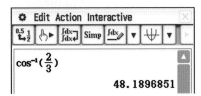

Open the **Keyboard** and use the **Trig** menu to access the inverse trigonometric functions.

TI-Nspire

Press **trig** to open the mini-palette to access the inverse trigonometry functions.

5 Answer to the nearest degree. $\theta = 48.19 \approx 48°$

EXERCISE 7.1 The trigonometric ratios

ANSWERS p. 421

Mastery

1. **WORKED EXAMPLE 1** For the right-angled triangle with the reference angle, state the value of each of the trigonometric ratios exactly and rounded to four decimal places, where appropriate. Interpret their meaning geometrically.

   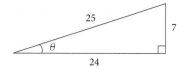

   a $\sin \theta$ b $\cos \theta$ c $\tan \theta$

2. **WORKED EXAMPLE 2**

   a State the value of β if $\sin 28° = \cos \beta$, where β is acute.

   b State the value of α if $\sin \alpha = \cos 73°$, where α is acute.

3. **WORKED EXAMPLE 3**

   a Draw a labelled right-angled triangle ΔABC such that $\angle ACB = \theta$, AC is the hypotenuse and $\tan \theta = \dfrac{9}{40}$.

   b Use Pythagoras' theorem to calculate the length of AC.

   c Hence, state the exact values of the ratios $\sin \theta$ and $\cos \theta$.

4. **WORKED EXAMPLE 4** Show the use of the trigonometric ratios to find the value of the pronumerals, correct to two decimal places, in each of the right-angled triangles.

   a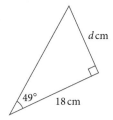

   b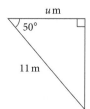

   c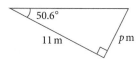

   d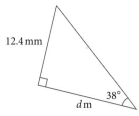

   e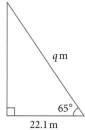

   f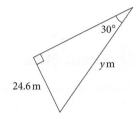

5. **Using CAS 1** Find the value of the pronumerals, correct to two decimal places, in each of the right-angled triangles.

   a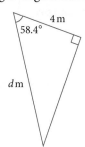

   b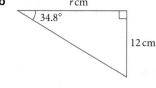

   c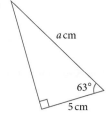

6 WORKED EXAMPLE 5 Find the value of the pronumeral in each triangle, correct to the nearest degree.

a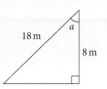

b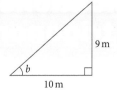

c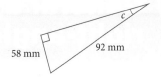

d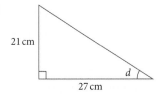

e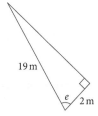

f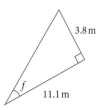

g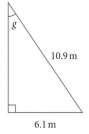

h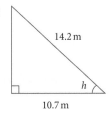

Calculator-free

7 (6 marks) Consider the right-angled triangle shown.

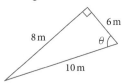

a State the simplified value of the trigonometric ratio, sin θ. (2 marks)

b State the simplified value of the trigonometric ratio, cos θ. (2 marks)

c Evaluate the fraction $\dfrac{\sin\theta}{\cos\theta}$ and compare your answer to tan θ. (2 marks)

8 (5 marks) For a right-angled triangle, it is known that $\sin\theta = \dfrac{5}{13}$.

a Draw a labelled diagram of this triangle, showing the known information. (2 marks)

b Hence, calculate the exact value of cos θ. (3 marks)

Calculator-assumed

9 (4 marks) Consider the triangle KLM shown, with ∠LKM = θ.

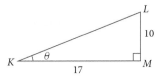

a Evaluate tan θ to four decimal places and interpret its significance for the sides LM and KM. (2 marks)

b Calculate the value of θ, correct to one decimal place. (2 marks)

10 (8 marks) A right-angled triangle ABC with ∠ABC = 90°, has BC = 16 cm and AC = 25 cm.

a Draw a clearly labelled diagram representing △ABC. (2 marks)

b Use Pythagoras' theorem to calculate the length of AB, correct to two decimal places. (2 marks)

c Determine all unknown angles in △ABC, correct to one decimal place. (4 marks)

Video playlist
General applications of right-angled trigonometry

7.2 General applications of right-angled trigonometry

Practical problems involving unknown lengths and angles

Wherever there is a right-angled triangle in the real-world, the trigonometric ratios can be used to calculate unknown lengths and angles. These calculations are important in many practical contexts, such as surveying, design, architecture, motion and forces.

WORKED EXAMPLE 6 Solving practical problems involving an unknown length

A wheelchair ramp is inclined at 18° to the horizontal. Determine the length of the ramp if it links two levels that are 8.7 m apart, horizontally. Round your answer correct to one decimal place.

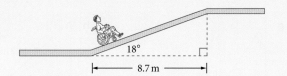

Steps	Working
1 Redraw the triangle using the given values. Represent the length of the ramp by x.	(triangle with x, 18°, 8.7 m)
2 Write the letters O, A and H on the right-angled triangle to show the opposite and adjacent sides and the hypotenuse.	(triangle labelled H = x, O, A = 8.7 m, 18°)
3 Identify whether the labelled sides are opposite, adjacent or hypotenuse and select the matching trigonometric ratio.	8.7 is the adjacent side. x is the hypotenuse. 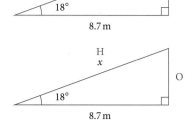
4 Substitute the known values and solve the equation for the unknown, using the **solve** function on CAS if necessary.	$\cos(18°) = \dfrac{8.7}{x}$ $x = \dfrac{8.7}{\cos(18°)} = 9.1477\ldots$

ClassPad

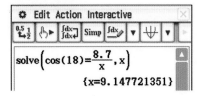

TI-Nspire

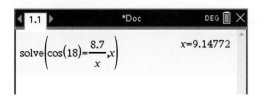

5 Write your answer in the required units and round to the required level of accuracy.

$x = 9.1$ m (to one decimal place)

WORKED EXAMPLE 7 — Solving practical problems involving an unknown angle

A playground slide is made up of a ladder and a metal slide. The top of the metal slide is 2.2 m from the ground, and the bottom of the slide is 4.8 m from a point directly under the top of the slide. Calculate the angle that the slide makes with the ground, correct to the nearest degree.

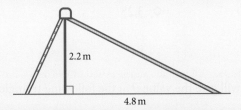

Steps	Working
1 Redraw the triangle using the given values. Represent the required angle by θ.	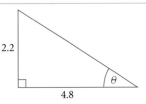
2 Write the letters O, A and H on the right-angled triangle to show the opposite and adjacent sides and the hypotenuse.	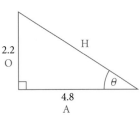
3 Identify whether the labelled sides are opposite, adjacent or hypotenuse and select the matching trigonometric ratio.	2.2 is the opposite side. 4.8 is the adjacent side. $\tan \theta = \dfrac{\text{opposite}}{\text{adjacent}}$
4 Substitute the known values and solve for θ with CAS, using the inverse trigonometric function and rounding to the nearest degree.	$\tan \theta = \dfrac{2.2}{4.8}$ $\theta = \tan^{-1}\left(\dfrac{2.2}{4.8}\right)$ $\theta = 24.624\ldots$

ClassPad

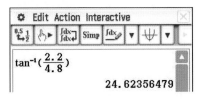

TI-Nspire

5 Write your answer in the required units and round to the required level of accuracy.

The slide makes an angle of 25° with the ground.

EXERCISE 7.2 General applications of right-angled trigonometry

ANSWERS p. 422

Recap

1 The value of the pronumeral, correct to two decimal places, in the diagram is

A 2.29 **B** 2.80 **C** 3.28
D 5.71 **E** 6.97

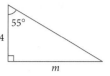

2 Identify the calculation that will give the value θ in the following triangle.

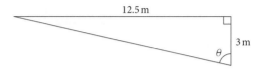

A $\theta = \tan^{-1}\left(\dfrac{12.5}{3}\right)$ **B** $\theta = \sin^{-1}\left(\dfrac{12.5}{3}\right)$ **C** $\theta = \cos^{-1}\left(\dfrac{3}{12.5}\right)$

D $\theta = \tan^{-1}\left(\dfrac{3}{12.5}\right)$ **E** $\theta = \sin^{-1}\left(\dfrac{3}{12.5}\right)$

Mastery

3 WORKED EXAMPLE 6 The pitch of a roof is 28° from the horizontal level of the gutter, as shown in the diagram. If the peak of the roof is 5.4 m from the gutter, determine the distance, h, between the peak of the roof and horizontal level of the gutter, correct to one decimal place.

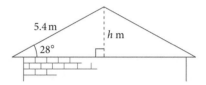

4 WORKED EXAMPLE 7 Selwa is on a swing of length 1.6 m. When she is 0.7 m away from the swing's frame, determine the angle that the swing makes with the vertical pole. Round your answer correct to one decimal place.

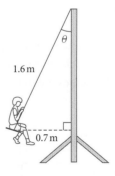

Calculator-free

5 (2 marks) Alexis is on a swing of length 1.4 m. She pulls back 0.6 m from the frame to commence swinging.

Write an expression for θ, the angle which the swing makes with the vertical pole.

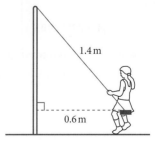

Calculator-assumed

6 (3 marks) Tom and Matt are having a competition to see who can kick a football the longest distance. They both kick from point *A*, and the length of the kick is measured from point *A* to the point where the ball first makes contact with the ground. Tom's kick lands 40 metres away at point *B*. Matt kicks the ball at an angle of 22° to the direction of Tom's kick and it lands at point *C*, such that *BC* makes a right angle with *AB* as shown in the diagram above.

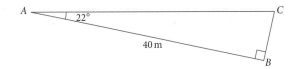

Determine the length of Matt's kick, rounded to the nearest metre.

7 (3 marks) Danielle flew in a helicopter from the ground at sea level to the summit of a mountain. The helicopter flew at a constant incline of 20° to the ground level. If the summit was 2.4 km above sea level, determine the total distance that the helicopter flew, rounded to the nearest kilometre.

8 (3 marks) A bungee jumper leaps off a bridge at an angle of 7° to the vertical. At the point where he is swinging 16.3 m off-centre as shown, calculate the vertical distance that he has dropped, measured to where his feet are tied, correct to one decimal place.

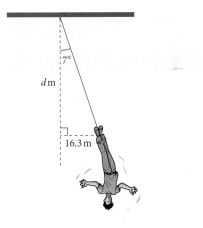

9 (2 marks) A vertical tree of height 20 m stands on horizontal ground. The tree is 100 m away from point *P*, as shown in the diagram.

Determine the value of *x*, as shown in the diagram, correct to one decimal place.

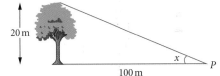

10 (2 marks) Find the angle of inclination, θ, of the ramp shown in the diagram, correct to the nearest degree.

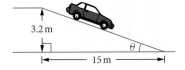

11 (6 marks) A staircase is inclined at an angle of 34° and has a horizontal length of 4.5 m. A bookcase, 1.45 m long, is placed under the stairs.

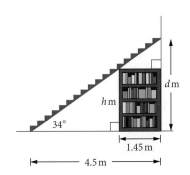

Find, correct to two decimal places,
  a the vertical height, *h*, of the tallest bookcase that will fit under the stairs. (2 marks)
  b the distance, *d*, of the top step of the staircase from the ground. (2 marks)
  c the length from the bottom of the staircase to the top. (2 marks)

12 (3 marks) A soccer goal is 7.4 metres wide. A rectangular region ABCD is marked out directly in front of the goal. In this rectangular region, AB = DC = 11.0 metres and AD = BC = 5.5 metres. The goal line XY lies on DC and M is the midpoint of both DC and XY.

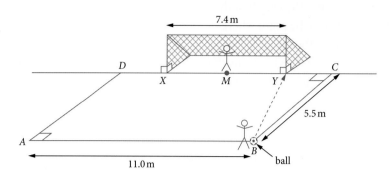

Ben kicks the ball from point B. It travels in a straight line to the base of the goal post at point Y on the goal line. Calculate ∠CBY, the angle that the path of the ball makes with the line BC, correct to one decimal place.

7.3 Angles of elevation and depression

Video playlist
Angles of elevation and depression

Worksheet
Angles of elevation and depression

Puzzles
Angles of elevation and depression

Trigonometric match-up

Angles of elevation

When we look up at an object, the angle that our line of sight makes with the horizontal is called the **angle of elevation**.

Angle of elevation

Angle of elevation – looking up from the horizontal

| WORKED EXAMPLE 8 | Solving practical problems involving an angle of elevation |

Amala stands 40 m from the base of a 25 m tower and looks up at the top of the tower. By first drawing a labelled diagram, find the angle of elevation θ of the top of the tower from Amala. Answer to the nearest degree.

Steps	Working
1 Draw a diagram with all the measurements, including the angle. 2 Write the letters O, A and H on the right-angled triangle to show the opposite and adjacent sides and the hypotenuse.	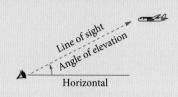
3 Identify the trigonometric ratio.	Use $\tan \theta = \dfrac{\text{opposite}}{\text{adjacent}}$
4 Substitute the known values and solve with CAS, rounding to the required level of accuracy.	$\tan \theta = \dfrac{25}{40}$ $\theta = \tan^{-1}\left(\dfrac{25}{40}\right)$ $\approx 32°$

Angles of depression

When we look down at an object, the angle that our line of sight makes with the horizontal is called the **angle of depression**.

Angle of depression

Angle of depression – looking down from the horizontal

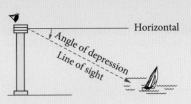

> 🔓 **Exam hack**
>
> If a question refers to angles of elevation or depression, make sure to label the angle from the horizontal and not from the vertical.

Alternate angles

The angle of elevation X to Y is equal to the angle of depression Y to X.

Recall that two angles drawn inside a 'Z shape' are called **alternate angles** and are always equal, due to the horizontals at X and Y being parallel lines.

WORKED EXAMPLE 9 Solving practical problems involving an angle of depression

Sergio is standing on a 220-metre-high vertical cliff looking down at a ship. The angle of depression of the ship is 28°. By first drawing a labelled diagram, find the distance d from the ship to the base of the cliff. Answer to the nearest metre.

Steps	Working
1 Draw a diagram with all the measurements including the angle. If necessary, use the fact that the angle of elevation X to Y = angle of depression Y to X.	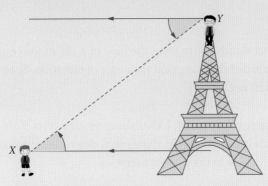
2 Write the letters O, A and H on the right-angled triangle to show the opposite and adjacent sides and the hypotenuse.	
3 Identify whether the labelled sides are opposite, adjacent or hypotenuse and select the matching trigonometric ratio.	d is the adjacent side. 220 is the opposite side. Use $\tan \theta = \dfrac{\text{opposite}}{\text{adjacent}}$
4 Substitute the known values and solve with CAS. Write your answer in the required units and round to the required level of accuracy.	$\tan(28°) = \dfrac{220}{d}$ $d = \dfrac{220}{\tan(28°)}$ $\approx 414 \text{ m}$

> 🔓 **Exam hack**
>
> Even if you are not asked to draw a diagram for elevation/depression problems, always draw the triangle you need to solve the problem!

EXERCISE 7.3 Angles of elevation and depression

ANSWERS p. 422

Recap

1. The height, h m, of the flagpole shown in the diagram is
   - **A** 20 m
   - **B** 21 m
   - **C** 19 m
   - **D** 16 m
   - **E** 30 m

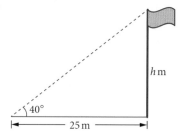

2. An escalator ramp has a length of 32 m and covers a horizontal distance of 29 m. The angle of inclination, θ, of the ramp is
   - **A** 0.9°
   - **B** 42°
   - **C** 47°
   - **D** 65°
   - **E** 25°

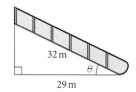

Mastery

3. **WORKED EXAMPLE 8** Lisbet stands 156 m from the base of a 180 m tower and looks up at the top of the tower. By first drawing a labelled diagram, find the angle of elevation, θ, of the top of the tower from Lisbet. Answer to the nearest degree.

4. **WORKED EXAMPLE 9** From her apartment, 130 m above ground level, Zahiya sights the park at an angle of depression of 50°. By first drawing a labelled diagram, determine the distance, d, to the park from the base of Zahiya's building. Answer to the nearest metre.

Calculator-assumed

5. (4 marks) A ferry, F, is 400 metres from point O at the base of a 50-metre-high cliff, OC.

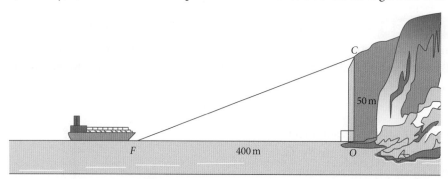

   a. Calculate the angle of elevation of point C from F, correct to one decimal place. (2 marks)
   b. Calculate the distance FC, in metres, correct to one decimal place. (2 marks)

6. (4 marks) For an observer on the ground at A, the angle of elevation of a weather balloon at B is 37°. C is a point on the ground directly under the balloon. The distance AC is 2200 m.
   a. Draw a labelled diagram to represent this situation. (2 marks)
   b. To the nearest metre, calculate the height of the weather balloon above the ground. (2 marks)

7. (4 marks) The angle of depression of a car sighted from the top of a tower is 23°. The tower is 34.6 metres tall and the base of the tower and the car are at the same level.
   a. Draw a labelled diagram to represent the situation. (2 marks)
   b. Hence, calculate the horizontal distance of the car from the base of the tower, correct to the nearest 0.1 m. (2 marks)

8 (4 marks) Milos observed a plane at a 33° angle of elevation. It flew at a constant height of 800 m until it was directly above him.

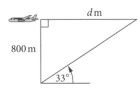

a Determine the direct distance, to the nearest metre, between the plane and Milos at the instance when he first saw it. (2 marks)

b Determine the horizontal distance, d, travelled by the plane, correct to the nearest metre. (2 marks)

9 (3 marks) The point Q on building B is visible from the point P on building A, as shown in the diagram.

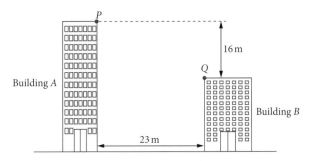

Building A is 16 metres taller than building B. The horizontal distance between point P and point Q is 23 metres. Calculate the angle of depression of point Q from point P, correct to the nearest 0.1 degree.

7.4 Bearings and navigation

True bearings

A **bearing** is the direction of a fixed point, or the path of an object, from a point of observation. Bearings are represented as angles, but these angles need to be described with reference to an established direction. For this, we use the **cardinal points** of a compass: north, east, south, west.

Bearings can be described in two ways. The first type of bearing is a **true bearing** (also called a **three-figure bearing**). These are bearings with angles measured from the north line, in a clockwise direction. As the name suggests, a three-figure bearing always has three figures. When talking about the direct lines, north, east, south and west, rather than the general direction, the word 'due' is often used. Due north has a true bearing of 000°, due east has a true bearing of 090°, due south has a true bearing of 180° and due west has a true bearing of 270°.

Other examples of true bearings include:

Three-figure bearing = 009°   Three-figure bearing = 056°   Three-figure bearing = 324°

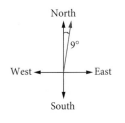 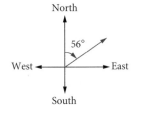 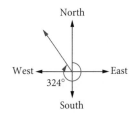

Three-figure bearings

A true bearing of Y from X is the angle of Y from X measured in a clockwise direction from north, and is always written as three-figures.

If the bearing of Y from X is θ, then the bearing of X from Y is 180° + θ, where θ < 180°.

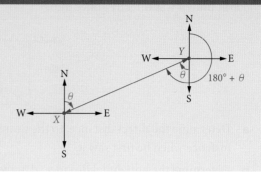

WORKED EXAMPLE 10 — Identifying true bearings

For each of the following, state the true bearing of the point A and draw a diagram showing the angle from north.

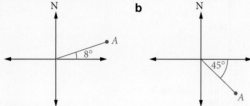

Steps	Working
For parts **a–d**: **1** Find the angle clockwise from north, adding or subtracting 90° angles if necessary. **2** Write your answer, making sure the answer has three figures. **3** Draw a diagram showing the angle.	**a** 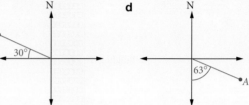 90° − 8° = 82° The three-figure bearing is 082°.

Exam hack

Watch for the word 'from' in a bearing question. This tells you where your starting point is.

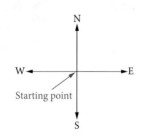

b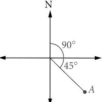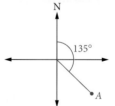

90° + 45° = 135°   The three-figure bearing is 135°.

c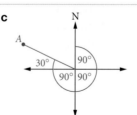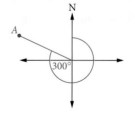

270° + 30° = 300°   The three-figure bearing is 300°.

d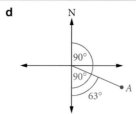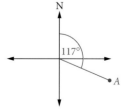

90° + (90° − 63°) = 117°   The three-figure bearing is 117°.

WORKED EXAMPLE 11 Solving practical problems involving true bearings

A hot air balloon travelled 1.5 km due east and then 1.8 km south.

a Determine the true bearing of the hot air balloon from its starting point, correct to the nearest degree.

b Calculate the balloon's distance from its starting point. Round your answer to one decimal place.

Steps	Working
a 1 Draw a diagram showing the information as a right-angled triangle. Include θ as an unknown angle in the triangle.	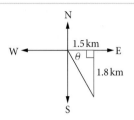
2 Identify whether the labelled sides are opposite, adjacent or hypotenuse and select the matching trigonometric ratio.	 Use $\tan(\theta) = \dfrac{\text{opposite}}{\text{adjacent}}$
3 Substitute the known values and solve with CAS, rounding to the nearest whole unit.	$\tan(\theta) = \dfrac{1.8}{1.5}$ $\theta = \tan^{-1}\left(\dfrac{1.8}{1.5}\right) \approx 50°$
4 Mark the three-figure bearing angle on the diagram. Find the angle clockwise from north, adding or subtracting 90° angles if necessary. Write the answer to the required level of accuracy.	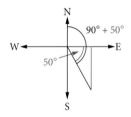 $90° + 50° = 140°$ The balloon is at a bearing of 140° from its starting point.
b 1 Use Pythagoras' theorem to find d, the distance from the starting point.	$d^2 = 1.5^2 + 1.8^2$ $= 5.49$ $d = 2.343...$
2 Write your answer in the required units and round to the required level of accuracy.	The balloon is 2.3 km from its starting point.

Compass bearings

An alternative method for describing direction is to use a **compass bearing**, which is an angle measurement taken from either north or south in the direction of east or west. For example, the compass bearing N 50° E can be read as 'start facing north, turn 50° in the direction of east', whereas the compass bearing S 50° E can be read as 'start facing south, turn 50° in the direction of east'.

Conventionally, these bearings always start from north or south, and never from east or west. Additionally, the angle given must be an acute angle, that is between 0° and 90°, because anything larger than 90° will end up being closer to the opposing cardinal point. For example, N 100° E would be better described as S 80° E because it is closer to south than it is to north.

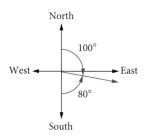

Compass bearings

A compass bearing of Y from X is written in the convention:

Closest starting point	Acute angle to the nearest whole degree	Direction
north or south	0 – 90°	east or west

WORKED EXAMPLE 12 | Identifying compass bearings

For each of the following, state the compass bearing of the point A.

a 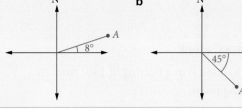 b

c 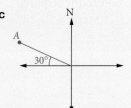 d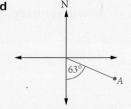

Steps	Working
For parts **a–d**: **1** Identify whether point A is closer to north or south. **2** Calculate the acute angle from north or south, in the direction of either east or west. **3** Express the bearing in the form N/S 0–90° E/W.	**a** Closer to north. 90 – 8 = 72 Easterly direction. N 72° E
	b Closer to south 90 – 45 = 45 Easterly direction. S 45° E
	c Closer to north. 90 – 30 = 60 Westerly direction. N 60° W
	d Closer to south Easterly direction. S 63° E

WORKED EXAMPLE 13 | Solving practical problems involving compass bearings

A ship sails 22 km due east and then 10 km due north.

a Calculate the compass bearing on which the ship travelled.

b On what compass bearing would the ship need to travel to return directly to its starting point?

Steps	Working
a 1 Draw a diagram showing the information as a right-angled triangle. Include θ as an unknown angle in the triangle.	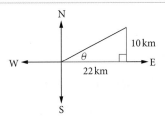
2 Identify whether the labelled sides are opposite, adjacent or hypotenuse and select the matching trigonometric ratio.	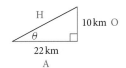 Use $\tan\theta = \dfrac{\text{opposite}}{\text{adjacent}}$.
3 Substitute the known values and solve with CAS, rounding to the nearest whole unit.	$\tan\theta = \dfrac{10}{22}$ $\theta = \tan^{-1}\left(\dfrac{10}{22}\right)$ $= 24.44\ldots$ $\approx 24°$
4 Find the angle clockwise from north, adding or subtracting 90° angles if necessary. Write the answer to the required level of accuracy.	$90 - 24 = 66°$ The ship travelled on a bearing of N 66° E.
b 1 At the end point of the ship's travel, draw a new compass plane. **2** Use alternate angles on parallel lines to locate the 66° angle between the south line and the direction of travel.	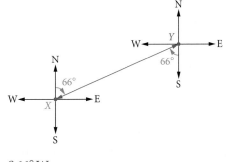
3 State the required bearing in the correct form.	S 66° W

EXERCISE 7.4 Bearings and navigation   ANSWERS p. 422

Recap

1 A zipline has been built with a 30° angle of elevation from the ground to the top of a tower. The height of the tower is 35 m. Assuming there is no slack in the zipline, calculate its length.

2 Jess is standing on her balcony, which is 60 m above street level. She spots her friend, Albie, on his balcony, which is 20 m above street level. If the two buildings are 35 m apart, calculate the angle of depression, θ, of Albie from Jess, correct to the nearest degree.

Mastery

3 WORKED EXAMPLE 10 For each of the following, state the true bearing of the point A and draw a diagram showing the angle from north.

a 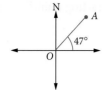   b 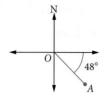   c 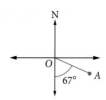   d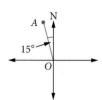

e 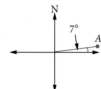   f 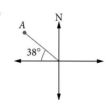   g 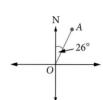   h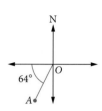

4 WORKED EXAMPLE 11 A ship sails due south for 20 km and due east for a further 15 km.

a Determine the true bearing of the ship from its starting point, correct to the nearest degree.

b Calculate the ship's distance from its starting point.

5 WORKED EXAMPLE 12 For each of the following, state the compass bearing of the point A.

a 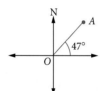   b 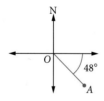   c 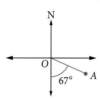   d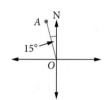

e 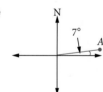   f 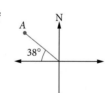   g   h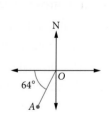

6 WORKED EXAMPLE 13 A ship sails due west for 7 km and due north for 5 km.

a Calculate the compass bearing on which the ship travelled, correct to the nearest degree.

b On what compass bearing would the ship need to travel to return directly to its starting point?

Calculator-free

7 (6 marks) The locations of three towns, Q, R and T, are shown in the diagram.

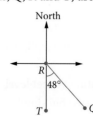

Town T is due south of town R. The angle TRQ is 48°.

a State the true bearing of town Q from town R. (2 marks)

b State the compass bearing of town R from town Q. (2 marks)

c Is it possible to determine the bearing of town Q from town T? Justify your answer. (2 marks)

8 (4 marks) A waterfall in a national park is 4 km due east of a camp site. A lookout tower is 4 km due south of the waterfall.

   a Draw a labelled diagram showing the positions of the waterfall, W, camp site, C, and the lookout tower, L. (2 marks)

   b State the bearing of the camp site from the lookout tower as a
      i compass bearing (1 mark)
      ii true bearing. (1 mark)

Calculator-assumed

9 (7 marks) Alan walked due east from his home and then due south. His final position was 100 m from his home on a bearing of 125°, as shown in the diagram.

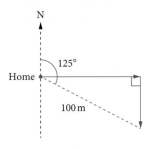

   a Calculate the distance, to the nearest metre, that Alan travelled east. (3 marks)
   b Calculate the distance, to the nearest metre, that Alan travelled south. (2 marks)
   c On what compass bearing would Alan need to travel if he wanted to walk home directly from his final position? (2 marks)

10 (4 marks) Salena practises golf at a driving range by hitting golf balls from point T. The first ball that Salena hits travels directly north, landing at point A. The second ball that Salena hits travels 50 m on a true bearing of 030°, landing at point B. The diagram shows the positions of the two balls after they have landed.

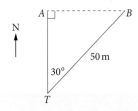

   a Given that the ball at point B is due east of the ball at point A, calculate how far apart the two golf balls are. (2 marks)

   b A fence is positioned at the end of the driving range. The fence is 16.8 m high and is 200 m from the point T.

   Calculate the angle of elevation of the top of the fence from point T, correct to the nearest 0.1 degrees. (2 marks)

11 (5 marks) A dairy farm has a farmhouse, a milking shed and a manufacturing building. The farmhouse is located due east of the milking shed. The manufacturing building is located due south of the farmhouse. The manufacturing building is 160 m from the milking shed, as shown.

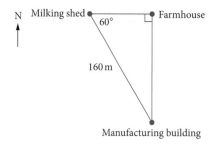

a How far east of the milking shed is the manufacturing building located? (2 marks)

b A storage facility is located 900 m due east and 400 m due north of the manufacturing building, as shown.

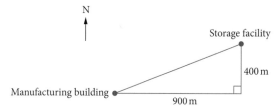

Determine the true bearing of the storage facility from the manufacturing building. Round your answer to the nearest degree. (3 marks)

12 (9 marks) A ship leaves from point A and sails on a bearing of 060° for 50 km to point B. Then it leaves point B and sails on a bearing of 150° for 12 km to point C.

  a Draw a labelled diagram showing the positions of points A, B and C, including all relevant angles and distances. (3 marks)
  b Explain why $\angle ABC = 90°$. (1 mark)
  c Calculate the length of AC, correct to the nearest kilometre. (2 marks)
  d Determine the true bearing required for the ship to travel directly back from point C to point A. (3 marks)

7.5 Area of a non-right-angled triangle

Video playlist
Area of a non-right-angled triangle

From our prior knowledge, we know that we can calculate the area of a triangle using the formula $A = \frac{1}{2}$ base × height, where height refers to the perpendicular height. For example, in the triangle KLM shown, the area is $A = \frac{1}{2}(17)(10) = 85$ units².

But what if we didn't know the perpendicular height of a triangle? For example, as seen in the triangles ABC and DEF below.

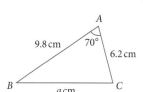 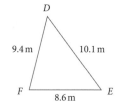

As a result, we need to two further approaches to calculate the areas of such triangles.

Using $A = \frac{1}{2}ab\sin C$

7.5

Consider a triangle ABC with the side lengths a and b known and the angle at C known. Let h be the perpendicular height of the triangle.

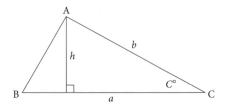

Knowing the length of two sides and the **included angle**, that is, the angle between them, is enough to calculate the area of ABC.

Using right-angled trigonometry, we can say that $\sin C = \dfrac{h}{b}$ and so $h = b \sin C$.

Remembering that the area of a triangle is given by $A = \dfrac{1}{2}$base × height, we can establish that

$A = \dfrac{1}{2}ah$

$A = \dfrac{1}{2}ab\sin C$

> **Area of a triangle using two sides and an angle**
>
> For a triangle with side lengths a and b, and the value of the included angle C, then the area is given by
>
> $A = \dfrac{1}{2}ab\sin C$
>
> This area formula is particularly useful for non-right-angled triangles, when two sides and the angle between them are known.

WORKED EXAMPLE 14 Calculating area using $A = \frac{1}{2}ab\sin C$

Calculate the area of triangle ABC, rounded to one decimal place.

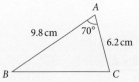

Steps	Working
1 Check the diagram to ensure two side lengths and the size of the angle between them are known.	
2 State the formula $A = \dfrac{1}{2}ab\sin C$.	$A = \dfrac{1}{2}ab\sin C$
3 Substitute the values $a = 9.8$, $b = 6.2$ and $C = 70°$ into the formula.	$A = \dfrac{1}{2}(9.8)(6.2)\sin(70°)$
4 Use CAS to evaluate the area, writing the answer in the correct units and to the required level of accuracy.	$A = 28.54...$ $A \approx 28.5\,\text{cm}^2$

Using Heron's rule

What if we wanted to calculate the area of a triangle when three side lengths, a, b and c, are known but no known angles? For example, consider triangle DEF shown. Once we have covered the cosine rule in section 7.6, we would be able to find an angle and then apply $A = \frac{1}{2}ab\sin C$, but it could become tedious. Instead, we can use a measurement formula called **Heron's rule**. The derivation and proof of Heron's rule is outside of the Applications 11 course due to the significant algebra involved, but it is important to understand its features and how to use it.

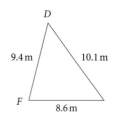

For a triangle with side lengths a, b and c, its perimeter is $P = a + b + c$. We can then define a new concept called the **semi-perimeter** of a triangle, which as the name suggests, is half of the value of the perimeter.

$$s = \frac{1}{2}P = \frac{a+b+c}{2}$$

> ### Heron's rule
>
> Heron's rule says that for a triangle with sides a, b and c, the area of the triangle is given by
>
> $$A = \sqrt{s(s-a)(s-b)(s-c)}$$
>
> where s is the semi-perimeter of the triangle,
>
> $$s = \frac{1}{2}P = \frac{a+b+c}{2}$$

WORKED EXAMPLE 15 — Calculating area using Heron's rule

Calculate the area of triangle DEF, rounded to one decimal place.

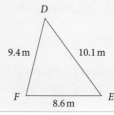

Steps	Working
1 Check the diagram to ensure all three side lengths are known.	
2 State Heron's rule and calculate the semi-perimeter, $s = \frac{a+b+c}{2}$.	$A = \sqrt{s(s-a)(s-b)(s-c)}$ $s = \frac{9.4 + 10.1 + 8.6}{2} = \frac{28.1}{2} = 14.05$
3 Substitute the values $a = 9.4$, $b = 10.1$, $c = 8.6$ and $s = 14.05$ into the formula.	$A = \sqrt{14.05(14.05 - 9.4)(14.05 - 10.1)(14.05 - 8.6)}$ $= \sqrt{14.05(4.65)(3.95)(5.45)}$
4 Use CAS to evaluate the area, writing the answer in the correct units and to the required level of accuracy.	$= 37.50\ldots$ $A \approx 37.5 \, \text{m}^2$

> ### 🔓 Exam hack
>
> In some cases, you may not be directed to using a specific area formula, but instead must choose the most appropriate formula. Just remember, if the triangle has
>
> - two known sides and the included angle, use $A = \frac{1}{2}ab\sin C$
> - three known sides, use Heron's rule.

CAS can be used to store and evaluate the formulas, which may be useful when asked to calculate one of the side lengths or an angle, when given the area of the triangle.

USING CAS 2 | Choosing and using area formulas

The area of the triangle shown is 216 cm^2. The side length labelled x is unknown.

a Identify which of the following area formulas would be the most appropriate to use in this situation. Justify your answer.

$A = \frac{1}{2}bh$ $\quad$ $A = \frac{1}{2}ab\sin C$ $\quad$ $A = \sqrt{s(s-a)(s-b)(s-c)}$

b Hence, determine the value of x, rounding to the nearest whole.

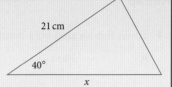

Steps	Working
a 1 Identify the features of the triangle. 2 Choose a rule and explain your choice using the features.	$A = \frac{1}{2}ab\sin C$ as two sides and the included angle between the two sides are known.

ClassPad

1. Open the **NumSolve** application.
2. In the **Equation** section, input the sine rule $A = \frac{1}{2}ab\sin(C)$ using the **Var** tab on the Keyboard for the variables. Press **EXE**.

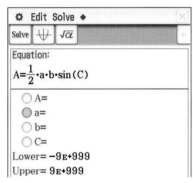

TI-Nspire

There is no equivalent **NumSolve** template in TI-Nspire but there is an **nSolve** function.

All letters in TI-Nspire are in lower case so unlike ClassPad, it will not distinguish between a and A, or b and B.

b 3 Make a unknown by ensuring the blue circle next to a is highlighted, representing the unknown x in the problem.

4 Input **A = 216**, **b = 21** and **C = 40°**.

5 Press **Solve** in the top left and read the solution from the pop-up window.

6 Write your answer in the required units, and round to the required level of accuracy.

b When finding a side length, use **nSolve** by substituting in the values.

Solve will give an exact answer whereas **nSolve** will give a decimal answer.

1 Press **menu > Algebra > nSolve**.

2 Enter **216 = $\frac{1}{2}$ a × 21 sin(40),a**.

3 Press **enter**.

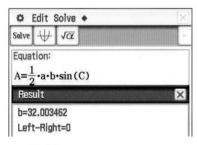

$x = 32.00...$
$x \approx 32 \text{ cm}$

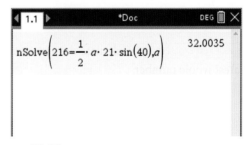

$x = 32.00...$
$x \approx 32 \text{ cm}$

Alternatively, the **solve** function in **Main** can be used to solve for the unknown side.

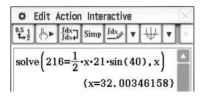

EXERCISE 7.5 Area of a non-right-angled triangle

ANSWERS p. 423

Recap

1 Town B is located on a bearing of 060° from Town A. The diagram that could illustrate this is

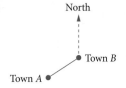 **A**

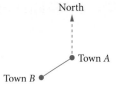 **B**

 **C**

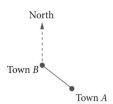 **D**

 **E**

2 Town A is due west of Town B. Town C is due south of town B. The bearing of Town A from Town C is

- **A** between 000° and 090°
- **B** between 090° and 180°
- **C** exactly 135°
- **D** between 180° and 270°
- **E** between 270° and 360°

Mastery

3 WORKED EXAMPLE 14   Calculate the area of the following triangles, rounding your answers to one decimal place.

a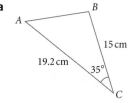

b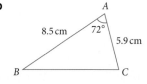

c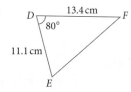

d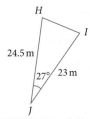

e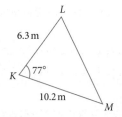

4 WORKED EXAMPLE 15   Calculate the area of the following triangles, rounding your answers to the nearest whole number.

a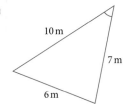

b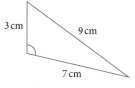

c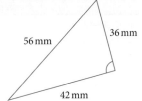

5 [Using CAS 2] The area of the triangle shown is 46 cm².

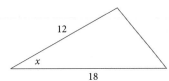

The acute angle labelled x is unknown.

a Identify which of the following area formulas would be the most appropriate to use in this situation. Justify your answer.

$$A = \frac{1}{2}bh \qquad A = \frac{1}{2}ab\sin C \qquad A = \sqrt{s(s-a)(s-b)(s-c)}$$

b Hence, determine the value of x, rounding to the nearest whole degree.

Calculator-free

6 (2 marks) Given that $\sin 90° = 1$, explain the relationship between the two area formulas below. Use a diagram of a right-angled triangle to support your explanation.

$$A = \frac{1}{2}bh \qquad A = \frac{1}{2}ab\sin C$$

7 (3 marks) Consider the triangle provided.

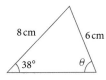

a Explain why the area formula $A = \frac{1}{2}ab\sin C$ cannot be used immediately to calculate the area of the triangle. (1 mark)

b Let $\theta = 55°$. Show that the area of the triangle can be calculated using $A = 24\sin(87°)$. (2 marks)

Calculator-assumed

8 (5 marks) A cross-section of a glass greenhouse is shown in the diagram.

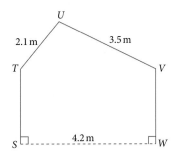

The sides of the glass panels TU and UV are 2.1 metres and 3.5 metres long respectively. The greenhouse is 4.2 metres wide. The walls ST and WV are vertical and are both 3 metres tall.

a Determine the cross-sectional area of the glass greenhouse, correct to one decimal place. (3 marks)

b If the greenhouse has a uniform cross-section as shown above, and the building is 11.4 metres long, determine the volume of the greenhouse to the nearest cubic metre. (2 marks)

9 (5 marks)

a Triangle ACB has an area of 1477 m².

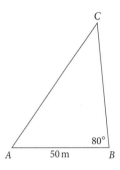

Determine the length of side BC, to the nearest metre. (2 marks)

b Triangle XYZ has an area of 7.1 km².

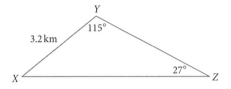

Determine the length of side XZ, to the nearest 0.1 km. (3 marks)

10 (6 marks) Consider the triangle shown.

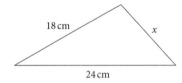

The side length labelled x is unknown.

a Express the perimeter of the triangle, P, in terms of x. (1 mark)
b Hence, express the semi-perimeter of the triangle, s, in terms of x. (1 mark)
c Write an equation for the area of the triangle, A, in terms of x. (2 marks)
d If the area of the triangle is 104.6 cm², determine the value of x to the nearest centimetre, given that it is the shortest side of the triangle. (2 marks)

7.6 The sine and cosine rules for non-right-angled triangles

The sine rule

In Sections 7.1 to 7.4, we used trigonometry to solve problems involving right-angled triangles, and in Section 7.5 we were introduced to some trigonometric results that can be used to calculate the area of non-right-angled triangles. However, what if we wanted to calculate unknown side lengths and angles of non-right-angled triangles?

One way of doing this is to break a non-right-angled triangle into two right-angled triangles using the perpendicular height of the triangle, as seen in the diagram showing $\triangle ABC$ with sides a, b and c.

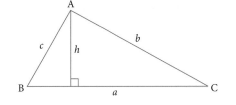

From here, there are two ways we can write a relationship for the sides b and c in terms of h.

$$\sin B = \frac{h}{c} \quad \text{[equation 1]} \quad \text{and} \quad \sin C = \frac{h}{b} \quad \text{[equation 2]}$$

From equation 1, we can multiply both sides by c and say that $h = c \sin B$.

From equation 2, we can multiply both sides by b and say that $h = b \sin C$.

If both equations are for h, then

$$c \sin B = b \sin C$$

which could be re-written as

$$\frac{c}{\sin C} = \frac{b}{\sin B}$$

This process could be done again to show that

$$\frac{a}{\sin A} = \frac{b}{\sin B}$$

and again to show that

$$\frac{a}{\sin A} = \frac{c}{\sin C}$$

This result is called the **sine rule**, which says that the proportion of each side length to the sine of its opposing angle is the same for all three sides.

Video playlist
The sine and cosine rules for non-right-angled triangles

Worksheets
Trigonometric calculations

Discovering the sine rule

The sine rule – Finding lengths of sides

The sine rule – Finding angles

The cosine rule – Angles and sides

Applying trigonometry

Finding an unknown side

Finding an unknown angle

Sine rule problems

Cosine rule problems

The sine rule

For the triangle ABC

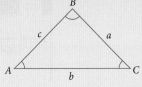

where $A°$, $B°$ and $C°$ are the angles at points A, B and C respectively and a, b and c are the sides opposite each respective angle, then

$$\frac{a}{\sin A} = \frac{b}{\sin B} = \frac{c}{\sin C}$$

The reciprocal of this rule is also true, such that

$$\frac{\sin A}{a} = \frac{\sin B}{b} = \frac{\sin C}{c}$$

Chapter 7 | Applications of trigonometry

While there are three parts to the sine rule, we only ever need two!

First, we will consider an example with two known angles and one opposing side. We can use the sine rule to calculate the unknown side.

WORKED EXAMPLE 16 — Solving for unknown side lengths using the sine rule

Find the value of x in the following triangle, rounded to the nearest whole unit.

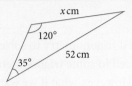

Steps	Working
1 Label the vertices A, B, C and label the sides opposite the angles a, b, c.	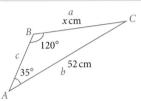
2 List the known values, and let the unknown value be x.	$A = 35°$, $B = 120°$, $a = x$, $b = 52\,\text{cm}$
3 Select the equality needed from the sine rule.	$\dfrac{a}{\sin A} = \dfrac{b}{\sin B}$
4 Substitute the values into the sine rule and rearrange appropriately to solve for x.	$\dfrac{x}{\sin 35°} = \dfrac{52}{\sin 120°}$ $x = \dfrac{52 \times \sin 35°}{\sin 120°}$
5 Evaluate x using CAS, write your answer in the required units, and round to the required level of accuracy.	$x = 34.44\ldots$ $x \approx 34\,\text{cm}$

> 🔒 **Exam hack**
>
> Always put the unknown side x as the numerator of the first fraction to make it an easy rearrangement.

> 🔒 **Exam hack**
>
> You can also use the **solve** function of CAS, like in Using CAS 1, to solve for the value of an unknown side length, but it is important to always show the line of working out involving the sine rule containing the substituted values before you do so.

ClassPad

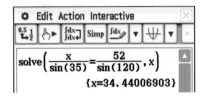

TI-Nspire

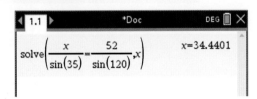

Now, we will consider an example where there are two known sides and one opposing angle. We can use the sine rule to calculate the unknown angle, as long as it is opposite a known side.

WORKED EXAMPLE 17 — Solving for unknown angles using the sine rule

Find the value of θ in the following triangle, rounded to the nearest degree.

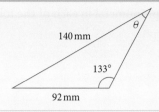

Steps	Working
1 Label the vertices A, B, C and label the sides opposite the angles a, b, c.	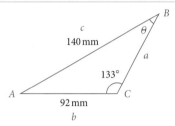
2 List the known values and let the unknown value be θ.	$B = \theta$, $C = 133°$, $b = 92$, $c = 140$
3 Select the equality needed from the sine rule.	$\dfrac{\sin B}{b} = \dfrac{\sin C}{c}$
4 Substitute the values into the sine rule and rearrange appropriately to solve for θ.	$\dfrac{\sin \theta}{92} = \dfrac{\sin 133°}{140}$ $\sin \theta = \dfrac{92 \times \sin 133°}{140}$ $\theta = \sin^{-1}\left(\dfrac{92 \sin 133°}{140}\right)$
5 Evaluate θ using a calculator, write your answer in the required units, and round to the required level of accuracy.	$\theta = 28.72…$ $\theta \approx 29°$

> **Exam hack**
> Use the version of the sine rule where the sine of the angles are on the numerator, as you are solving for an unknown angle.

ClassPad

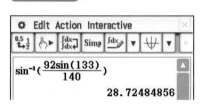

TI-Nspire

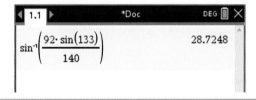

While we do not cover it in this course, when solving for an angle using the sine rule, there are technically two possible answers we would need to consider: an acute angle (between 0° and 90°) or an obtuse angle (between 90° and 180°). We will only ever deal with the acute solution here.

As a result, we could also use CAS to solve for the unknown angle, but we would need to put a restriction on the possible values of θ such that $0° < \theta < 90°$.

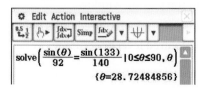

USING CAS 3 — Solving for a side length or an angle using the sine rule

Find the value of the pronumeral in each of the following triangles. Answer to the nearest whole unit.

a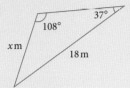

b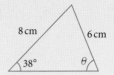

ClassPad

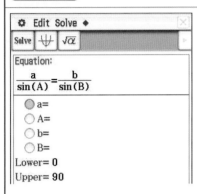

1 Open the **NumSolve** application.

2 In the Equation section, input the sine rule $\dfrac{a}{\sin(A)} = \dfrac{b}{\sin(B)}$ using the **Var** tab on the Keyboard for the variables. Press **EXE**.

a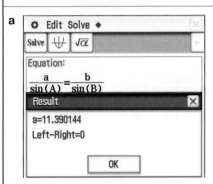

b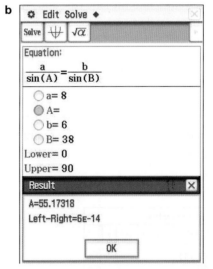

1 Leave a unknown and the blue circle next to a highlighted, representing the unknown x in the problem.

2 Input **A = 37°**, **b = 18** and **B = 108°**.

3 Press **Solve** in the top left and read the solution from the pop-up window.

4 Write your answer in the required units, and round to the required level of accuracy.

1 Change the blue circle to A to represent the unknown angle, θ. Input the values of **a = 8**, **b = 6** and **B = 38**.

2 Now that you are solving for an angle, change the **Lower=** setting to 0 and **Upper=** setting to 90, to represent the acute angle.

3 Press **Solve** in the top left and read the solution from the pop-up window.

4 Write your answer in the required units, and round to the required level of accuracy.

a $x = 11.39\ldots$
$x \approx 11$ m

b $\theta = 55.17\ldots$
$\theta \approx 55°$

> **TI-Nspire**
>
> There is no equivalent **NumSolve** template in TI-Nspire but there is an **nSolve** function.
>
> All letters in TI-Nspire are in lower case so unlike ClassPad, it will not distinguish between a and A, or b and B.
>
> **a** When finding a side length, use **nSolve** by substituting in the values.
>
>   **Solve** will give an exact answer, whereas **nSolve** will give a decimal answer.
>
> **b** When finding an angle, use **nSolve** by substituting in the values.
>
>   Without restricting the domain, **Solve** gives a general solution, whereas **nSolve** gives a decimal angle between 0° and 180°.
>
>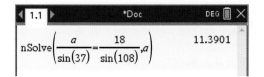
>
>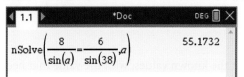
>
> 1 Press **menu** > **Algebra** > **nSolve**.
>
> 2 Enter $\left(\dfrac{a}{\sin(37)} = \dfrac{18}{\sin(108)}, a\right)$.
>
> 3 Press **enter**.
>
> 1 Press **menu** > **Algebra** > **nSolve**.
>
> 2 Enter $\left(\dfrac{8}{\sin(a)} = \dfrac{6}{\sin(38)}, a\right)$.
>
> 3 Press **enter**.
>
> **a** $x = 11.39\ldots$
>
> $x \approx 11\,\text{m}$
>
> **b** $\theta = 55.17\ldots$
>
> $\theta \approx 55°$

The cosine rule

The **cosine rule** is another way to find unknown sides and angles in non-right-angled triangles. The derivation of the cosine rule involves some algebra beyond the scope of our course, so we can take the rule for granted.

If two sides and the angle between them are known, use the cosine rule to find the unknown side:

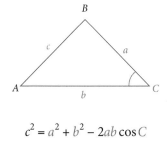

$$c^2 = a^2 + b^2 - 2ab\cos C$$

If three sides are known, use the rearranged cosine rule to find any angle:

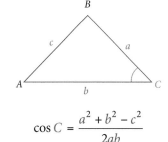

$$\cos C = \dfrac{a^2 + b^2 - c^2}{2ab}$$

It is important to recognise that when we are solving for an unknown side using the cosine rule, it has to be the matching side opposite the angle we are given!

WORKED EXAMPLE 18 — Solving for unknown side lengths using the cosine rule

Find the value of x in the following triangle, rounded to the nearest whole unit.

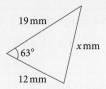

Steps | Working

1 Label the known angle C, then label the other angles A and B and the sides opposite the angles a, b, c.

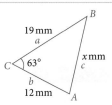

2 List the known values, with the value that needs to be found labelled x.

$C = 63°$, $c = x$, $b = 12$, $a = 19$

3 State the cosine rule used to find an unknown side length and substitute all the information.

$c^2 = a^2 + b^2 - 2ab \cos C$
$x^2 = 19^2 + 12^2 - 2(19)(12) \cos 63°$

4 Evaluate the right-hand side and solve for x by taking the square root.

$x^2 = 297.98\ldots$
$x = 17.26\ldots$

5 Write your answer in the required units and round to the required level of accuracy.

$x \approx 17$ mm

WORKED EXAMPLE 19 — Solving for unknown angles using the cosine rule

Find the value of θ in the following triangle, rounded to the nearest whole degree.

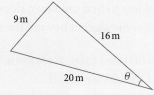

Steps | Working

1 Label the unknown angle C, then label the other angles A and B and the sides opposite the angles a, b, c.

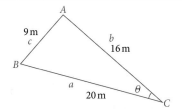

2 List the known values, with the value that needs to be found labelled θ.

$C = \theta°$, $c = 9$, $b = 16$, $a = 20$

3 State the cosine rule used to find an unknown angle and substitute all the information.

$$\cos C = \frac{a^2 + b^2 - c^2}{2ab}$$

$$\cos \theta = \frac{20^2 + 16^2 - 9^2}{2(16)(20)}$$

4 Evaluate the right-hand-side and solve for θ by taking the inverse of the cosine ratio.

$\cos \theta = 0.898\ldots$
$\theta = \cos^{-1}(0.898\ldots)$
$\theta = 26.046\ldots$

5 Write your answer in the required units and round to the required level of accuracy.

$\theta \approx 26°$

The cosine rule

The cosine rule for triangle ABC can be used in two forms.

- When finding an unknown side
$$c^2 = a^2 + b^2 - 2ab \cos C$$

- When finding an unknown angle
$$\cos C = \frac{a^2 + b^2 - c^2}{2ab}$$

where angle C is the angle between sides a and b.

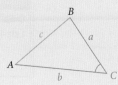

You might notice that the start of the cosine rule looks familiar. In fact, if you use your calculator to evaluate $\cos 90°$, you will see that it gives a value of 0. This means that for a right-angled triangle, the cosine rule becomes

$$c^2 = a^2 + b^2 - 2ab \times 0$$
$$c^2 = a^2 + b^2$$

which is Pythagoras' theorem!

USING CAS 4 Solving for a side length or an angle using the cosine rule

Find the value of the pronumeral in each of the following triangles. Answer to the nearest whole unit.

a Triangle ABC with $AB = c$ cm, $BC = 15$ cm, $AC = 19.2$ cm, and angle $C = 35°$.

b Triangle with sides 9.4 m, 10.1 m, 8.6 m, and angle θ between the 9.4 m and 10.1 m sides.

ClassPad

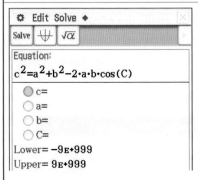

1. Open the **NumSolve** application.

2. In the **Equation** section, input the cosine rule $c^2 = a^2 + b^2 - 2ab \cos(C)$ using the **Var** tab on the Keyboard for the variables. Press **EXE**.

a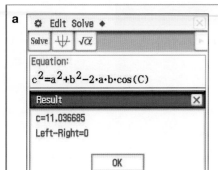

b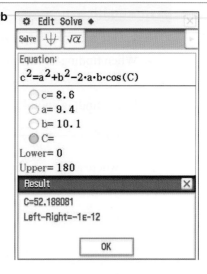

1. Leave c unknown and the blue circle next to c highlighted, representing the unknown x in the problem.
2. Input **a = 15, b = 19.2** and **C = 35°**.
3. Press **Solve** in the top left and read the solution from the pop-up window.
4. Write your answer in the required units, and round to the required level of accuracy.

1. Change the blue circle to to represent the unknown angle, θ. Input the values of **c = 8.6, a = 9.4, b = 10.1**.

   Note that it is really important that the c is the 8.6 as it is the side length opposite the unknown angle. The values of a and b do not matter.
2. When solving for an angle, change the **Lower=** setting to **0** and **Upper=** setting to **180**, to represent an angle inside a triangle.
3. Press **Solve** in the top left and read the solution from the pop-up window.
4. Write your answer in the required units, and round to the required level of accuracy.

TI-Nspire

There is no equivalent **NumSolve** template in TI-Nspire but there is an **nSolve** function. All letters in TI-Nspire are in lower case so unlike ClassPad, it will not distinguish between c and C.

a When finding a side length, use **nSolve** by substituting in the values.

Solve gives both a positive and negative answer, whereas **nSolve** gives only the positive answer.

b When finding an angle, use **nSolve** by substituting in the values.

Without restricting the domain, **Solve** gives a general solution, whereas **nSolve** gives a decimal angle between 0° and 180°.

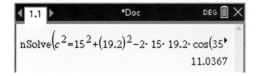

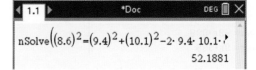

1. Press **menu > Algebra > nSolve**.
2. Enter **($c^2 = 15^2 + 19.2^2 − 2 × 15 × 19.2 × \cos(35)$,c)**.
3. Press **enter**.

1. Press **menu > Algebra > nSolve**.
2. Enter **($8.6^2 = 9.4^2 + 10.1^2 − 2 × 9.4 × 10.1 × \cos c$, c)**.
3. Press **enter**.

a $c = 11.036…$
$c \approx 11$ cm

b $\theta = 52.18…$
$\theta \approx 52°$

Practical problems involving non-right-angled triangles

Now that we have multiple skills in the topic of trigonometry, it is important to distinguish the type of problem we are solving. A good first step is to check whether the triangle is right-angled or not.

Selecting the method to find unknowns in a triangle

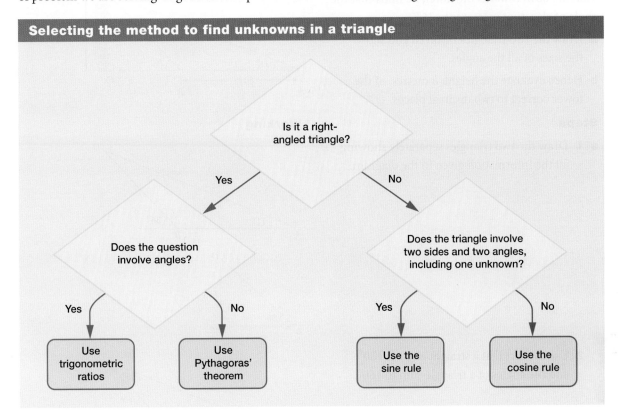

The practical problems that we may encounter could include general applications, as well as the contexts of angles of elevation and depression (from Section 7.3) and bearings (from Section 7.4). Some practical problems may even revisit the area of a triangle (from Section 7.5)!

> **Exam hack**
>
> When you're solving problems involving non-right-angled triangles, you may need to remember that the angles inside a triangle add to 180° and that a straight-line angle is 180°.

WORKED EXAMPLE 20 Solving a practical problem involving angles of elevation

Ryan observes a tower at an 11° angle of elevation. Walking 80 m towards the tower, he finds that the angle of elevation increases to 36°.

a Draw the two triangles involved and show the sizes of all the angles.

b Hence evaluate the height, h metres, of the tower correct to two decimal places.

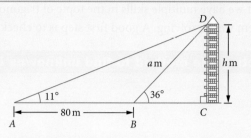

Steps	Working
a 1 Draw the two triangles separately showing all the information given in the diagram.	 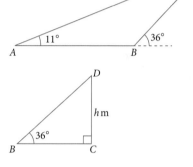
2 Use the fact that a straight angle is 180° and the angles of a triangle sum to 180°.	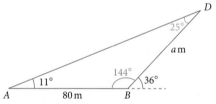 $180° - 36° = 144°$ $180° - 11° - 144° = 25°$ 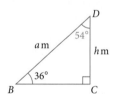 $180° - 36° - 90° = 54°$
b 1 Select the equality needed from the sine rule to find the unknown in the first triangle: $$\frac{a}{\sin A} = \frac{b}{\sin B} = \frac{c}{\sin C}$$ Solve using CAS.	$$\frac{a}{\sin(11)} = \frac{80}{\sin(25)}$$ $a = 36.119…$
2 Use the unknown length found and the sine rule to calculate the required value from the second triangle. Solve using CAS.	$$\frac{36.119…}{\sin(90)} = \frac{h}{\sin(36)}$$ $h = 21.230…$
3 Write your answer in the required units and round to the required level of accuracy.	The height of the tower is 21.23 m.

WORKED EXAMPLE 21 Solving a practical problem involving bearings

Putri is going on a three-day bushwalking adventure. She starts at camp site X and walks for 15 km on a bearing of 150° to camp site Y. On the second day, she walks 20 km due north to camp site Z. On the third day, Putri plans to return to camp site X.

a Find the distance, to the nearest kilometre, that Putri needs to travel to return to camp site X.
b Calculate the true bearing that Putri needs to travel on in order to return to camp site X. Express your answer correct to the nearest degree.
c Calculate the area of land encompassed within the triangle defined by Putri's bushwalking adventure.

Steps	Working
a 1 Draw a diagram showing the information as a triangle, including the unknown we are asked to find. Let this unknown be d.	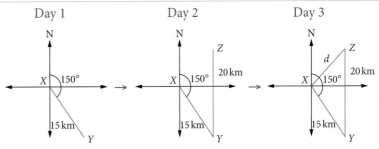
2 Calculate an angle in the triangle using the bearings and the angle sum of a triangle.	150° − 90° = 60° 90° − 60° = 30° 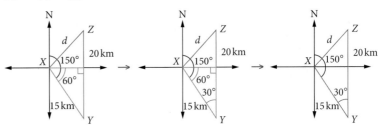
3 Is it a right-angled triangle?	No.
4 Does the triangle involve two sides and two angles, including one unknown?	No, it involves three sides and an angle. Use the cosine rule.
5 Redraw the triangle. Label the shown angle C, then label the other angles A and B, and the sides opposite the angles a, b, c.	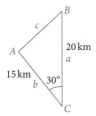
6 List the known values, with the value that needs to be found labelled d.	$C = 30°$, $c = d$, $b = 15$, $a = 20$
7 State the cosine rule used to find an unknown side length and substitute all the information.	$c^2 = a^2 + b^2 - 2ab \cos C$
8 Solve using CAS.	$d^2 = 20^2 + 15^2 - 2(20)(15) \cos(30°)$ $d = 10.265\ldots$
9 Write your answer in the required units and round to the required level of accuracy.	The distance that Putri needs to travel to return to camp site X is 10 km.

b 1 Add north to the bearing starting point on the redrawn triangle, include the unrounded distance found in part *a*, and show the angle that needs to be found.

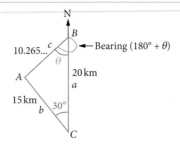

2 Is it a right-angled triangle?

No.

3 Does the triangle involve two sides and two angles, including one unknown?

Yes. Use the sine rule.

Note that here the cosine rule could also be used as all three sides are now known.

4 State the sine rule used to find an unknown angle and substitute all the information.

$c = 10.265...$, $C = 30°$, $b = 15$, $B = \theta$

$$\frac{\sin(\theta)}{15} = \frac{\sin(30°)}{10.265...}$$

$\theta = 46.93...$

5 Solve using CAS for $0 < \theta < 90°$, rounding to the nearest degree.

$\theta \approx 47°$

6 Answer the question by calculating the true bearing.

$180° + 47° = 227°$

The true bearing is 227°.

c 1 Choose an appropriate area formula for a non-right-angled triangle:

$A = \frac{1}{2}ab \sin C$

or

$A = \sqrt{s(s-a)(s-b)(s-c)}$

$A = \frac{1}{2}ab \sin C$

2 Substitute the known values and calculate the area.

$A = \frac{1}{2}(20)(15) \sin 30°$

$A = 75 \text{ km}^2$

The area of land within the triangle is 75 km^2.

EXAMINATION QUESTION ANALYSIS

Calculator-assumed **(12 marks)**

A tree is growing near a triangular block of land, *PST*. The base of the tree, *T*, is on the same horizontal plane as the corners, *P* and *S*, of the block of land.

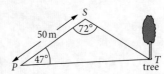

a Explain why $\angle PTS = 61°$. (1 mark)

b Show that the distance *ST* is approximately 42 metres. (2 marks)

c Calculate the area of the block of land, *PST*, correct to the nearest whole square metre. (2 marks)

d From point *S*, the angle of elevation to the top of the tree is 22°. Calculate the height of the tree, correct to the nearest whole metre. (2 marks)

e Calculate the angle of depression from the top of the tree to the point *P*, correct to the nearest degree. (5 marks)

Reading the question

- Recognise that 'triangular' and 'same horizontal plane' are there for you to be able to apply trigonometric concepts.
- Highlight the command words such as *calculate, show, explain why*.
- Highlight all rounding and accuracy commands, so that you do not lose marks for not meeting the requirements of the question.

Thinking about the question

- Think about what needs to be shown in the 'show' question to demonstrate full working out.
- You will need to add a third dimension to this problem when using the angles of elevation and depression. Think about how you can effectively represent this in a diagram.

Worked solution (✓ = 1 mark)

a Given that the sum of the interior angles of a triangle add to 180°,
$\angle PTS = 180 - 72 - 47 = 61°$

uses angle sum of triangle to explain $\angle PTS$ ✓

b $\dfrac{x}{\sin(47°)} = \dfrac{50}{\sin(61°)}$

establishes the sine rule correctly, with *ST* unknown ✓

$x = 41.80...$
$x = 42\,\text{m}$

calculates *ST* and shows appropriate rounding ✓

c $A = \dfrac{1}{2}(50)(41.80...)\sin(72°)$

establishes the area formula correctly ✓

$A = 994.08...$
$A \approx 994\,\text{m}^2$ ✓

d $\tan(22°) = \dfrac{h}{41.81}$

establishes trigonometric ratio correctly for height of the tree ✓

$h = 16.89...$
$h \approx 17\,\text{m}$ ✓

e $PT^2 = 50^2 + 41.81^2 - 2(50)(41.81)\cos(72°)$

establishes the cosine rule correctly, with *PT* unknown ✓

$PT = 54.36...$ ✓

$\tan(\theta) = \dfrac{16.89...}{54.37...}$

establishes trigonometric ratio correctly for angle of depression ✓

$\theta = \tan^{-1}\left(\dfrac{16.89...}{54.37...}\right) = 17.25...$ ✓

$\theta \approx 17°$ ✓

 Exam hacks

- It is often useful to draw multiple two-dimensional diagrams involving triangles rather than attempting a three-dimensional diagram.
- Be sure to keep the full values in CAS and use those when solving further problems instead of the rounded values.

EXERCISE 7.6 The sine and cosine rules for non-right-angled triangles   ANSWERS p. 423

Recap

1 The area of the triangle shown is

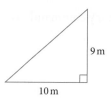

A $90\,m^2$ **B** $45\,cm^2$ **C** $90\,cm^2$ **D** $45\,m^2$ **E** $9.5\,m^2$

2 The distances from a kiosk to points A and B on opposite sides of a pond are found to be 12.6 m and 19.2 m respectively. The angle between the lines joining these points to the kiosk is 63°.

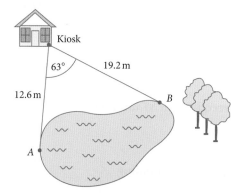

The area covered by the triangle enclosed between the Kiosk, and points A and B is found by evaluating

A $\dfrac{19.2 \times \sin 63°}{12.6}$

B $\dfrac{1}{2} \times 12.6 \times 19.2 \times \sin 63°$

C $\sqrt{12.6^2 + 19.2^2}$

D $\sqrt{12.6^2 + 19.2^2 - 2 \times 12.6 \times 19.2 \times \cos 63°}$

E $\sqrt{s(s-12.6)(s-19.2)(s-63)}$, where $s = \dfrac{1}{2}(12.6 + 19.2 + 63)$

Mastery

3 WORKED EXAMPLE 16   Find the value of the pronumeral in the following triangles, rounded to two decimal places.

a 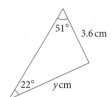   **b** 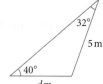   **c**

4 **WORKED EXAMPLE 17** Find the acute value of θ in the following triangle, rounded to the nearest whole degree.

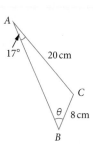

5 **Using CAS 3** Find the value of the pronumeral in each of the following triangles. Answer to the nearest whole unit. Note: the value of θ that you find should be an acute angle!

a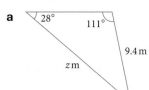

b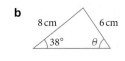

6 **WORKED EXAMPLE 18** Find the value of the pronumeral in each of the following triangles, rounded to one decimal place.

a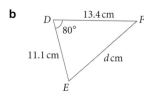

b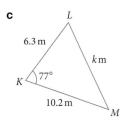

c

7 **WORKED EXAMPLE 19** Find the value of θ in each of the following triangles, rounded to the nearest whole degree.

a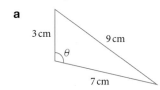

b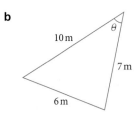

8 **Using CAS 4** Find the value of the pronumeral in each of the following triangles. Answer to the nearest whole unit.

a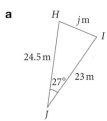

b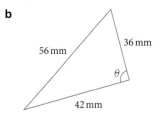

9 **WORKED EXAMPLE 20** Reshan observes the top and bottom of a mountain resort to be at angles of elevation of 57° and 43° respectively. He is 400 m from the bottom of the resort.

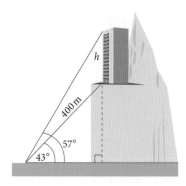

Calculate the height, h metres, of the resort correct to two decimal places.

10 **WORKED EXAMPLE 21** Adib is going on a long-distance run to practice for a marathon. He starts at point X and runs for 12 km on a bearing of 140° to point Y. Then he runs 16 km due north to point Z. He then decides to head back to his starting point, X.

  a Find the distance, to the nearest kilometre, that Adib needs to travel to return to his starting point X.
  b Calculate the true bearing that Adib needs to run on in order to return to X. Express your answer correct to the nearest degree.
  c Calculate the area of land encompassed within the triangle defined by Adib's long-distance run, correct to the nearest square kilometre.

Calculator-assumed

11 (2 marks) The diagram shows the route of a cross-country race.

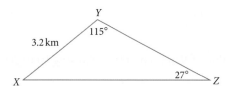

Point X lies due west of point Z. Given that the length XY is 3.2 km, calculate the length of XZ, correct to one decimal place.

12 (4 marks) In the triangle ACB, $\angle CAB = 60°$ and $\angle ABC = 80°$. The length of side $AB = 50$ m.
  a Draw a diagram to represent this information. (2 marks)
  b Calculate the length of side AC, correct to the nearest metre. (2 marks)

13 (3 marks) Marcus is on the opposite side of a large lake from a horse and its stable. The stable is 150 m due east of the horse. Marcus is on a bearing of 170° from the horse and on a bearing of 205° from the stable.

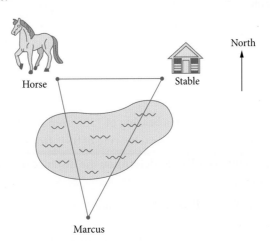

Calculate the direct distance between Marcus and the horse, correct to the nearest metre.

14 (6 marks) A yacht, Y, is 7 km from a lighthouse, L, on a bearing of 210°. A ferry can also be seen from the lighthouse. The ferry is 3 km from L on a bearing of 135°.
  a Draw a labelled diagram representing this situation. (3 marks)
  b State the angle between LY and LF. (1 mark)
  c Hence, calculate the length of FY, correct to two decimal places. (2 marks)

15 (3 marks) The cross-section of a warehouse is shown in the diagram below.

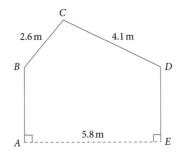

The sides of the roof, BC and CD, are 2.6 metres and 4.1 metres long respectively. The warehouse is 5.8 metres wide. The walls AB and ED are vertical and equal in height. Calculate the size of ∠BCD, correct to one decimal place.

16 (3 marks) In a game of beach soccer, a player at point F is attempting to kick a goal. She is 6 metres from one goal post at G and 8 metres from the other goal post at H. From where she is standing, the angle through which she can shoot at the goal and still score a goal is 19°, as shown.

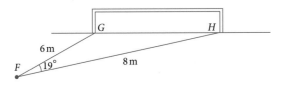

Calculate the distance between the goal posts, GH, to the nearest metre.

17 (4 marks) During a game of golf, Salena hits a ball twice, from P to Q and then from Q to R. The path of the ball after each hit is shown in the diagram below.

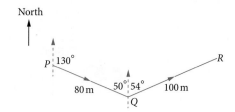

After Salena's first hit, the ball travelled 80 m on a bearing of 130° from point P to point Q.
After Salena's second hit, the ball travelled 100 m on a bearing of 054° from point Q to point R.

a If Salena was to hit a ball directly from P to R, determine the distance travelled by this ball, correct to the nearest metre. (2 marks)

b Determine the true bearing of R from P, correct the nearest degree. (2 marks)

18 (8 marks) A helicopter trip started in Rome and travelled directly to Genzano di Roma, as shown in the diagram.

Genzano di Roma is 37 km from Rome on a bearing of 153°.

a How far South of Rome is Genzano di Roma? Round your answer to the nearest kilometre. (2 marks)

From Genzano di Roma, the helicopter then followed a route to Tivoli, then Calcata and then back to Rome, as shown in the diagram.

Tivoli is 25 km from Genzano di Roma on a bearing of 016°. Calcata is 42 km from Tivoli on a bearing of 309°.

b Calculate the direct distance between Genzano di Roma and Calcata. Round your answer to the nearest kilometre. (3 marks)

c Calcata is 40.3 km north and 11.8 km west of Rome. Calculate the bearing of Rome from Calcata as a

  i compass bearing, to the nearest degree. (2 marks)

  ii bearing, to the nearest degree. (1 mark)

19 (6 marks) A tree, 12 m tall, is growing at point *T* near the shed. The distance, *CT*, from corner *C* of the shed to the centre base of the tree is 13 m.

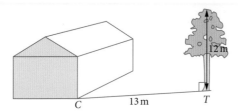

a Calculate the angle of elevation of the top of the tree from point *C*. Write your answer, in degrees, correct to one decimal place. (2 marks)

N and *C* are two corners at the base of the shed. *N* is 10 m due north of *C*. The angle, *TCN*, is 65°.

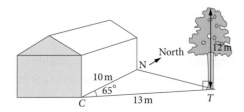

b Show that the distance, *NT*, is 12.6 m, correct to one decimal place. (2 marks)

c Calculate the angle, *CNT*, correct to the nearest degree. (2 marks)

20 (8 marks) Two ships are observed from point *O*. At a particular time, their positions *A* and *B* are as shown in the diagram.

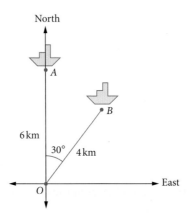

a Calculate the distance between the ships at this time, correct to two decimal places. (2 marks)

b Calculate the compass bearing of the ship at *B* from the ship at *A* at this point in time, correct to the nearest degree. (4 marks)

c Hence, state the true bearing of the ship at *A* from the ship at *B* at this point in time, correct to the nearest degree. (2 marks)

21 (8 marks) There are plans to construct a series of straight paths on the flat top of the mountain. A straight path will connect the cable car station at C to a communications tower at T, as shown in the diagram.

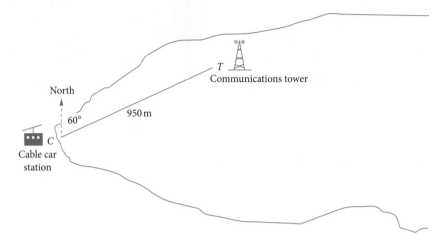

The true bearing of the communications tower from the cable car station is 060°. The length of the straight path between the communications tower and the cable car station is 950 m.

a How far north of the cable car station is the communications tower? (2 marks)

Paths will also connect the cable car station and the communications tower to a camp site at E, as shown.

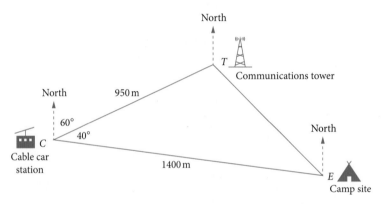

The length of the straight path between the cable car station and the camp site is 1400 m. The angle TCE is 40°.

b **i** Determine the length of the straight path between the communications tower and the camp site, correct to the nearest metre. (2 marks)

ii Show use of the cosine rule to find the true and compass bearings of the camp site from the communications tower, correct to the nearest degree. (4 marks)

⑦ Chapter summary

The trigonometric ratios

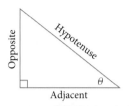

Trigonometric ratio	Notation	Meaning	Ratio
sine of θ	$\sin \theta$	compares how long the opposite side is when compared to the hypotenuse.	$\sin \theta = \dfrac{\text{opposite}}{\text{hypotenuse}}$
cosine of θ	$\cos \theta$	compares how long the adjacent side is when compared to the hypotenuse.	$\cos \theta = \dfrac{\text{adjacent}}{\text{hypotenuse}}$
tangent of θ	$\tan \theta$	compares how long the opposite side is when compared to the adjacent side.	$\tan \theta = \dfrac{\text{opposite}}{\text{adjacent}}$

The complementary relationship between sine and cosine

$\sin \alpha = \cos \beta$ if $\alpha + \beta = 90°$.

The inverse trigonometric ratios

Inverse trigonometric ratios can be used to find unknown angles in right-angled triangles, when we have the appropriate ratio of two sides.

If $\sin \theta = \dfrac{\text{opposite}}{\text{hypotenuse}}$, then $\theta = \sin^{-1}\left(\dfrac{\text{opposite}}{\text{hypotenuse}}\right)$.

If $\cos \theta = \dfrac{\text{adjacent}}{\text{hypotenuse}}$, then $\theta = \cos^{-1}\left(\dfrac{\text{adjacent}}{\text{hypotenuse}}\right)$.

If $\tan \theta = \dfrac{\text{opposite}}{\text{adjacent}}$, then $\theta = \tan^{-1}\left(\dfrac{\text{opposite}}{\text{adjacent}}\right)$.

When using CAS, make sure to set it to degree mode.

Angles of elevation and depression

Angle of elevation – looking up

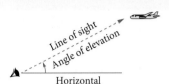

Angle of depression – looking down

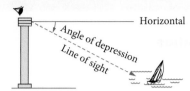

- angle of elevation X to Y = angle of depression Y to X

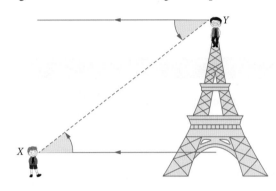

Bearings and navigation

A **true (three-figure) bearing** of Y from X is the angle of Y from X measured in a clockwise direction from north.

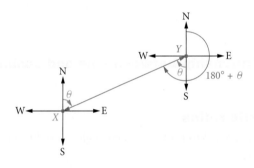

If the bearing of Y from X is θ, then the bearing of X from Y is $180° + \theta$, where $\theta < 180°$.

A **compass bearing** of Y from X is written in the convention:

Closest starting point	Acute angle to the nearest whole number	Direction
north or south	0–90°	east or west

Area of a triangle

The area of a triangle can be found using:

- $A = \dfrac{1}{2}bh$ if the base length and the perpendicular height are known.

- $A = \dfrac{1}{2}ab\sin C$ if two sides, a and b, and the included angle, C, are known.

- **Heron's rule**, $A = \sqrt{s(s-a)(s-b)(s-c)}$, if all three sides a, b and c are known, where s is the semi-perimeter, $S = \dfrac{1}{2}P = \dfrac{a+b+c}{2}$.

The sine and cosine rules

For the triangle ABC

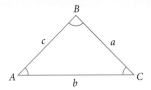

where A, B and C are the angles at points and respectively and a, b and c are the sides opposite each respective angle, then the **sine rule** says

$$\frac{a}{\sin A} = \frac{b}{\sin B} = \frac{c}{\sin C}$$

The reciprocal of this rule is also true, such that

$$\frac{\sin A}{a} = \frac{\sin B}{b} = \frac{\sin C}{c}$$

The **cosine rule** for triangle ABC can be used in two forms.

When finding an unknown side:

$$c^2 = a^2 + b^2 - 2ab \cos C$$

When finding an unknown angle:

$$\cos C = \frac{a^2 + b^2 - c^2}{2ab}$$

where angle C is the angle between sides a and b.

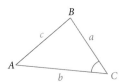

Practical problems involving non-right-angled triangles

- Select the method to find unknowns in a triangle.

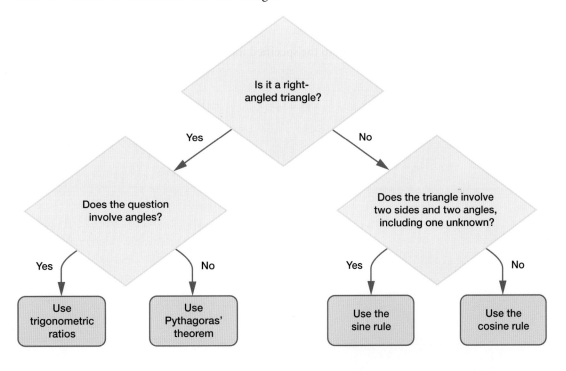

Cumulative examination: Calculator-free

Total number of marks: 14   Reading time: 2 minutes   Working time: 14 minutes

1 (4 marks) The first row of a spreadsheet is shown in the table below.

	A	B	C	D
1	150	200	400	50
2				

   a Determine the value of the cell A2 if it is defined by the formula A2 = 2B1 + 4D1. (2 marks)

   b Write an appropriate formula if the cell B2 is 150% of the difference between C1 and A1. (2 marks)

2 (2 marks) Researchers conducted a survey of 403 school leavers who had recently entered the workforce. The aim was to determine whether the type of work they undertook was gender related. Work type was classified as 'trade', 'clerical', 'manual' or 'professional'. Classify the variables *work type* (trade, clerical, manual or professional) and *gender* (male or female) as categorical or numerical variables.

3 (4 marks) For the stem plot, find the

```
Stem | Leaf
  1  | 2 5 6 7
  2  | 6 6 7 7 8 9 9
  3  | 0 3 5 7 8 8
  4  | 3
  5  |
  6  | 0
```
Key: 1 | 2 means 12

   a median (1 mark)

   b lower quartile (Q_1) (1 mark)

   c upper quartile (Q_3) (1 mark)

   d interquartile range (IQR). (1 mark)

4 (4 marks) Convert the following bearings to the specified form.

   a 260° as a compass bearing. (2 marks)

   b S 34° E as a true bearing. (2 marks)

Cumulative examination: Calculator-assumed

Total number of marks: 36   Reading time: 4 minutes   Working time: 36 minutes

1 (3 marks) Magda sells books for $56 each. Each year, she receives 1% commission on the first 4000 copies and 2.5% commission on the remaining copies of books she sells.

Calculate the commission Magda receives in a year when she sells

  a 3650 books (1 mark)

  b 7000 books. (2 marks)

2 (6 marks) The number of individual points scored by Rhianna (R), Suzy (S), Tina (T), Ursula (U) and Vicki (V) in five basketball matches (F, G, H, I, J) is shown in matrix P below.

$$P = \begin{bmatrix} 2 & 0 & 3 & 1 & 8 \\ 4 & 7 & 2 & 5 & 3 \\ 6 & 4 & 0 & 0 & 5 \\ 1 & 6 & 1 & 4 & 5 \\ 0 & 5 & 3 & 2 & 0 \end{bmatrix} \begin{matrix} R \\ S \\ T \\ U \\ V \end{matrix} \text{ Player}$$

with columns labelled $F\ G\ H\ I\ J$ (Match).

  a Identify the player who scored the highest number of points and the match in which this occurred. (2 marks)

  b State the number of players who scored no points in some of their matches. (1 mark)

  c Let $U = \begin{bmatrix} 1 \\ 1 \\ 1 \\ 1 \\ 1 \end{bmatrix}$. Evaluate the matrix product PU and interpret the significance of this matrix in context of the question. (3 marks)

3 (3 marks) A rectangular swimming pool is enclosed by a brick wall on one side and a fence along the other three sides. The pool is 8 m long and 5 m wide. The fence is 1.2 m from the pool on each side. The area between the pool and the fence will be concreted, as shown shaded in the diagram.

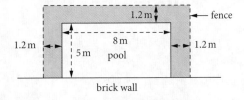

  a Determine the total area that will be concreted. (2 marks)

  b If it costs $25/m² for the concrete, determine the total cost to concrete the shaded section. (1 mark)

4 (4 marks) For the stem plot below, determine if there is a possible outlier? Justify your answer.

Stem	Leaf
1	2 5 6 7
2	6 6 7 7 8 9 9
3	0 3 5 7 8 8
4	3
5	
6	0

Key: 1 | 2 means 12

5 (7 marks) Calculate the value of the pronumeral, correct to one decimal place, in each of the following triangles.

a

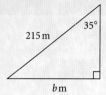

(2 marks)

b

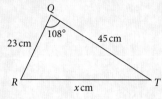

(2 marks)

c

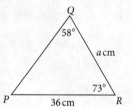

(3 marks)

6 (4 marks) A section of a rollercoaster is shown in the diagram provided.

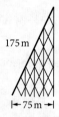

a Determine the angle of elevation of this section to the nearest degree. (2 marks)

b Hence, or otherwise, determine the height of this portion of the rollercoaster to the nearest metre. (2 marks)

7 (9 marks) An 80 m high lookout tower, T, stands in the centre of town. Two landmarks, on the same horizontal plane, are visible from the top of the lookout tower.

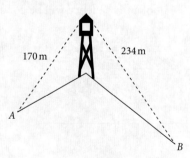

The direct distance from the top of the lookout tower to the base of Landmark A is 170 m.

The direct distance from the top of the lookout tower to the base of Landmark B is 234 m.

The bearing of Landmark B from Landmark A is 105°.

The bearing of Landmark B from the lookout tower is 142°.

a Calculate the direct distance along the ground, to the nearest metre, between Landmark A and Landmark B. (7 marks)

b Calculate the area of land enclosed in the triangle formed between Landmarks A, B and the lookout tower. Answer to the nearest square metre. (2 marks)

LINEAR EQUATIONS AND GRAPHS

CHAPTER 8

Syllabus coverage

Nelson MindTap chapter resources

8.1 Solving linear equations
Using CAS 1: Solving linear equations

8.2 Solving problems using linear equations

8.3 Linear equations in the form $y = ax + b$
Linear graphs and tables of values
Using CAS 2: Drawing linear graphs and generating tables of values
Determining if a point lies on a line
Identifying the slope
The slope formula
Identifying the y-intercept

8.4 Interpreting linear equations in the form $y = ax + b$
Constant rate of change and initial value

Examination question analysis

Chapter summary

Cumulative examination: Calculator-free

Cumulative examination: Calculator-assumed

Syllabus coverage

TOPIC 2.3: LINEAR EQUATIONS AND THEIR GRAPHS

Linear equations

2.3.1 identify and solve linear equations (with the aid of technology where complicated manipulations are required)

2.3.2 develop a linear formula from a word description and solve the resulting equation

Straight-line graphs and their applications

2.3.3 construct straight-line graphs both with and without the aid of technology

2.3.4 determine the slope and intercepts of a straight-line graph from both its equation and its plot

2.3.5 construct and analyse a straight-line graph to model a given linear relationship; for example, modelling the cost of filling a fuel tank of a car against the number of litres of petrol required.

2.3.6 interpret, in context, the slope and intercept of a straight-line graph used to model and analyse a practical situation

Mathematics Applications ATAR Course Year 11 syllabus p. 13 © SCSA

Video playlists (5):
- 8.1 Solving linear equations
- 8.2 Solving problems using linear equations
- 8.3 Linear equations in the form $y = ax + b$
- 8.4 Interpreting linear equations in the form $y = ax + b$

Examination question analysis Linear equations and graphs

Skillsheet (1):
- 8.3 Graphing linear equations

Worksheets (11):
- 8.3 General equation of a straight line • Graphing linear functions 1 • Graphing linear functions 2 • A page of number planes • Number plane grid paper • Drawing gradients • Identifying the slope • Finding the gradient between 2 points on a line • Gradient and y-intercept • Finding the equation of a line
- 8.4 Practical applications

Nelson MindTap

To access resources above, visit
cengage.com.au/nelsonmindtap

8.1 Solving linear equations

An equation in mathematics is made up of algebraic expressions and numerical values, which may appear on either side of the equal sign. The equal sign is what makes it an equation rather than just an expression.

For example, $2x - 6$ is an expression but $2x - 6 = 10$ is an equation.

Equations can be solved by determining the value of an unknown, for example, x, that satisfies the equation. x is a variable and it can take on any numerical values.

> **Exam hack**
>
> The variable doesn't have to be represented by the letter 'x'; it can be represented by any letter or symbol. It is simply a placeholder. Any other letter or symbol can be used.

A **linear equation** is an equation where the highest power of the variable is 1. For example, $2x - 6 = 10$ is a linear equation since the highest power of x is 1, but $2x^2 - 6 = 10$ is not a linear equation because x has a power greater than 1.

Linear equations can contain multiple variables. For example, an equation such as $2x + 4y = 7$ is still considered linear because all variables have a power of 1.

We can solve linear equations with only one variable (or unknown) by isolating the variable, on one side of the equation.

Video playlist
Solving linear equations

WORKED EXAMPLE 1 Solving linear equations with one variable on one side of the equation

Solve $2x - 6 = 10$.

Steps	Working
1 Add 6 to both sides to isolate the unknown on the left-hand side (LHS).	$2x - 6 + 6 = 10 + 6$ $2x = 16$
2 Divide both sides by 2 to obtain the value of x.	$\dfrac{2x}{2} = \dfrac{16}{2}$ $x = 8$

WORKED EXAMPLE 2 Solving linear equations with the variable on both sides of the equation

Solve $-4x + 3 = 2x + 21$.

Steps	Working
1 Subtract 3 from both sides to eliminate the constant from the LHS.	$-4x + 3 - 3 = 2x + 21 - 3$ $-4x = 2x + 18$
2 Subtract $2x$ from both sides to isolate the unknown on the LHS.	$-4x - 2x = 2x + 18 - 2x$ $-6x = 18$
3 Divide both sides by -6 to obtain the value of x.	$\dfrac{-6x}{-6} = \dfrac{18}{-6}$ $x = -3$

In some cases, an equation may contain a term or multiple terms expressed as fractions. When this occurs, it is necessary to determine the **lowest common denominator (LCD)** so that the common denominator can be cancelled out. This involves finding the smallest common multiple of the denominators of all the fractions in the equation. By multiplying both sides of the equation by the LCD, the denominators are cancelled out, allowing the equation to be solved as normal.

WORKED EXAMPLE 3 — Solving linear equations with fraction

Solve $\dfrac{2x}{5} - \dfrac{6x}{7} = 2$.

Steps	Working
1 Determine the LCD.	$5 \times 7 = 35$ The LCD is 35.
2 Convert the two fractions and re-write the equation with the fractions with the LCD.	$\dfrac{2x}{5}\left(\dfrac{7}{7}\right) - \dfrac{6x}{7}\left(\dfrac{5}{5}\right) = 2$ $\dfrac{14x}{35} - \dfrac{30x}{35} = 2$
3 Multiply both sides by the LCD to eliminate the fractions.	$35 \times \left(\dfrac{14x}{35} - \dfrac{30x}{35}\right) = 35 \times 2$ $14x - 30x = 70$
4 Solve the equation. Remember to simplify the fraction (it is acceptable to leave the answer as a simplified improper fraction).	$-16x = 70$ $x = -\dfrac{70}{16}$ $x = -\dfrac{35}{8}$

We can also use CAS to solve equations.

USING CAS 1 — Solving linear equations

Use CAS to solve $\dfrac{x}{5} - \dfrac{6}{5} = -\dfrac{7x}{3}$.

ClassPad

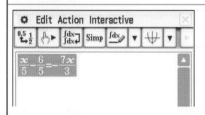
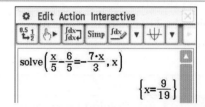

1 Open **Main**.

2 Enter and highlight the equation
$\dfrac{x}{5} - \dfrac{6}{5} = -\dfrac{7x}{3}$.

3 Tap **Interactive** > **Equation/Inequality** > **solve**.

4 In the dialogue box, keep the default variable x and tap **OK**.

TI-Nspire

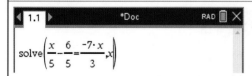

1 Add a **Calculator** page.

2 Press **menu** > **Algebra** > **Solve**.

3 Enter the equation $\dfrac{x}{5} - \dfrac{6}{5} = -\dfrac{7x}{3}$ followed by **,x**.

4 Press **enter**.

$x = \dfrac{9}{19}$

An equation may contain brackets, such as 2(x − 3) = 3x. In this case, it's necessary to apply the **distributive law** by multiplying the term outside the bracket (in this case, the number 2) by each term inside the bracket. This expands the expression within the bracket, resulting in an equation that can be solved using the usual methods. Once the brackets have been expanded, the equation can be simplified and the variable can be isolated and solved.

WORKED EXAMPLE 4 — Solving linear equations with bracket

Solve $2(x − 3) = 3x$.

Steps	Working
1 Expand expression by multiplying each term by 2.	$2 \times x − 2 \times 3 = 3x$ $2x − 6 = 3x$
2 Solve for x.	$2x − 3x = 6$ $−x = 6$ $x = −6$

WORKED EXAMPLE 5 — Solving linear equations with brackets on both sides of the equation

Solve $5(x − 1) = 3(x + 2)$.

Steps	Working
1 Expand the expressions.	$5 \times x − 5 \times 1 = 3 \times x + 3 \times 2$ $5x − 5 = 3x + 6$
2 Solve for x.	$5x − 3x = 6 + 5$ $2x = 11$ $x = \dfrac{11}{2}$

Again, we can use CAS to solve these equations.

🔒 Exam hack

Note that it's not always necessary to express your answer in decimal form, unless it is specifically required. In fact, leaving the answer in the form of an improper fraction is perfectly acceptable as long as it is simplified. This means that the numerator and denominator of the fraction cannot be divided by any common factors other than 1.

EXERCISE 8.1 Solving linear equations   ANSWERS p. 424

Mastery

1 WORKED EXAMPLE 1 Solve the following equations without CAS.

     a $-4x - 6 = 2$ b $7 - x = -8$ c $12 - 3x = -15$

     d $4 = 6x - 20$ e $-5x + 10 = -10$ f $2x + 4 = 11$

2 WORKED EXAMPLE 2 Solve the following equations without CAS.

     a $2x = 3x + 5$ b $6x + 4 = 4x - 2$ c $3x + 16 = 8 + 5x$

     d $3x - 42 = -48 - x$ e $-4 + x = -3x + 10$ f $7x + 12 = -13 + 2x$

3 WORKED EXAMPLE 3 Solve the following equations without CAS.

     a $\dfrac{x}{2} + \dfrac{3}{4} = 1$ b $-\dfrac{x}{3} + \dfrac{x}{2} = 2$ c $-\dfrac{1}{2} + \dfrac{x}{4} = -\dfrac{1}{6} + \dfrac{x}{3}$

     d $\dfrac{2x}{3} + \dfrac{1}{2} = -\dfrac{2}{3} + \dfrac{5x}{6}$ e $\dfrac{2x}{3} - \dfrac{3x}{4} = \dfrac{1}{4} - \dfrac{1}{6}$ f $\dfrac{4x}{5} - \dfrac{x}{2} = -\dfrac{2}{3} + \dfrac{1}{4}$

4 **Using CAS 1** Use CAS to solve the following equations, rounding to two decimal places where appropriate.

a $\dfrac{5x}{6} - \dfrac{2}{3} = \dfrac{1}{2}$

b $-\dfrac{3x}{4} + \dfrac{7}{8} = 2$

c $\dfrac{5x}{6} - \dfrac{2}{3} = \dfrac{2x}{3} + \dfrac{1}{2}$

d $0.5x + 2.4 = 3.1x - 1.8$

e $0.9x + 1.5 = 2.4x - 3.1$

f $1.2 + 2.5x = 3.8x + 2.4$

g $3.5x - 2.1 = 1.2x + 1.9$

h $\dfrac{x}{4} - 0.3 = \dfrac{2x}{3} + \dfrac{1}{2}$

i $-\dfrac{2}{3} + 1.5x = \dfrac{3x}{4} + \dfrac{1}{2}$

5 **WORKED EXAMPLE 4** Solve the following equations without CAS first, and then use CAS to check your answer.

a $3(x + 4) = 2x + 5$

b $2(x - 3) + 4 = 5x$

c $5(x + 2) - 2(x - 1) = 6$

d $5(x + 3) - 2(x - 1) = 4x + 7$

e $2(4x - 3) - 3(x + 2) = 8$

f $3(x - 1) - 2(x + 2) = 4$

6 **WORKED EXAMPLE 5** Solve the following equations without CAS.

a $4(2x + 1) = 5(3x - 2)$

b $2(x - 1) + 3(2x + 4) = 5x$

c $4(3x + 2) - 2(x - 1) = 2x + 10$

d $3(x + 2) + 2(x - 3) = 4x - 1$

e $2(3x - 5) + 4(x + 1) = 3(x + 2) + 10$

f $5(x + 3) - 2(3x - 1) = 4(2x + 1) + 1$

g $5(2x - 1) = 2(3x + 2) + 3(x - 1) - 2$

h $5(2x + 3) = 3(x + 2) + 2(4x - 1) + 1$

7 Solve the equations in Question 6 using CAS.

 Exam hack

It is important to keep in mind that the coefficient of the variable in a linear equation could be a decimal number. To solve such an equation, you can follow the same steps as outlined previously, or you can use CAS to help with the calculations.

Calculator-free

8 (5 marks) Solve for the unknown in the following linear equations.

a $8x + 25 = 41$ (1 mark)

b $7(p + 3) = 4(3p - 1)$ (2 marks)

c $\dfrac{3a - 4}{4} = 8$ (2 marks)

Calculator-assumed

9 (5 marks) Solve for the unknown in the following linear equations. Give your answers to two decimal places where appropriate.

a $\dfrac{x - 2}{7} = \dfrac{x - 4}{6}$ (1 mark)

b $0.32x - \dfrac{4x}{5} = 11$ (2 marks)

c $\dfrac{4(x + 3)}{3} - 1 = 0.78x + 6$ (2 marks)

8.2 Solving problems using linear equations

It is important to realise that linear equations can be used to solve practical problems in real life. For instance, they can be used to solve problems involving age differences, the dimensions of objects, and unknown numbers in various scenarios.

> **Solving worded problems**
>
> To solve worded problems:
> 1. Assign a **pronumeral** to the unknown quantity, if it is not given.
> 2. Translate the problem into an equation in terms of the assigned pronumeral.
> 3. Solve the problem by using the techniques learned.
> 4. State the solutions to the problem clearly.

Video playlist
Solving problems using linear equations

WORKED EXAMPLE 6 — Solving worded problems involving consecutive numbers

Determine three consecutive even numbers whose sum is 78.

Steps	Working
1 Assign a pronumeral to the unknown.	Let x = the first number. Therefore, the three consecutive even numbers are x, $x + 2$ and $x + 4$.
2 Write a linear equation.	$x + (x + 2) + (x + 4) = 78$
3 Solve the linear equation.	$3x + 6 = 78$ $3x = 72$ $x = \dfrac{72}{3} = 24$
4 State the solution to the problem.	The three consecutive numbers are 24, 26 and 28.

WORKED EXAMPLE 7 — Solving worded problems involving mathematical operations

If two times a number is increased by 3 and the result is multiplied by −2, the final product is −22. Find the value of the number.

Steps	Working
1 Assign a pronumeral to the unknown.	Let x = the number.
2 Write a linear equation.	
Two times the number.	$2x$
Increase by three.	$2x + 3$
Multiply by −2.	$-2(2x + 3)$
Equate to −22.	$-2(2x + 3) = -22$
3 Solve the linear equation.	$-4x - 6 = -22$ $-4x = -16$ $x = \dfrac{-16}{-4} = 4$
4 State the solution to the problem.	The number is 4.

WORKED EXAMPLE 8 | Solving worded problems involving dimensions

A fence is to be built on the boundary of a rectangular paddock where the length is twice the width. The perimeter of the paddock is 96 metres.

a Determine the length and width of the paddock.
b Determine the area of the paddock.

Steps	Working
a 1 Assign a pronumeral to the unknown.	Let W = the width of the paddock in metres. Therefore, the length $l = 2W$ and the perimeter = $2(W + l)$.
2 Write a linear equation.	$2(W + 2W) = 96$
3 Solve the linear equation.	$2W + 4W = 96$ $6W = 96$ $W = \dfrac{96}{6} = 16$
4 State the solution to the problem.	The width of the paddock is 16 m and the length of the paddock is 32 m.
b Substitute the length and width from part a into the area formula, $A = l \times w$.	Area = $16 \times 32 = 512$ The area of the paddock is $512 \, m^2$.

 Exam hack

It is important to always include the unit of measurement when stating a value or answer.

WORKED EXAMPLE 9 | Solving worded problems involving proportions

At a local fish and chips shop, Nick only bought chips while Pippa bought fish and chips. Pippa spent $10 more than Nick, and it was three times the amount Nick spent. Find the amount each of them spent.

Steps	Working
1 Assign a pronumeral to the unknown.	Let N = the amount of money Nick spent. Therefore, the amount of money Pippa spent is $N + 10$.
2 Write a linear equation.	$3N = N + 10$
3 Solve the linear equation.	$2N = 10$ $N = 5$
4 State the solution to the problem.	Nick spent $5 and Pippa spent $15.

WORKED EXAMPLE 10 Solving worded problems involving ages of persons

David is 30 years older than his youngest son, Mark. In two years, David's age will be four times Mark's age. What are their ages now?

Steps	Working
1 Assign a pronumeral to the unknown.	Let D = David's current age. Therefore, Mark's current age is $D - 30$.
2 Write a linear equation. In two years: David's age Mark's age David's age = four times Mark's age	 $D + 2$ $(D - 30) + 2 = D - 28$ $D + 2 = 4(D - 28)$
3 Solve the linear equation.	$D + 2 = 4D - 112$ $3D = 114$ $D = 38$
4 State the solutions to the problem.	David is currently 38 years old and Mark is 8 years old.

 Exam hack

Decide if it's faster to use CAS to solve the equation in the calculator-assumed section of the exam.

EXERCISE 8.2 Solving problems using linear equations   ANSWERS p. 424

Recap

1 Solve the following linear equations without CAS.

   a $3x + 4 = 19$

   b $2(x - 3) + 5 = 17 - x$

2 Use CAS to solve the following equations.

   a $\dfrac{x}{2} - \dfrac{4}{3} = -\dfrac{x}{3} + \dfrac{1}{6}$

   b $\dfrac{2}{3}(4x - 1) = 10$

   c $3(4x - 2) - 2(2x + 5) = 14 - 3x$

▶ **Mastery**

For questions 3 to 11, write a linear equation and solve for the unknown.

3 WORKED EXAMPLE 6 The sum of three consecutive numbers is 39. Determine the three numbers.

4 The sum of four consecutive odd numbers is 72. Determine the four numbers.

5 WORKED EXAMPLE 7 If twice a number is increased by 1 and the result is multiplied by 4, the final product is −28. Find the value of the number.

6 If four times a number is decreased by 5 and the result is multiplied by 6, the final product is −54. Find the value of the number.

7 WORKED EXAMPLE 8 A rectangular block of land has a perimeter of 120 m, with a length that is three times its width. Find the area of the rectangular block.

8 WORKED EXAMPLE 9 Stephen went to the cinema with a friend and bought 5 ice creams and 2 medium popcorns to share. The medium popcorn cost $3 more than the ice cream. In total, they spent $30.50. What was the cost of an ice cream and a medium popcorn?

9 Mrs. White's mathematics class has a total of 32 students. If the number of female students is two more than the number of male students, how many male students are there?

10 WORKED EXAMPLE 10 Martha is currently 24 years older than her son Alex. In two years, Martha's age will be twice that of her son. What are their present ages?

11 A father is currently two times as old as his daughter. Ten years ago, the father was three times as old as his daughter. Determine their current ages.

Calculator-free

12 (6 marks) Solve the following by first writing an equation. Clearly define the variable you use.

   a The sum of three consecutive even numbers is 42. Determine the three numbers. (3 marks)

   b When three is deducted from a number and the result is multiplied by a third, the final product is 39. Determine the number. (3 marks)

Calculator-assumed

13 (8 marks) Solve the following by first writing an equation. Clearly define the variable you use.

   a Mark allocated a quarter of his weekly pay to groceries and a third of his weekly pay to rent. Afterward, he had $210 remaining. Find the total amount of Mark's weekly pay. (4 marks)

   b If one-half of Heather's current age two years from now, combined with one-third of her current age three years ago, equals to 20 years, what is her current age? (4 marks)

8.3 Linear equations in the form $y = ax + b$

Let us now shift our attention to linear equations with two variables, which are typically expressed in the form $y = ax + b$. In this equation, both a and b are constants, and x and y are the variables. In this form, we usually refer to it as a linear equation.

The graph of a linear equation is a straight line. The **coordinates** of points on the line are found by substituting x values into the equation to calculate corresponding y values.

Linear graphs and tables of values

Equations can be drawn using a table of values.

WORKED EXAMPLE 11 Drawing linear graphs from tables of values

For the linear equation $y = -3x + 5$

a find the y values for each of the x values: $-2, -1, 0, 1, 2$
b use the y values to construct a table of values for the linear equations
c use the table of values to draw the linear equation by hand.

Steps	Working
a Use the linear equation to calculate the y value for each x value.	When $x = -2$, $y = -3 \times (-2) + 5 = 6 + 5 = 11$. When $x = -1$, $y = -3 \times (-1) + 5 = 3 + 5 = 8$. When $x = 0$, $y = -3 \times (0) + 5 = 0 + 5 = 5$. When $x = 1$, $y = -3 \times (1) + 5 = -3 + 5 = 2$. When $x = 2$, $y = -3 \times (2) + 5 = -6 + 5 = -1$.
b Set up a table of values for the linear equation to show the x values and their corresponding y values.	<table><tr><th>x</th><td>-2</td><td>-1</td><td>0</td><td>1</td><td>2</td></tr><tr><th>y</th><td>11</td><td>8</td><td>5</td><td>2</td><td>-1</td></tr></table>
c 1 Draw a Cartesian plane and plot the points from the table. 2 Draw a line through the points, extending it and placing arrows on both ends of the line. 3 Label the graph with its equation.	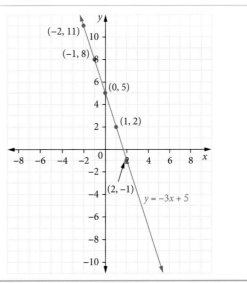

🔒 Exam hack

Note that the linear equation in two variables, $y = ax + b$, is also commonly expressed in the form $y = mx + c$. You will also often see linear equations written in a different order. For example, $y = 5 - 3x$ rather than $y = -3x + 5$.

Video playlist
Linear equations in the form $y = ax + b$

Skillsheet
Graphing linear equations

Worksheets
General equation of a straight line

Graphing linear functions 1

Graphing linear functions 2

A page of number planes

Number plane grid paper

USING CAS 2 — Drawing linear graphs and generating tables of values

For the linear equation $y = 2x - 1$, draw the graph and generate a table of values for $x = -2, -1, 0, 1, 2$.

ClassPad

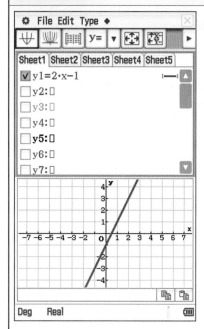
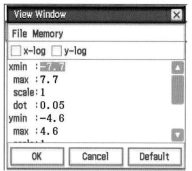
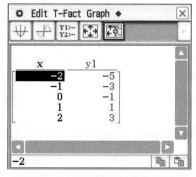
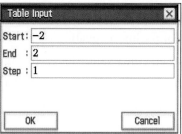

1. Tap **Menu** and then tap **Graph&Table**.
2. Clear all equations and enter **2x – 1** into the line for **y1**.
3. Press **EXE** to select the equation.
4. Tap the **Graph** tool to graph it in the lower window.
5. If the graph does not display, tap the **View Window** tool.
6. In the **View Window** screen, tap **Default**.
7. Tap **OK**.
8. Tap the **Table** tool to view the table of values.
9. A table of values will be displayed.
10. Tap the **Table Input** tool.
11. In the **Table Input** dialogue box, enter the values as shown above.
12. Tap **OK**.

TI-Nspire

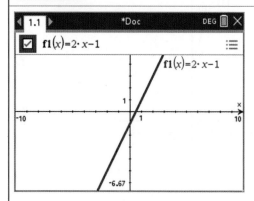
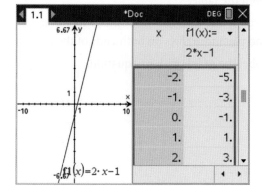

1. Start a new document and add a **Graphs** page.
2. In the **Graph Entry Line** at the top of the page, enter **2x–1**.
3. Press **enter**. The graph of the equation will be displayed.

   Note: after pressing **enter**, the Graph Entry Line will be hidden.
4. Press **menu** > **Table** > **Split-screen Table**. The table of values will be displayed in the right-hand window.
5. Scroll up to view the table for $x = -2, -1, 0, 1$ and 2.

Determining if a point lies on a line

To determine whether a point lies on a line, substitute the coordinates into the equation of the line and check if the result is true or false. If it's true, then the point lies on the line, and if it's false, then the point doesn't lie on the line.

WORKED EXAMPLE 12	**Determining if a point lies on a line**
Determine whether the point $(-2, 3)$ lies on each of the following lines, showing a calculation to justify your answer.	
a $y = 7 + 2x$	**b** $y = 3x - 8$
Steps	**Working**
a 1 Substitute the coordinates into the equation of the line, evaluate, and state whether the equation is true or false.	$x = -2, y = 3$ $y = 7 + 2x$ Does $3 = 7 + 2 \times (-2)$? $3 = 7 - 4$ true
2 Write the answer.	$(-2, 3)$ lies on the line $y = 7 + 2x$ as $3 = 7 - 4$.
b 1 Substitute the coordinates into the equation of the line, evaluate, and state whether the equation is true or false.	$x = -2, y = 3$ $y = 3x - 8$ Does $3 = 3 \times (-2) - 8$? $3 \neq -6 - 8$ false
2 Write the answer.	$(-2, 3)$ doesn't lie on the line $y = 3x - 8$ as $3 \neq -6 - 8$.

Identifying the slope

When a straight line is written in the form $y = ax + b$, a represents the **slope** (or **gradient**) of the line. Slopes can be positive, negative, zero or not defined.

Positive slope

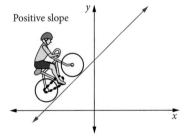

Line sloping up from left to right

Negative slope

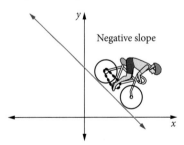

Line sloping down from left to right

Worksheets
Drawing gradients

Identifying the slope

The slope of a linear equation

For a linear equation $y = ax + b$:

$$a = \text{slope} = \frac{\text{rise}}{\text{run}}$$

$$= \frac{\text{vertical distance between two points}}{\text{horizontal distance between the same two points}}$$

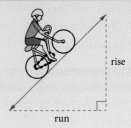

Positive slope	Negative slope	Zero slope for a horizontal line; rise = 0	Slope not defined for a vertical line; run = 0	Parallel lines have the same slope

WORKED EXAMPLE 13 — Finding the slope using $\frac{\text{rise}}{\text{run}}$

For each of the lines shown below
  i state whether the slope is positive, negative, zero or not defined.
  ii if it is positive or negative, calculate the slope of the line using $\frac{\text{rise}}{\text{run}}$ for the two points shown.

a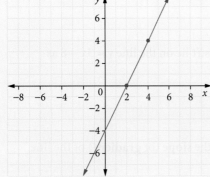

b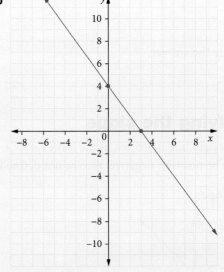

c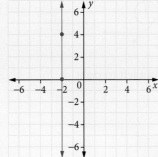

d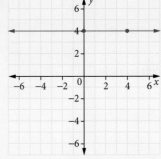

Steps	Working
a i Is the line sloping up or down from left to right?	The line is sloping up from left to right, so the slope is positive.
ii 1 Draw in a right-angled triangle using the two points and find the rise and run between them.	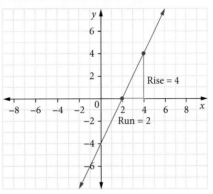
2 Use slope = $\frac{\text{rise}}{\text{run}}$ and simplify.	slope = $\frac{4}{2} = 2$
b i Is the line sloping up or down from left to right?	The line is sloping down from left to right, so the slope is negative.
ii 1 Draw in a right-angled triangle using the two points and find the rise and run between them.	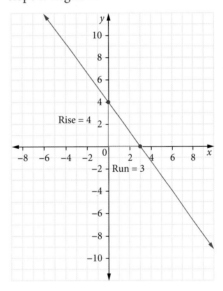
2 Use slope = $\frac{\text{rise}}{\text{run}}$ and simplify.	slope = $-\frac{4}{3}$
c i The line is vertical.	The slope is not defined.
d i The line is horizontal.	The slope is 0.

The slope formula

If we know any two points on a straight line, the slope of a line can be found without drawing a graph.

> **The slope formula**
>
> If two points on a straight line are (x_1, y_1) and (x_2, y_2),
> then the slope $= \dfrac{y_2 - y_1}{x_2 - x_1}$.

 Exam hack

If you have a choice of two points on a straight line to select from, choose the two that will make the calculation the easiest. This will often be the points that have a zero coordinate or lie on a grid line.

Worksheets
Finding the gradient between 2 points on a line

Gradient and y-intercept

WORKED EXAMPLE 14 — Finding the slope from two points

Calculate the slope of the line for each of the following.

a a straight line through the points $(2, 5)$ and $(4, 13)$

b
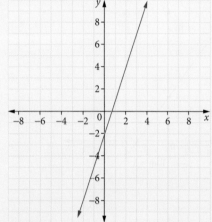

c a straight line drawn from the following table of values

x	−2	0	2	4	6
y	16	6	−4	−14	−24

Steps	Working
a Use slope $= \dfrac{y_2 - y_1}{x_2 - x_1}$ for (x_1, y_1) and (x_2, y_2), and simplify.	slope $= \dfrac{13 - 5}{4 - 2} = \dfrac{8}{2} = 4$
b 1 Select two points on the line that can be clearly read from the graph.	$(0, -2)$ and $(2, 4)$
2 Use slope $= \dfrac{y_2 - y_1}{x_2 - x_1}$ for (x_1, y_1) and (x_2, y_2), and simplify.	slope $= \dfrac{4 - (-2)}{2 - 0} = \dfrac{6}{2} = 3$
c 1 Select two points from the table.	$(0, 6)$ and $(2, -4)$
2 Use slope $= \dfrac{y_2 - y_1}{x_2 - x_1}$ for (x_1, y_1) and (x_2, y_2), and simplify.	slope $= \dfrac{-4 - 6}{2 - 0} = \dfrac{-10}{2} = -5$

Identifying the y-intercept

When a straight line is written in the form $y = ax + b$, b is the **y-intercept**, which is the point where the graph crosses the y-axis.

> **Straight lines in the form $y = ax + b$**
>
> For a straight line in the form $y = ax + b$:
>
> a = slope
>
> b = y-intercept

8.3

WORKED EXAMPLE 15 | Finding the straight line equation from the intercept and slope

Find the equation of each of the following straight lines by determining the y-intercept and slope.

a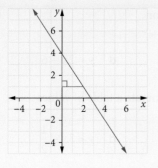

b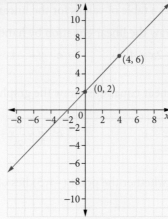

c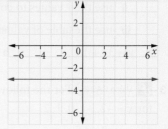

Steps	Working
a 1 Where does the graph cross the y-axis?	y-intercept = 4
2 Is the line sloping up or down from left to right?	The line is sloping down from left to right, so the slope is negative.
3 Use slope = $\frac{\text{rise}}{\text{run}}$ and simplify.	slope = $-\frac{3}{2}$
4 Identify the values for a and b in the equation $y = ax + b$.	$a = -\frac{3}{2}$ $b = 4$
5 Write the equation of the line.	The equation of the line is $y = -\frac{3}{2}x + 4$.
b 1 Where does the graph cross the y-axis?	y-intercept = 2
2 Use slope = $\frac{y_2 - y_1}{x_2 - x_1}$ for (x_1, y_1) and (x_2, y_2) and simplify.	slope = $\frac{6 - 2}{4 - 0} = \frac{4}{4} = 1$
3 Identify the values for a and b in the equation $y = ax + b$.	$a = 1$ $b = 2$
4 Write the equation of the line.	The equation of the line is $y = x + 2$.
c 1 Where does the graph cross the y-axis?	y-intercept = −3
2 Is the line sloping up or down from left to right?	The line is neither sloping up or down. It's a horizontal line. slope = 0
3 Identify the values for a and b in the equation $y = ax + b$.	$a = 0$ and $b = -3$
4 Write the equation of the line.	The equation of the line is $y = -3$.

Worksheets
Gradient and y-intercept

Finding the equation of a line

EXERCISE 8.3 Linear equations in the form $y = ax + b$

ANSWERS p. 425

Recap

1 Chris plans to paint two identical walls in his room, where the width of each wall is 1.4 m longer than its height. The perimeter of one of the walls is 12.4 m. Find the total area of the two walls.

Mastery

2 WORKED EXAMPLE 11 For each of the following linear equations

    i find the y values for each of the x values: $-2, -1, 0, 1, 2$

    ii use the y values to construct a table of values for the linear equation

    iii use the table of values to draw the linear equation by hand.

  a $y = 3x + 1$ **b** $y = -2x + 4$ **c** $y = -4x + 3$ **d** $y = 2x - 6$

3 Using CAS 2 For the following linear equations, draw the graph and generate a table of values for $x = -2, -1, 0, 1, 2$.

  a $y = -2x + 5$ **b** $y = x - 1$ **c** $y = \frac{1}{2}x + 2$ **d** $y = 3x - \frac{1}{2}$

4 WORKED EXAMPLE 12 Determine whether the point $(-1, 5)$ lies on each of the following lines, showing a calculation to justify your answer.

  a $y = 15x - 5$ **b** $y = -5x - 1$ **c** $y = -10x - 5$

5 WORKED EXAMPLE 13 For each of the lines shown

    i state whether the slope is positive, negative, zero or not defined.

    ii if it is positive or negative, calculate the slope of the line using $\frac{\text{rise}}{\text{run}}$ for the two points shown.

a

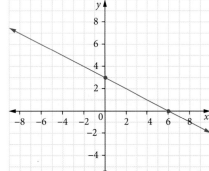

b

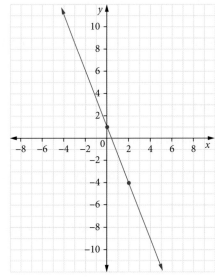

c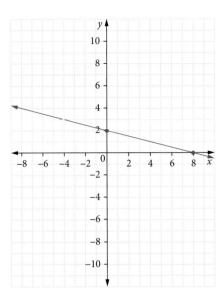

d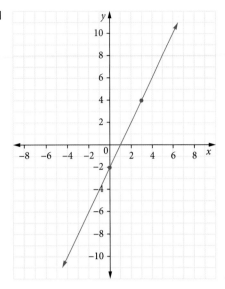

e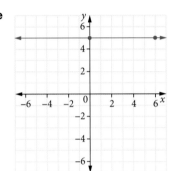

f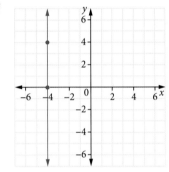

6 WORKED EXAMPLE 14 Calculate the slope of the line for each of the following.

a a straight line through the points $(5, 3)$ and $(2, 9)$

b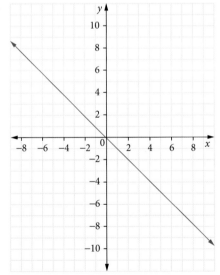

c a straight line drawn from the following table of values

x	0	4	7	10
y	−1	11	20	29

7 **WORKED EXAMPLE 15** Find the equation of each of the following straight lines by determining the slope and y-intercept.

a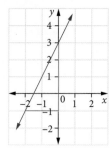

b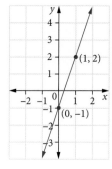

c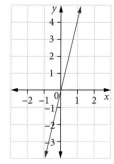

d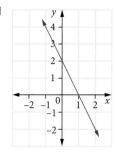

e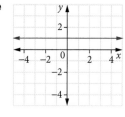

Calculator-free

8 (1 mark) Give a reason why $y = \dfrac{5}{6}x - \dfrac{3}{8}$ is linear.

9 (2 marks) On the graph shown, the line passes through the origin $(0, 0)$ and the point $(2, 1)$. Determine the slope of this line.

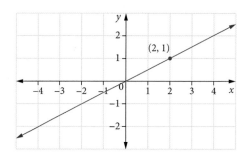

10 (1 mark) State the equation of the line that passes through the point $(1, 2)$ plotted on the graph, as shown below.

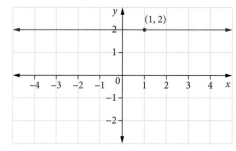

11 (2 marks)

   a State the gradient and y-intercept of $y = 2 + \frac{1}{2}x$.   (1 mark)

   b Copy the Cartesian plane below and sketch the graph from part **a**.   (1 mark)

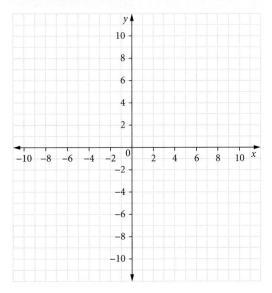

Calculator-assumed

12 (4 marks) A straight line is plotted on a Cartesian plane as shown on the diagram below.

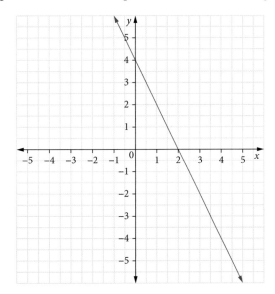

   a Determine the slope of this line.   (2 marks)

   b State the coordinates of the x-intercept and y-intercept.   (1 mark)

   c State the equation of this line.   (1 mark)

13 (2 marks) Determine the equation of the line that passes through the points $(8, 0)$ and $(8, 3)$.

14 (3 marks) Determine which one of the following points does **not** lie on the straight line with equation $y = 1 - 3x$. Justify your answer.

Point $A\ (0, 1)$, Point $B\ (-1, 4)$ and Point $C\ (-4, -11)$.

Video playlist
Interpreting linear equations in the form $y = ax + b$

Worksheet
Practical applications

8.4 Interpreting linear equations in the form $y = ax + b$

Constant rate of change and initial value

Linear equations in the form $y = ax + b$ can be used to model real-world situations, where a is the **constant rate of change** (or **rate**) and b is the **initial value**. The initial value is the value of y at the start. The constant rate of change measures the change in y as x changes. For example:

A computer support service company charges a $50 call-out fee plus $80 per hour.

This can be modelled by the linear equation

$C = 80n + 50$

where C is the total charge
n is the number of hours.

The initial value is $50, the charge before any time has passed. The constant rate of change is $80 per hour. For each hour n increases, the charge C increases by $80.

> **Linear models using initial value and constant rate of change**
>
> For a linear equation in the form $y = ax + b$:
>
> a = constant rate of change
>
> b = initial value

WORKED EXAMPLE 16 Working with the initial value and constant rate of change

The TruBlu Energy Company charges its customers a combination of an upfront cost (in dollars) and a rate per megajoule (MJ) of electricity used, according to the graph shown.

Find the

a vertical intercept of the graph

b slope of the graph

c equation of the graph using C for the total charge and m as the amount of electricity used in megajoules.

From the equation, find the

d upfront cost

e rate of change per megajoule

f total cost of using 2200 megajoules of electricity

g number of megajoules of electricity that have a total cost of $90, rounding your answer to the nearest megajoule.

Steps	Working
a Read from the graph.	The vertical intercept is 10.
b 1 Select two points on the line that can be clearly read from the graph.	(0, 10) and (5000, 160)
2 Use slope = $\dfrac{y_2 - y_1}{x_2 - x_1}$ for (x_1, y_1) and (x_2, y_2) and simplify.	slope = $\dfrac{160 - 10}{5000 - 0}$ = $\dfrac{150}{5000}$ = 0.03

c	1 Identify a and b in the equation $y = ax + b$.	$a = 0.03$ $b = 10$
	2 Write the equation $y = ax + b$ using the variables given.	$C = 0.03m + 10$
d	Find the initial value (vertical intercept).	The upfront cost is $10.
e	Find the constant rate of change (slope).	The rate of change is $0.03/MJ or 3 c per megajoule.
f	Substitute the value of m into the equation and solve for C, using CAS if necessary.	$m = 2200$ $C = 0.03 \times 2200 + 10 = 76$ The total cost of using 2200 megajoules of electricity is $76.
g	1 Substitute the value of C into the equation and solve for m, using CAS if necessary.	$C = 90$ $90 = 0.03m + 10$ $0.03m = 80$ $m = \dfrac{80}{0.03} = 2666.666\ldots$
	2 Write the answer rounding to the nearest megajoule.	2667 megajoules of electricity have a total cost of $90.

 Exam hack

When modelling practical situations, make sure to use the pronumerals given in the question, rather than x and y. To identify the constant rate of change, look for the word 'per'.

We can use linear equations to make predictions about real-world situations. However, in real life not every value makes sense. We need to watch out for answers that are mathematically correct but impossible in real life. Linear models may only apply to a specific range of values.

WORKED EXAMPLE 17 Dealing with the unrealistic solutions in real-life problems

Sabine is organising the annual Year 11 dance. The total cost for the event will include $500 venue hire, $280 for the DJ and $25 per head for food. The maximum capacity of the venue is 200.

a Find the linear equation in the form $C = an + b$ for the total cost of the event, C, for n students.
b Use the equation to find the total cost of the dance if 2 students attend.
c Explain why your answer to part **b** makes no sense in real life.
d Use the equation to find the total cost of the dance if 1000 students attend.
e Explain why your answer to part **d** makes no sense in real life.

Steps	Working
a 1 Identify the initial value and constant rate of change.	The initial value = $500 + $280 = $780 So, $b = 780$. The constant rate of change is $25. So, $a = 25$.
2 Write the total cost equation.	$C = 25n + 780$
b Substitute the value of n into the equation and evaluate. Write the answer in words.	$n = 2$ $C = 25 \times 2 + 780$ = 830 The total cost if 2 students attended is $830.

c	What would happen in real life?	The dance would be cancelled if so few students were attending.
d	Substitute the value of n into the equation and evaluate. Write the answer in words.	$n = 1000$ $C = 25 \times 1000 + 780 = 25\,780$ The total cost if 1000 students attended is $25\,780.
e	What would happen in real life?	The venue has a maximum capacity of 200, so Sabine would have to find another venue and the cost equation would be different.

Linear equations can often be used to model business **cost** (money going out) and **revenue** (money coming in). We can combine these to form the linear equation

profit = revenue − cost

If the profit is negative according to this equation, then the business will have made a loss.

> **Exam hack**
>
> You will have to know the **break-even point** at which revenue begins to exceed the cost.

WORKED EXAMPLE 18 — Modelling profit and loss

Snow domes are sold for $6.50 each, and the cost C of making n snow domes is given by the equation
$C = 5n + 130$.

a Find the revenue equation in terms of n.
b Find the profit equation in terms of n.
c How much profit would be made if 100 snow domes were sold?
d How much profit would be made if 80 snow domes were sold?
e How many snow domes need to be sold to make at least $1000 profit? Explain why you need to round up for this calculation.

Steps	Working
a Use the price of one item to calculate the revenue from selling n items.	revenue = 6.50 × number of snow domes sold = $6.5n$
b Use the profit equation and simplify.	profit = revenue − cost = $6.5n - (5n + 130)$ = $1.5n - 130$
c Substitute $n = 100$ into the profit equation and write the answer.	profit = $1.5 \times 100 - 130$ = 20 100 snow domes would make a profit of $20.
d Substitute $n = 80$ into the profit equation and write the answer.	profit = $1.5 \times 80 - 130$ = −10 80 snow domes would make a loss of $10.
e 1 Let profit equal 1000 and solve for n, using CAS if necessary.	$1.5n - 130 = 1000$ $n \approx 753.33$
2 Round the answer according to the question. The answer needs to be a whole number as n represents the number of snow domes.	The profit needs to be at least $1000 and selling 753 snow domes won't quite make that profit. So, the number of snow domes that need to be sold is 754.

> **Exam hack**
>
> Always look at the context of the question when rounding. Sometimes the real-world context means that it makes sense to round up to the nearest whole number, even if the decimal indicates to round down.

EXAMINATION QUESTION ANALYSIS

Calculator-assumed (5 marks)

Charlie's weekly profit from his bookshop is given by the equation $P = 35n - 525$, where P is the profit earned in dollars and n is the number of books that he sells.

a Explain the meaning of the values 35 and 525 in the context of the question. (2 marks)

b How much profit will Charlie earn if he sells 100 books in a given week? (1 mark)

c What is the minimum number of books that Charlie must sell in order to cover the costs of running his bookshop? (2 marks)

Reading the question

- Note the pronumeral representing profit and the number of books sold.
- Recognise that the equation is a linear equation in the form $y = ax + b$.
- Understand that the cost of running the bookshop is the constant of the linear equation.

Thinking about the question

- What does the slope of the equation represent?
- What does the constant of the equation represent?
- When calculating the profit, what is the unit?
- In part **c**, what does 'cover the costs of running his bookshop' mean?

Video
Examination question analysis: Linear equations and graphs

Worked solution (✓ = 1 mark)

a 35: Each book is sold for $35.

correctly interprets the meaning of the value of 35 in context ✓

−525: the cost of running the shop is $525 per week.

correctly interprets the meaning the value of −525 in context ✓

b Substitute $n = 100$ into the profit equation.

$P = 35(100) - 525 = 2975$

Charlie's profit is $2975 in a given week.

determines the profit using the equation ✓

c The cost for running the bookshop is $525. To cover the cost, Charlie must sell a certain amount of book so that the profit = $0. Therefore, we need to determine n when $P = 0$.

$35n - 525 = 0$
$35n = 525$
$n = 15$

Charlie must sell at least 15 books to cover his costs of running his bookshop.

substitutes $P = 0$ into the equation ✓

states the minimum number of books Charlie must sell ✓

Chapter 8 | Linear equations and graphs

EXERCISE 8.4 Interpreting linear equations in the form $y = ax + b$

ANSWERS p. 427

Recap

1 Determine the equation of the straight line passing through the points $(-2, 0)$ and $(0, 2)$, as shown in the diagram.

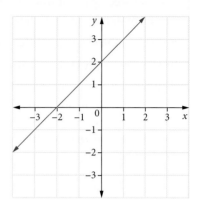

2 For the graph of $y = 5x - 2$
   a State the slope and the y-intercept.
   b Determine if $(7, 32)$ lies on the line.

Mastery

3 WORKED EXAMPLE 16   The United Water Company charges its commercial customers a combination of an upfront cost (in dollars) and a rate per kilolitre (kL) used, according to the graph shown.

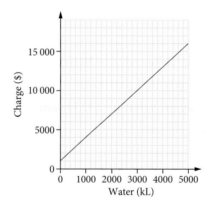

Find the
   a vertical intercept of the graph
   b slope of the graph
   c equation of the graph using C for the total charge and k as the amount of water used in kilolitre.

From the equation, find the
   d upfront cost
   e rate of charge per kilolitre
   f total cost of using 4200 kilolitres of water
   g number of kilolitres of water that have a total cost of $12 000, rounding your answer to the nearest kilolitre.

4 **WORKED EXAMPLE 17** During Australian summers, crickets chirp faster at night if the temperature is higher. The linear equation for calculating the chirp rate, R (chirps/min), for a given temperature, t (°C), is $R = 8t - 24$.

   a Use the equation to find the chirp rate of crickets when the temperature is 26°C.
   b Use the equation to find the chirp rate of crickets when the temperature is 4°C.
   c Explain why your answer to part **b** makes no sense in real life.

5 **WORKED EXAMPLE 18** Key rings are sold for $5.50 each, and the cost C of making n key rings is given by the equation $C = 4n + 110$.

   a Find the revenue equation in terms of n.
   b Find the profit equation in terms of n.
   c How much profit would be made if 200 key rings were sold?
   d How much profit would be made if 50 key rings were sold?
   e How many key rings need to be sold to make at least $600 profit? Explain why you need to round **up** for this calculation.

Calculator-free

6 (2 marks) Steven is a wedding photographer. He charges his clients a fixed fee of $500, plus $250 per hour of photography. Find the equation that represents the total amount Steven charges, denoted as $C, for t hours of photography.

7 (2 marks) Initially there are 5000 litres of water in a tank. Water starts to flow out of the tank at a constant rate of 2 litres per minute until the tank is empty. Determine the equation that represents the total volume of water in the tank, V, after t minutes.

Calculator-assumed

8 (3 marks) A full tank holds 2000 litres of water. Water is pumped out of the tank at a constant rate. The graph shows how the volume of water in the tank, V, changes with time, t.

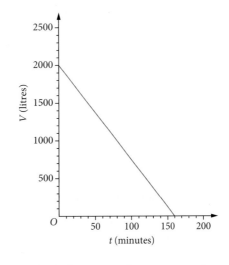

   a Determine the constant rate, in litres per minute, at which the water is being pumped out of the tank. (2 marks)
   b Hence or otherwise, state the equation of this relationship. (1 mark)

9 (3 marks) Justin makes and sells electrical circuit boards. He has one fixed cost of $420 each week. Each circuit board costs $15 to make. The selling price of each circuit board is $27. Determine the weekly profit if Justin makes and sells 200 circuit boards per week.

10 (3 marks) To raise funds, a club plans to sell lunches at a weekend market. The club will pay $190 to rent a stall. Each lunch will cost $12 to prepare and will be sold for $35. Determine the minimum number of lunches that must be sold to make a profit of at least $1000.

11 (3 marks) In one month, an energy company charges a $30 service fee, plus a supply charge of two cents per megajoule (MJ) of energy used. State the linear equation which represents this relationship, denoting $C as the cost and E as the energy used in MJ. Copy the Cartesian plane below and sketch the graph of this relationship.

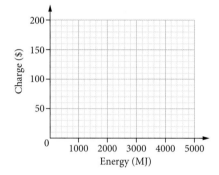

12 (2 marks) Ben will use a currency exchange agency to buy some Japanese yen (the Japanese currency unit). The graph shows the relationship between Japanese yen and Australian dollars on a particular day. This graph can be used to calculate a conversion between dollars and yen on that day.

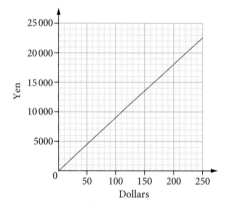

    a Ben converts his dollars into yen using this graph. How many yen does he receive for $200? (1 mark)

    b The slope of this graph is the exchange rate for converting dollars into yen on that particular day. How many yen will Ben receive for each dollar? (1 mark)

13 (3 marks) The weight of gold can be recorded in either grams or ounces. The following graph shows the relationship between *weight in grams* and *weight in ounces*.

The relationship between weight measured in grams and weight measured in ounces is shown in the equation

weight in grams = M × *weight in ounces*

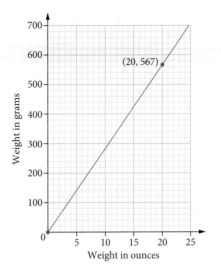

a Show that $M = 28.35$. (1 mark)

b Robert found a gold nugget weighing 0.2 ounces. Using the equation above, calculate the weight, in grams, of this gold nugget. (1 mark)

c Last year Robert sold gold to a buyer at $55 per gram. The buyer paid Robert a total of $12 474. Using the equation above, calculate the weight, in ounces, of this gold. (1 mark)

14 (2 marks) A rock-climbing activity will be offered to students at a camp on one afternoon. Each student who participates will pay $24. The organisers have to pay the rock-climbing instructor $260 for the afternoon. They also have to pay an insurance cost of $6 per student. Let n be the total number of students who participate in rock climbing.

a Write an expression for the profit that the organisers will make in terms of n. (1 mark)

b The organisers want to make a profit of at least $500. Determine the minimum number of students who will need to participate in rock climbing. (1 mark)

15 (6 marks) This table shows the cost, C cents, of mobile phone calls under the Oz-Zone Budget Plan, for calls of length t minutes.

Length of call, t (min)	1	2	5	10	15
Cost, C (cents)	102	182	422	822	1222

a Find the linear relationship in the form $C = at + b$. (1 mark)

b If this rule was graphed on a number plane, which variable would be shown on the vertical axis? (1 mark)

c Use the relationship you found in part **a** to calculate the cost of an 18-minute call. (1 mark)

d What is the vertical axis intercept of the graph and what does it represent? (1 mark)

e If a phone call is extended by 3 minutes, by how much would the cost of the call increase by? (1 mark)

f How long is a phone call under this plan if it costs $5.82? (1 mark)

8 Chapter summary

Solving linear equations
- A **linear equation** is an equation where the highest power of the variable is 1. An example is $2x - 6 = 10$.
- A linear equation with one unknown can be solved by isolating the variable, which is the unknown quantity, on one side of the equation.

Solving problems using linear equations
To solve worded problems:
1. Assign a pronumeral to the unknown quantity, if it is not given.
2. Translate the problem into an equation in terms of the assigned pronumeral.
3. Solve the problem by using the techniques learned.
4. State the solutions to the problem clearly.

Linear equations in the form $y = ax + b$
- A linear equation is an equation that can be written in the form $y = ax + b$, where x and y are variables and a and b are constants.
- The graph of a linear equation is a straight line.
- The **coordinates** of points on the line are found by substituting x values into the equation to calculate corresponding y values.
- Equations can be drawn by plotting values from a table of values onto a Cartesian plane.
- To determine whether a point lies on a line, substitute the coordinates into the equation of the line and check if the result is true or false.
- For a straight line in the form $y = ax + b$:

  $a = \textbf{slope} = \dfrac{\text{rise}}{\text{run}} = \dfrac{\text{vertical distance between two points}}{\text{horizontal distance between two points}}$

  $b = \textbf{y-intercept} = y$ value when $x = 0$.

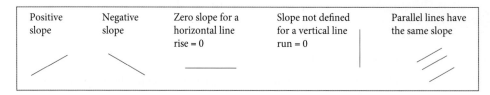

Positive slope | Negative slope | Zero slope for a horizontal line rise = 0 | Slope not defined for a vertical line run = 0 | Parallel lines have the same slope

Interpreting linear equations
- When interpreting linear equations in the form $y = ax + b$:

  $a = $ **constant rate of change**

  $b = $ **initial value**
- We can use linear equations to make predictions about practical situations, but not all calculated values make sense in the real world.

Cumulative examination: Calculator-free

Total number of marks: 13 marks   Reading time: 2 minutes   Working time: 13 minutes

1 (2 marks) A trapezium has parallel lengths of 5 cm and 8 cm with a perpendicular height of 3 cm. Determine the area of the trapezium.

2 (2 marks) The box plot shows the annual amounts earned by workers in a fast-food franchise.

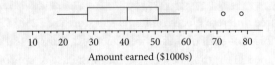

a What percentage of workers earn less than $28 000 per year? (1 mark)

b Describe the shape of the box plot and note any outliers. (1 mark)

3 (9 marks)

a Determine the equation of the following linear graphs.

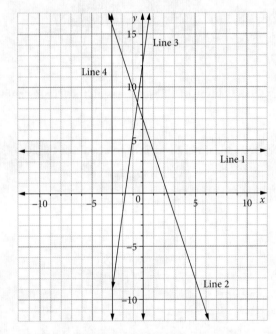

Copy and complete the following.

Line 1: (1 mark)

Line 2: (2 marks)

Line 3: (2 marks)

Line 4: (1 mark)

b Determine the equation of the straight line that passes through the points $(-2, 10)$ and $(1, -8)$. (3 marks)

Cumulative examination: Calculator-assumed

Total number of marks: 35 marks   Reading time: 4 minutes   Working time: 35 minutes

1 (2 marks) Renuta is a self-employed antique furniture auctioneer. On every item she sells, she charges 15% commission on the first $2000 of the sale price and 12.5% of the amount over $2000. How much will Renuta charge for selling an antique dining room suite that sold for $22 600?

2 (3 marks) The cost prices of three different electrical items in a store are $230, $290 and $310 respectively. The selling price of each of these three electrical items is 130% of the cost price plus a fixed commission of $20 for the salesperson. Show the use of appropriate matrix operations to determine the final selling price of each of the items.

3 (2 marks) Determine the volume, in cubic centimetres, of the witches' hat shown.

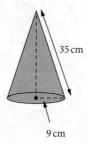

4 (2 marks) A rectangular prism with a volume of $1200 \, cm^3$ has all the lengths halved. What is the volume of the smaller prism?

5 (4 marks) The plan of a one bedroom apartment is drawn to a scale of 1 : 120.

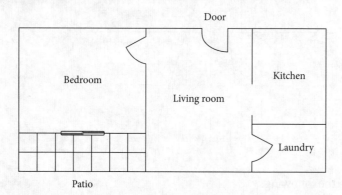

   a Calculate the dimensions of the living room in metres. (2 marks)

   b If new tiles are to be installed in the living room at a cost of $55/m^2$, determine the cost to install the tiles. (2 marks)

6 (3 marks) The dot plot shows the number of vehicles driving past Westvale High School every minute for a 20-minute period.

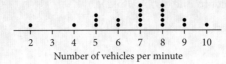

    a Find the mean. (1 mark)

    b Calculate the standard deviation, correct to two decimal places. (1 mark)

    c How many data values were within one standard deviation from the mean? (1 mark)

7 (4 marks) The box plot shows the distribution of the forearm *circumference*, in centimetres, of 252 people.

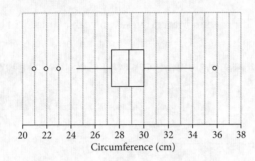

    a Determine the percentage of these 252 people with a forearm *circumference* of less than 30 cm. (1 mark)

    b Determine the five-number summary for the forearm *circumference* of these 252 people. (3 marks)

8 (4 marks)

    a If $\sin 30° = \dfrac{1}{2}$, determine the exact area of the following triangle. (2 marks)

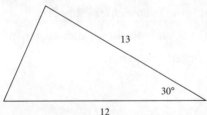

    b Use Heron's rule to show that the area of this triangle is between 11 and 12 units². (2 marks)

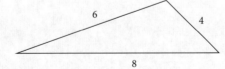

9 (2 marks) A flying fox is constructed between the top of a tree and a pile of rocks. It has a length of 34 m and a 38° angle of elevation. Calculate the (horizontal) distance between the tree and the pile of rocks, correct to one decimal place.

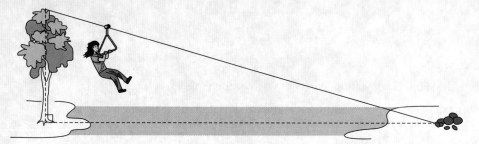

10 (9 marks)

a At present, Grace is 24 years older than Marcia. In seven years, Grace will be three times older than Marcia. Determine how old Grace and Marcia are now. (4 marks)

b The graph below shows the mass, W grams, of chicken pellets in a feed bowl after t hours.

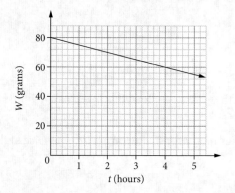

 i Determine the equation of the linear relationship connecting W and t. (2 marks)
 ii What mass of chicken pellets is in the bowl after 7.2 hours? (1 mark)
 iii If 25 grams of chicken pellets are left in the bowl, after how many hours did this occur? (1 mark)
 iv How long will it take for the pellets to run out? (1 mark)

CHAPTER 9
SIMULTANEOUS EQUATIONS AND OTHER LINEAR GRAPHS

Syllabus coverage
Nelson MindTap chapter resources

9.1 Simultaneous equations
   Solving simultaneous linear equations graphically
   Using CAS 1: Solving simultaneous equations graphically

9.2 Solving simultaneous linear equations algebraically
   Solving simultaneous equations by substitution
   Solving simultaneous equations by elimination
   Using CAS 2: Solving simultaneous equations

9.3 Modelling with simultaneous equations

9.4 Other linear graphs
   Piece-wise linear graph
   Step graph

Examination question analysis
Chapter summary
Cumulative examination: Calculator-free
Cumulative examination: Calculator-assumed

Syllabus coverage

TOPIC 2.3: LINEAR EQUATIONS AND THEIR GRAPHS

Simultaneous linear equations and their applications

2.3.7 solve a pair of simultaneous linear equations graphically or algebraically, using technology when appropriate

2.3.8 solve practical problems that involve determining the point of intersection of two straight-line graphs; for example, determining the break-even point where cost and revenue are represented by linear equations

Piece-wise linear graphs and step graphs

2.3.9 sketch piece-wise linear graphs and step graphs, using technology when appropriate

2.3.10 interpret piece-wise linear and step graphs used to model practical situations; for example, the tax paid as income increases, the change in the level of water in a tank over time when water is drawn off at different intervals and for different periods of time, the charging scheme for sending parcels of different weights through the post

Mathematics Applications ATAR Course Year 11 syllabus p. 13 © SCSA

Video playlists (5):
9.1 Simultaneous equations
9.2 Solving simultaneous linear equations algebraically
9.3 Modelling with simultaneous equations
9.4 Other linear graphs
Examination question analysis Simultaneous equations and other linear graphs

Worksheets (5):
9.1 Sketching simultaneous equations
   • Intersection of lines
9.2 Solving simultaneous equations
9.3 Simultaneous equations problems
9.4 Step graphs

Nelson MindTap

To access resources above, visit
cengage.com.au/nelsonmindtap

9.1 Simultaneous equations

Simultaneous equations are a set of two or more algebraic equations that involve the same unknown variables. These equations have a common solution, which means there is a set of values for the variables that satisfy all the equations simultaneously. One example of a set of simultaneous equations is $y = 4 + x$ and $2x + y = 7$.

Simultaneous equations can be solved using different methods, which involve graphical or algebraic techniques. The choice of method depends on the specific problem and the preferred approach.

Graphical method: In this approach, the equations are graphed on a Cartesian plane, and the solution is determined by finding the point of **intersection** of the graphs. The coordinates of the intersection point represent the solution to the simultaneous equations.

Algebraic method: **Algebraic techniques** involve manipulating the equations algebraically to solve for the variables. This can be done using either **substitution** or **elimination**.

Video playlist
Simultaneous equations

Worksheets
Sketching simultaneous equations

Intersection of lines

Solving simultaneous linear equations graphically

If two linear equations are graphed on the same axes, the intersection of the two lines is the point that lies on both lines. This point is the solution to the two simultaneous equations. Simultaneous linear equations can be in the form $y = ax + b$ or $Ax + By = C$.

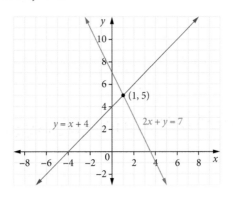

For example, the solution to the two simultaneous equations $y = x + 4$ and $2x + y = 7$ is shown as the point $(1, 5)$ in the graph, meaning $x = 1$ and $y = 5$.

WORKED EXAMPLE 1 | Solving simultaneous equations graphically

Graph and solve the pair of simultaneous equations $y = 2x + 4$ and $-2x - y = 8$.

Steps | **Working**

1 Clearly label the scale and the axes on a Cartesian plane.

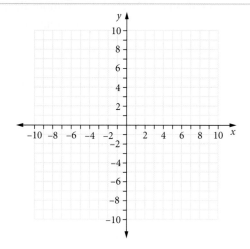

2 On the axes, plot $y = 2x + 4$.

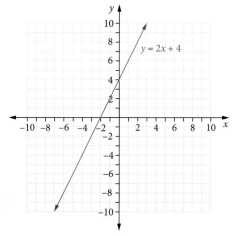

3 Plot $-2x - y = 8$ on the *same set of axes*.

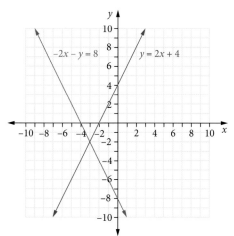

4 Identify the point of intersection of the two graphs. This is the solutions of the two simultaneous equations.

The point of intersection is $(-3, -2)$.

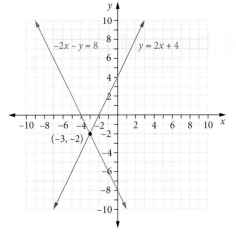

The solution is $x = -3$ and $y = -2$.

We can solve simultaneous equations graphically using CAS.

USING CAS 1 Solving simultaneous equations graphically

Graph and solve the pair of simultaneous equations $y = 3x - 2$ and $x + 2y = 10$.

ClassPad

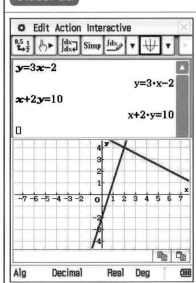
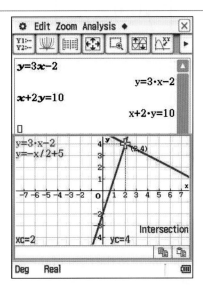

1 Tap **Main** and clear all equations.
2 Enter the equations $y = 3x - 2$ and $x + 2y = 10$.
3 Tap the **Graph** tool to add a graph window.
4 Drag the equations into the graph window.

Note: if the graphs do not display, tap the **View Window** tool and tap **Default**, then **OK**.

5 With the graph window highlighted, tap **Analysis > G-Solve > Intersection**.
6 The coordinates of the point of intersection will be displayed.

The solution is $x = 2$ and $y = 4$ or $(2, 4)$.

TI-Nspire

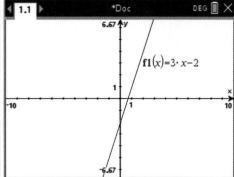
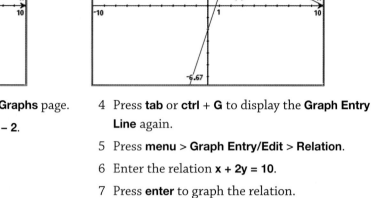

1. Start a new document and add a **Graphs** page.
2. In the **Graph Entry Line**, enter **3x − 2**.
3. Press **enter** to graph the line.

4. Press **tab** or **ctrl + G** to display the **Graph Entry Line** again.
5. Press **menu > Graph Entry/Edit > Relation**.
6. Enter the relation **x + 2y = 10**.
7. Press **enter** to graph the relation.

8. Press **menu > Analyze > Intersection**.
9. When prompted for the **lower bound**, use the arrow keys to move to the left of the point of intersection and press **enter**.
10. When prompted for the **upper bound**, use the right arrow key to move to the right of the point of intersection and press **enter**.
11. The coordinates of the point of intersection will be displayed on the screen.

🔒 **Exam hack**

It is important to keep in mind that the coefficient of the variable in a pair of simultaneous equations could be a decimal number. To solve such a situation, use CAS to determine the solutions.

The solution is $x = 2$ and $y = 4$ or $(2, 4)$.

EXERCISE 9.1 Simultaneous equations   ANSWERS p. 428

Mastery

1 ⚙ WORKED EXAMPLE 1 Graph and solve the following pairs of simultaneous equations.

    a $y = 2x - 1$ **b** $y = 8 - x$ **c** $2x + y = 8$ **d** $4 = y - x$
         $y = -5x - 1$ $2y - x = -2$ $3y + x = 9$ $2 = 2x - y$

2 ⚙ Using CAS 1 Graph and solve the following pairs of simultaneous equations.

    a $y = x + 3$ **b** $y = -x + 2$ **c** $2x + y = 3$ **d** $3x - y = 4$
         $2x - y = -4$ $y = x + 6$ $4x - y = 3$ $2x - y = 2$

▶ **Calculator-free**

3 (7 marks) The graph of $y = 4 - x$ is plotted on the Cartesian plane below.

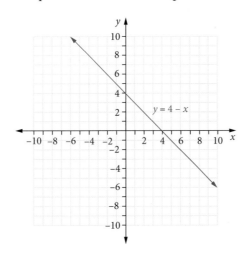

a Copy the above graph and plot $y = x - 6$ on the same set of axes. State the point of intersection between these two simultaneous equations. (3 marks)

b Plot $y = -4x + 4$ on the same set of axes from part **a**. (2 marks)

c Hence or otherwise, state the solutions to the simultaneous equations of

  i $y = 4 - x$ and $y = -4x + 4$ (1 mark)

  ii $y = -4x + 4$ and $y = x - 6$ (1 mark)

Calculator-assumed

4 (2 marks) Solve the following pairs of simultaneous equations by first plotting the graphs using CAS and state the solutions to the simultaneous equations.

a $y = 4 + \dfrac{x}{2}$ and $3x - 2y = 1$ (1 mark)

b $-2 = 2y + 3x$ and $y + \dfrac{5x}{2} = 1$ (1 mark)

> 🔒 **Exam hack**
>
> Always express your answer in exact form (fraction) unless you are specifically instructed to round your answer to a certain number of decimal places.

9.2 Solving simultaneous linear equations algebraically

Video playlist
Solving simultaneous linear equations algebraically

Worksheet
Solving simultaneous equations

It is important to note that while graphing can provide a quick and visual way to solve simultaneous equations, it may not always give precise solutions, especially for non-whole numbers. Algebraic methods, on the other hand, allow for more accurate calculations, particularly when dealing with fractions or decimals. So when accuracy is important, algebraic methods are preferred over graphical methods.

The two methods for solving simultaneous equations algebraically are substitution and elimination. These methods involve manipulating the equations to eliminate one variable and solve for the other, ultimately finding the values that satisfy both equations simultaneously.

Solving simultaneous equations by substitution

When solving simultaneous equations using the substitution method, we solve one equation for one variable and substitute that expression into the other equation. This allows us to eliminate one variable and solve for the remaining variable.

WORKED EXAMPLE 2	Solving simultaneous equations by substitution
Solve the pair of simultaneous equations $y = x + 4$ and $y - 3x = 16$.	
Steps	**Working**
1 Label the two equations.	$y = x + 4$ [1] $y - 3x = 16$ [2]
2 Substitute equation [1] into equation [2].	$(x + 4) - 3x = 16$
3 Solve for x.	$x + 4 - 3x = 16$ $-2x = 12$ $x = -6$
4 Substitute the value of x into equation [1].	$y = (-6) + 4$ $y = -2$
5 State the solution.	The solution is $x = -6$, $y = -2$ or $(-6, -2)$.

 Exam hack

It is important to always determine the other unknown variable by substituting the one you have solved into one of the original equations.

Solving simultaneous equations by elimination

When solving simultaneous equations using the elimination method, we manipulate the two equations by adding or subtracting them to eliminate one of the variables. This results in a new equation with only one unknown, making it easier to solve for the variable.

WORKED EXAMPLE 3	Solving simultaneous equations by elimination
Solve the pair of simultaneous equations $y - x = 1$ and $3x - y = 7$.	
Steps	**Working**
1 Label the two equations.	$y - x = 1$ [1] $3x - y = 7$ [2]
2 Since the variable y in both equations has the same coefficient but opposite signs, we can add the two equations to eliminate y.	[1] + [2]: $(y - x) + (3x - y) = 1 + 7$ $2x = 8$ $x = 4$
3 Substitute the value of x into equation [1].	$y - 4 = 1$ $y = 5$
4 State the solution.	The solution is $x = 4$, $y = 5$ or $(4, 5)$.

In the elimination method, it is important to ensure that one of the variables in the equations has the same coefficient. If the coefficients are different, one or both the equations can be multiplied by a suitable number(s) to make the coefficients equal before applying the elimination process. This step is necessary to ensure that when the equations are added or subtracted, one of the variables will be eliminated.

WORKED EXAMPLE 4 — Solving simultaneous equations by elimination (multiplying one equation by a suitable number)

Solve the pair of simultaneous equations $-3x + y = 7$ and $5x + 2y = 3$.

Steps	Working
1 Label the two equations.	$-3x + y = 7$ [1] $5x + 2y = 3$ [2]
2 Multiply equation [1] by 2 so that the coefficients for the variable y for both equations are the same.	$2 \times [1]$: $-6x + 2y = 14$ [3]
3 Subtract equation [2] from equation [3] to eliminate y and solve for x.	$[3] - [2]$: $(-6x + 2y) - (5x + 2y) = 14 - 3$ $-6x + 2y - 5x - 2y = 11$ $-11x = 11$ $x = -1$
4 Substitute the value of x into equation [1].	$-3(-1) + y = 7$ $y = 4$
5 State the solution.	The solution is $x = -1$, $y = 4$ or $(-1, 4)$.

WORKED EXAMPLE 5 — Solving simultaneous equations by elimination (multiplying both equations by two suitable numbers)

Solve the pair of simultaneous equations $4x + 5y = 14$ and $-5x + 3y = 1$.

Steps	Working
1 Label the two equations.	$4x + 5y = 14$ [1] $-5x + 3y = 1$ [2]
2 Multiply equation [1] by 5 and equation [2] by 4 so that the coefficient of x for both equations is the same but with opposite signs.	$5 \times [1]$: $20x + 25y = 70$ [3] $4 \times [2]$: $-20x + 12y = 4$ [4]
3 Add equations [3] and [4] to eliminate x and then solve for y.	$[3] + [4]$: $(20x + 25y) + (-20x + 12y) = 70 + 4$ $20x - 20x + 25y + 12y = 74$ $37y = 74$ $y = 2$
4 Substitute the value of y into equation [1].	$4x + 5(2) = 14$ $4x = 4$ $x = 1$
5 State the solution.	The solution is $x = 1$, $y = 2$ or $(1, 2)$.

> **Exam hack**
>
> When one of the equations is already in the form of $y = ...$ or $x = ...$, it is generally more efficient to use the substitution method to solve the simultaneous equations. Otherwise, you can choose the method you are familiar with.

USING CAS 2 Solving simultaneous equations

Solve the pair of simultaneous equations $y = 3x - 2$ and $x + 2y = 10$.

ClassPad

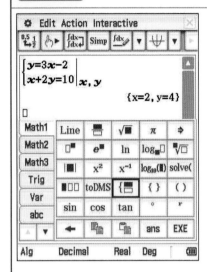

TI-Nspire

ClassPad steps	TI-Nspire steps
1 Tap **Main**.	1 Insert a **Calculator** page.
2 Open the **Keyboard** > **Math1**.	2 Press **menu** > **Algebra** > **Solve System of Equations** > **Solve System of Equations**.
3 Select the **simultaneous equations** template.	3 On the next screen, keep the default values and select **OK**.
4 Enter the equations into the template.	4 Enter the two equations into the template and press **enter**.
5 Enter **x,y** in the lower right corner of the template.	5 The coordinates of the point of intersection will be displayed.
6 Press **EXE**.	
7 The coordinates of the point of intersection will be displayed.	

The solution is $x = 2$, $y = 4$ or $(2, 4)$.

EXERCISE 9.2 Solving simultaneous linear equations algebraically   ANSWERS p. 430

Recap

1 Plot the graph of $y = -3x - 9$ and $x + 4y = 8$ on a Cartesian plane. Hence, state the point of intersection of both lines.

2 Using CAS, plot $y = 8 + 2x$, $2x + y = 5$ and $y = \dfrac{x}{2} - 3$. State the three points of intersection between the three lines.

Mastery

3 **WORKED EXAMPLE 2** Solving simultaneous equations by substitution.

a $y = 5 - x$ $2x - y = 4$	**b** $x = 8 - 2y$ $3x - y = -4$	**c** $2x + 3y = 13$ $x = 2y - 4$	**d** $y = 13 - 3x$ $4x = 3y$
e $x = 2 + y$ $2x + y = 4$	**f** $y = 2x - 6$ $x + y = 12$	**g** $4x - 3y = 9$ $x = 12 + y$	**h** $x + y = 19$ $y = 3x - 5$

4 **WORKED EXAMPLE 3** Solve the following simultaneous equations by elimination.

a $2x + y = 10$
   $x - y = 5$

b $2y + x = 8$
   $-2y + 4x = 2$

c $2x + 3y = 13$
   $6x + 3y = 9$

d $7x + 4y = 21$
   $4y + 8x = 16$

e $7y + 2x = 3$
   $-6x - 7y = 18$

f $3x + 4y = 18$
   $6 = -3x + 4y$

5 **WORKED EXAMPLE 4** Solve the following simultaneous equations by elimination by first multiplying one of the equations by a suitable number.

a $y - 2x = 8$
   $5y + 3x = 14$

b $3x + 2y = 14$
   $5x - y = 6$

c $5x + 4y = -7$
   $-x - 3y = 8$

d $4x - 2y = -6$
   $6x - y = -11$

e $2y + 2x = 22$
   $-x + 3y = 17$

f $2x + 7y = -12$
   $5y + x = -9$

6 **WORKED EXAMPLE 5** Solve the following simultaneous equations by elimination by multiplying both equations by two suitable numbers.

a $-2x + y = -8$
   $3x + 2y = 19$

b $7y + 3x = 25$
   $8y + 5x = 27$

c $2x + 3y = 12$
   $x + y = 6$

d $-4y + 5x = 10$
   $3y + 2x = 4$

e $9x - 7y = 21$
   $10x + 8y = -24$

f $x + 2y = 3$
   $2x - 3y = -1$

7 **Using CAS 2** Use CAS to solve the following simultaneous equations.

a $x + y = 3$
   $x - y = 1$

b $-y + 3x = 7$
   $3y + 2x = 10$

c $5x + 2y = 13$
   $4x - 3y = 8$

d $7y + 9x = 46$
   $-5y + 10x = 0$

e $2x + 3y = 12$
   $3x + 2y = 15$

f $x = -4 + 2y$
   $3x + y = 12$

g $4 = 2y + x$
   $x + 3y = 12$

h $x - y = 10$
   $6x = 5y$

i $y = 2x + 3$
   $6x - 2y = 13$

j $y = 2x + 16$
   $4x - 3y = 9$

k $5y = 12 + \dfrac{2x}{3}$
   $7x - \dfrac{3y}{4} = 9$

l $8x + 6y = 12$
   $3x + 2.5y = 4$

Calculator-free

8 (7 marks) Solve the following simultaneous equations algebraically.

a $y = 3x - 12$ and $3x + y = 8$ (3 marks)

b $5x + 8y = 27$ and $3x + 7y = 25$ (4 marks)

Calculator-assumed

9 (4 marks) Solve the following simultaneous equations, stating your answer to two decimal places.

a $\dfrac{22y}{3} + 4x = 12$ and $3y - 5.2x = 1$ (2 marks)

b $3.6y - 5x = 1$ and $4x = 12 - 2y$ (2 marks)

Video playlist
Modelling with simultaneous equations

Worksheet
Simultaneous equations problems

9.3 Modelling with simultaneous equations

Simultaneous equations can be used to solve real-world problems where there is more than one unknown.

Steps for solving worded problems using simultaneous equations

To solve worded problems using simultaneous equations:
1. Identify the unknowns and assign a pronumeral for each of them.
2. Set up two equations by converting the given information into mathematical language.
3. Solve the simultaneous equations using CAS.
4. State the solution to the problem in sentence form.

WORKED EXAMPLE 6 Solving problems using simultaneous equations

Eleanor spent $16.40 on pens and pencils. She purchased a total of 8 items. If the pens cost $2.80 each and the pencils cost 80 cents each, how many pens and pencils did Eleanor purchase?

Steps	Working
1 Identify the unknowns and assign a pronumeral for each of them.	Let p = number of pens and q = number of pencils
2 Set up two equations by converting the given information into mathematical symbols. Convert units where necessary.	There are 8 items, so $p + q = 8$ Pens cost $2.80 each and pencils cost $0.80 each. cost of pens = $2.8 \times p$ cost of pencils = $0.8 \times q$ Total cost of pens and pencils = $2.8p + 0.8q$ Eleanor spent $16.40.
3 Solve the simultaneous equations using CAS.	$2.8p + 0.8q = 16.4$

 ClassPad

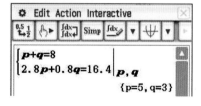

TI-Nspire

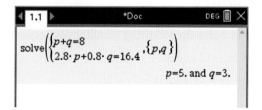

| 4 Answer the question in sentence form. | Eleanor purchased 5 pens and 3 pencils. |

EXERCISE 9.3 Modelling with simultaneous equations

ANSWERS p. 431

Recap

1. Determine the solution to the pair of simultaneous equations $2x - 3y = 7$ and $3x = 5 - y$.

2. Consider the linear graph on the right.
   a. Determine the equation of the linear relationship.
   b. Copy the graph and on the same set of axes, plot $y = x - 6$. Determine the coordinates of the point of intersection.

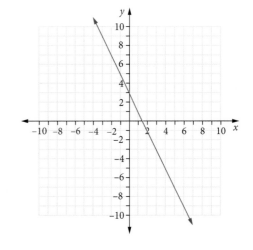

Mastery

3. **WORKED EXAMPLE 6** In a souvenir shop, Mark bought three magnets and a tea towel for $11.65. His brother, Thomas, bought two magnets and three tea towels for $17.10. By writing a pair of simultaneous equations, determine the price of a magnet and a tea towel.

4. A farm has cows and chickens. Altogether there are 110 heads and 290 legs. How many chickens are there on the farm?

5. At a school concert, the cost of the adult tickets was different from the cost of tickets for children. The Smith family bought three adult tickets and four child tickets for $67.00. The Chan family bought two adult tickets and five child tickets for $57.50. Let x be the ticket cost for one adult. Let y be the ticket cost for one child.
   a. Write a pair of simultaneous equations that could be used to represent the situation.
   b. Hence, determine the price of one adult ticket and one child ticket.

6. At the local bakery, the cost of four doughnuts and six buns is $14.70. The cost of three doughnuts and five buns is $11.90. Determine the cost of one doughnut and two buns from this bakery.

7 Robert wants to hire a geologist to help him find potential gold locations. One geologist, Jennifer, charges a flat fee of $600 plus 25% commission on the value of gold found. The following graph displays Jennifer's total fee in dollars.

Another geologist, Kevin, charges a total fee of $3400 for the same task.
   a Copy the above graph and add a graph of the line representing Kevin's fee.
   b For what value of gold found will Kevin and Jennifer charge the same amount for their work?

8 The graph shows the growth of two different types of trees, measured from the time they were seedlings.

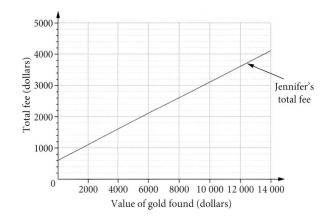

   a Which tree was taller as a seedling and what was the difference in height?
   b After how many months do the trees reach the same height?
   c From part **b**, what is that height?
   d Find an equation for the height of the
      i gum tree
      ii wattle tree.
   e Are these equations good models for the height of the trees after 20 years? Why or why not?

Calculator-free

9 (4 marks) Jack and Jill each have a number of Star Trek trading cards. Jill has 20 more cards than Jack and in total, they have 180 cards.
   a Write a pair of simultaneous equations to represent this situation. (2 marks)
   b Determine the number of Star Trek trading cards each of them have. (2 marks)

10 (5 marks) The sum of two numbers is 44 and their difference is 16. By first writing a pair of simultaneous equations, determine the two numbers.

Calculator-assumed

11 (5 marks) Ben went to a sports store and bought some items of clothing. He bought a number of t-shirts costing $10 each and a number of shorts costing $20 each. If he spent $120 in total and bought twice as many t-shirts as shorts, how many of each did he buy?

12 (10 marks) Martin is planning a holiday in New Zealand. He considers hiring a car from two different companies. Cars-are-cheap Rental charges a base fee of $80 and then an additional $0.15 per kilometre driven. Rent-a-bomb Car Rental does not charge a base fee but charges $0.55 per kilometre driven.

  a Let c = the rental cost and d = the distance driven.

  i Write an equation for the rental cost for Cars-are-cheap Rental using c and d. (1 mark)

  ii Write an equation for the rental cost for Rent-a-bomb Car Rental using c and d. (1 mark)

  b Copy the graph below and plot the graph for the rental cost charged by Cars-are-cheap Rental and Rent-a-bomb Car Rental. Clearly label each line with the name of the rental company. (4 marks)

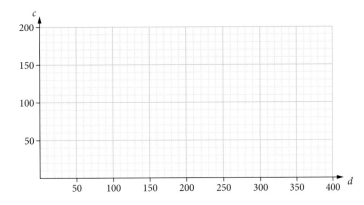

  c At what distance will the rental cost charged by the two rental companies be the same and what will the cost be? (2 marks)

  d Martin plans on driving a total of 350 km holidaying in New Zealand. By comparing the rental cost charged by both rental companies, which rental company should Martin choose? (2 marks)

9.4 Other linear graphs

There are two other linear graphs which we will explore:

- **piece-wise linear graph**, and
- **step graph**.

Piece-wise linear graph

A piece-wise linear graph consists of one or more non-overlapping line segments. It is sometimes called a **line segment graph**. For example:

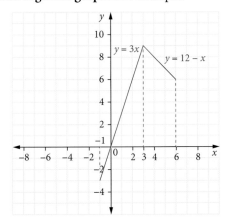

The equations for this graph are:

$y = 3x$ $-1 \leq x < 3$ ← This means for x values between -1 and 3, including -1.

$y = 12 - x$ $3 \leq x < 6$ ← This means for x values between 3 and 6, including 3.

Piece-wise graphs are useful for displaying information that involves rates. The rate for each line segment is the slope of the line.

WORKED EXAMPLE 7 — Plotting a piece-wise linear graph

Copy the Cartesian plane and plot the following piece-wise graph on it.

$$y = \begin{cases} x - 1 & x \leq -2 \\ 2x + 1 & -2 \leq x < 2 \\ -x + 7 & x \geq 2 \end{cases}$$

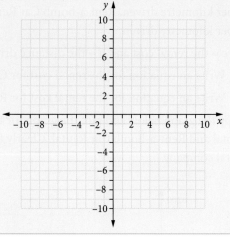

Steps

1 Start by plotting $y = x - 1$ on the Cartesian plane and stop at the point when $x = -2$.

> You can use any suitable method to plot this line but do not draw the line past $(-2, -3)$.

Working

Calculate the y-coordinate when $x = -2$.

$y = -2 - 1 = -3$

Plot the line $y = x - 1$ on the Cartesian plane and stop at this point $(-2, -3)$. Remember to include an arrow at the other end of the line.

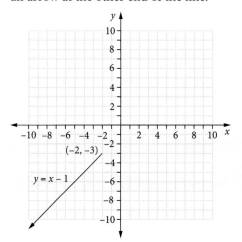

2. Plot the graph of $y = 2x + 1$ from $x = -2$ to $x = 2$.

Calculate the y-coordinate when $x = 2$.

$y = 2(2) - 1 = 5$

The coordinate $(2, 5)$ forms the last point of the graph of $y = 2x + 1$ in this case. Draw a straight line between $(-2, -3)$ and $(2, 5)$ to represent the line segment $y = 2x + 1$.

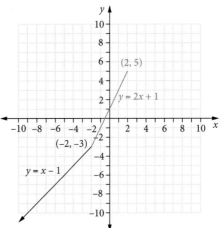

3. Plot the graph of $y = -x + 7$ from $x = 2$ onward.

From the point $(2, 5)$, plot the graph of $y = -x + 7$ using any method you are familiar with. Do not extend the line beyond $(2, 5)$. Remember to include an arrow at the end of the line.

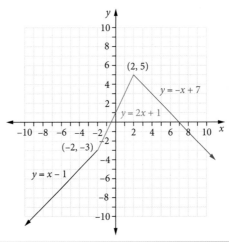

WORKED EXAMPLE 8 | Interpreting piece-wise graphs

A rainwater tank holds 50 000 litres. The line segment linear graph shows the rate at which the rainwater tank fills with water from the start of winter.

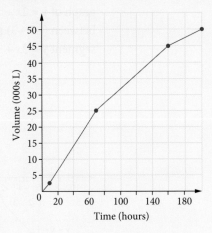

a Explain how we know the rainwater tank fills at four different rates.
b Approximately how many litres does the rainwater tank hold after 160 hours?
c After how many hours from the start of winter is the rainwater tank filled to capacity?
d Approximately during which times does the tank fill the fastest?
e What is the rate, in litres per hour, that the tank is filling in the last 20 hours?

Steps	Working
a Refer to the line segments and their slopes.	The graph is made up of four different line segments and each line segment has a different slope. This means the rainwater tank fills at four different rates.
b Read from the graph and note how the vertical scale is written.	After 160 hours, the rainwater tank holds 45 000 litres.
c Read from the graph.	The rainwater tank filled to capacity after 200 hours.
d Identify which of the line segments has the greatest slope.	The second time period has the greatest slope, so the tank fills the fastest from approximately 10 hours to 70 hours after the start of winter.
e Find the slope of the last line segment. Convert the rate to litres per hour.	Slope of last line segment = $\dfrac{\text{rise} \uparrow}{\text{run} \rightarrow}$ $= \dfrac{5}{40}$ $= 0.125$ thousand litres per hour. The tank is filling at a rate of $0.125 \times 1000 = 125$ litres per hour in the last 20 hours.

Piece-wise graphs can also be used to display the relationship between distance and time. In this situation, the slope of the line segments measures speed, which is the rate at which distance is travelled.

WORKED EXAMPLE 9 — Interpreting distance-time graphs

The distance-time graph shows how Celine walked to Nicky's house, stayed there for a while, then borrowed Nicky's bike and cycled home.

a The vertical intercept of this graph is 2 km. What does this mean?

b Explain how we know Celine started walking slower after 30 minutes.

c Explain how we know the horizontal line segment represents the time Celine is at Nicky's house.

d How long did Celine stay at Nicky's house?

e What is the distance between Celine's house and Nicky's house?

f Calculate Celine's cycling speed in kilometres/hour during the last section of her journey.

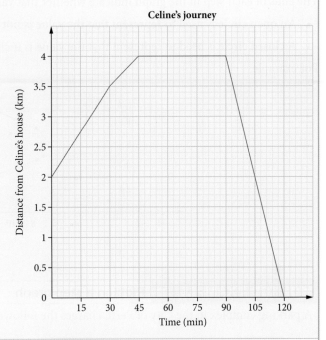

Celine's journey

Steps	Working
a The vertical intercept is the initial value. What is the vertical scale measuring?	Celine started her journey 2 km from her own house.
b Refer to the slopes of the line segments, which measure speed.	The slope of the line segment after 30 minutes is less than the slope of the line segment before 30 minutes, so Celine started walking slower after 30 minutes.
c Refer to the vertical axis value.	The horizontal line indicates that Celine is the same distance (4 km) from her house during this time, so she is at Nicky's house.
d Read from the graph.	Celine stayed at Nicky's house from the 45-minute mark to the 90-minute mark, so Celine stayed at Nicky's house for 45 minutes.
e Read the distance using the vertical scale of the graph.	The distance between Celine's house and Nicky's house is 4 km.
f Calculate the slope from the graph in kilometres/minute and convert to kilometres/hour.	Cycling speed = $\dfrac{\text{rise} \rightarrow}{\text{run} \downarrow} = \dfrac{4 \text{ km}}{30 \text{ min}}$ $= \dfrac{4 \text{ km}}{\frac{1}{2} \text{ h}}$ $= 8 \text{ km/h}$

Exam hack

Always check to see if all the information is given in the same units. Convert units where necessary.

Step graph

A step graph is a linear graph that has only horizontal straight-line pieces. Because of this, the graph doesn't change gradually but has distinct 'steps'. This kind of graph is common for fee structures, like the tariff structures in household bills, where a certain amount is charged for the first block of energy and then a different amount is charged for the next block and so forth.

The ends of each step in the graph indicate whether that value is included or not.

- An open circle is used to represent that the value is not included in the interval.
- A closed circle is used to represent that the value is included in the interval.

For example:

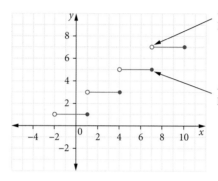

An open circle means this point is **not** included.

A closed circle means this point is included.

WORKED EXAMPLE 10 — Plotting a step graph

A parking complex in the City of Perth charges the following weekday rates:

Duration	Cost ($)
0 to less than 1 hour	$8.00
1 to less than 2 hours	$15.00
2 to less than 3 hours	$25.00
3 to exactly 5 hours	$27.50
More than 5 hours	$32.50

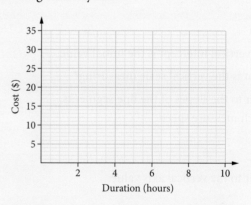

Copy the axes on the right and on it represent the information in the above table in a step graph.

Steps	Working
1 We know from the table that the cost is $8 from entry into the carpark to a duration of just under one hour.	Draw a horizontal straight line $y = 8$ from $x = 0$ and $x = 1$. Both circles are open circles. 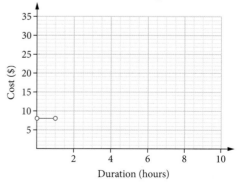

2 The cost for the duration after the first hour to just under 2 hours is $15.

Draw a horizontal straight line $y = 15$ from $x = 1$ and $x = 2$. The circle at the start is closed and the circle at the end is open.

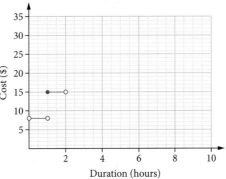

3 The cost for the duration from the second hour to just under 3 hours is $25.

Draw a horizontal straight line $y = 25$ from $x = 2$ and $x = 3$. The circle at the start is closed and the circle at the end is open.

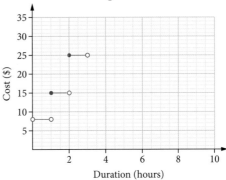

4 The cost for the duration from the third hour to exactly 5 hours is $27.50.

Draw a horizontal straight line $y = 25$ from $x = 3$ and $x = 5$. The circles at both the start and the end are closed.

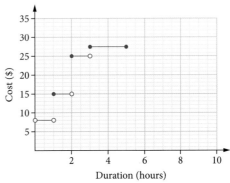

5 The cost is $32.50 for the duration over 5 hours.

Draw a horizontal straight line $y = 25$ from $x = 3$ and $x = 5$. The circle at the start is open, while there is an arrow at the end of the straight line.

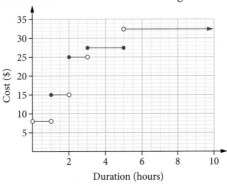

WORKED EXAMPLE 11 — Interpreting step graphs

This step graph shows the daily parking charges at a car park.

a Find the charge for parking for a duration of
  i 55 minutes
  ii 3 hours
  iii $3\frac{1}{2}$ hours
  iv 4 hours 20 minutes.

b What range of times can a driver park for $25?

c What does the arrow on the $40 step mean?

Steps	Working
a Read from the graph. 	i Parking for a duration of 55 minutes will cost $5. ii Parking for a duration of 3 hours will cost $15. iii Parking for a duration of $3\frac{1}{2}$ hours will cost $25. iv Parking for a duration of 4 hours 20 minutes will cost $35.
b 1 Find the relevant step on the graph. Remember, a closed circle means the value is included and an open circle means it's not included. 	
2 Write the answer.	For $25 the driver can park for a time that is more than 3 hours up to and including 4 hours.
c The arrow indicates the line continues.	After 6 hours, the parking charge remains constant at $40, so the maximum daily charge is $40.

EXAMINATION QUESTION ANALYSIS

Calculator-free (8 marks)

FareJet Airlines offers air travel between destinations in regional Western Australia. The table shows the fares for some distances travelled.

Distance (km)	Fare
$0 < distance \leq 100$	$100
$100 < distance \leq 250$	$160
$250 < distance \leq 400$	$220

a What is the maximum distance a passenger could travel for $160? (1 mark)

The fares for the distances travelled in the table are graphed below.

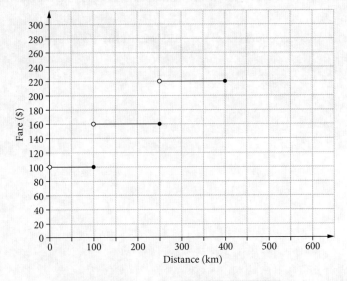

b The fare for a distance longer than 400 km, but not longer than 550 km, is $280. Copy the graph above and draw this information on it. (1 mark)

FareJet Airlines is planning to change its fares. A new fare will include a service fee of $40, plus 50 cents per kilometre travelled. An equation used to determine this new fare is given by

$fare = 40 + 0.5 \times distance$

c A passenger travels 300 km. How much will this passenger save on the fare calculated using the equation above compared to the fare shown in the table? (2 marks)

d At a certain distance between 250 km and 400 km, the fare, when calculated using either the new equation or the table, is the same. What is this distance? (2 marks)

e An equation connecting the maximum distance that may be travelled for each fare in the table can be written as

$fare = a + b \times maximum\ distance$

Determine a and b. (2 marks)

Reading the question

- Note the words 'longer than' and 'not longer than', and the mathematical symbols they represent.
- Part **c** is asking how much the passenger 'saves' on the fare, not how much they pay.
- The maximum distances for each fare are indicated both in the table and on the graph.

Thinking about the question

- How do the words 'longer than' and 'not longer than' appear on the graph?
- Part **d** involves calculating using two different methods.
- In part **e**, what do a and b represent in the equation of a straight line?

Worked solution (✓ = 1 mark)

a Read from the vertical axis and then down to the horizontal axis, noting that the dot means the point is included.

The maximum distance a passenger could travel for $160 is **250 km**. ✓

b
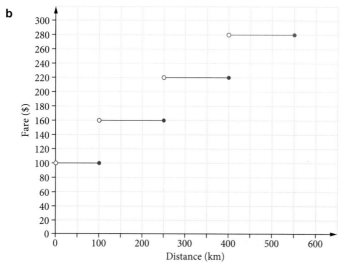

correctly draws the graph ✓

c fare = 40 + 0.5 × distance

If distance = 300 km, then fare = 40 + 0.5 × 300 = 190.

So, the cost using this equation is $190.

The fare shown in the table for travelling 300 km is **$220**.

So, the passenger will save **$30** on the fare calculated using the equation compared to the fare shown in the table.

calculates the fare based on the formula ✓

states the difference in fare ✓

d From the graph, the fare for the distance between 250 km and 400 km is $220.

The equation is fare = 40 + 0.5 × distance

To find the distance when fare = 220, solve using CAS if necessary.

40 + 0.5 × distance = 220

$$\text{distance} = \frac{220 - 40}{0.5} = 360$$

The distance when the two fares are the same is **360 km**.

- calculates the fare based on the formula ✓
- calculates and states the distance ✓

e Draw a line connecting the maximum distance that may be travelled for each fare.

Find the slope using $\frac{\text{rise}}{\text{run}}$ and find the vertical axis intercept.

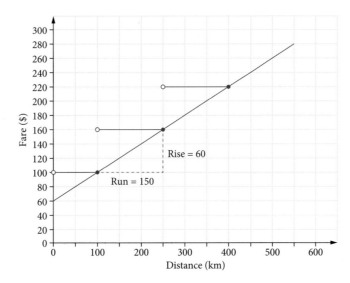

Slope = $\frac{60}{150} = \frac{2}{5}$ = **0.4** and vertical intercept from the graph is **60**.

The equation of the line is fare = $a + b$ × maximum distance, so a is the vertical axis intercept and b is the slope.

Therefore, a = **60** ✓ and b = **0.4**. ✓

EXERCISE 9.4 Other linear graphs   ANSWERS p. 432

Recap

1 Determine the solution to the simultaneous equations $7x - y = -8$ and $-5x - y = 16$.

2 One afternoon at the beach Mr Smith bought four ice creams and three drinks for his family at a cost of $21.40. Mrs Brown bought five of the same ice creams and two of the same drinks for $20.80. Based on these prices, determine the cost of one drink.

Mastery

3 WORKED EXAMPLE 7 Copy the Cartesian plane below and plot the following piece-wise linear graph:

$$y = \begin{cases} -x + 3 & x < -2 \\ -2x + 1 & -2 \leq x < 1 \\ 3x - 4 & x \geq 1 \end{cases}$$

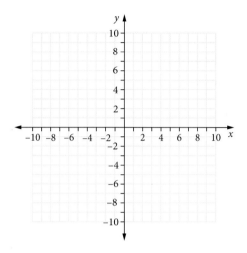

4 WORKED EXAMPLE 8 A rainwater tank holds 20 000 litres. The line segment graph shows the rate at which the rainwater tank fills with water from the start of the rainy season.

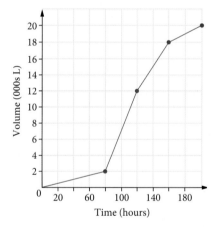

a Explain how we know the rainwater tank fills at four different rates.
b Approximately how many litres does the rainwater tank hold after 160 hours?
c After how many hours from the start of the rainy season is the rainwater tank filled to capacity?
d Approximately during which times does the tank fill the fastest?
e What is the rate, in litres per hour, at which the tank is filling in the first 80 hours?

5 **WORKED EXAMPLE 9** The distance-time graph shows how Volker and Herb walked from the park to Herb's house. Volker stayed there a while and then jogged home.

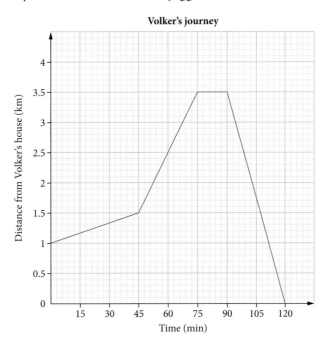

a The vertical intercept of this graph is 1 km. What does this mean?
b Explain how we know Volker and Herb started walking faster after 45 minutes?
c Explain how we know the horizontal line segment represents the time Volker and Herb were at Herb's house.
d How long did Volker stay at Herb's house?
e What is the distance between Volker's house and Herb's house?
f Calculate Volker's jogging speed in kilometres/hour during the last section of his journey.

6 **WORKED EXAMPLE 10** A t-shirt store offers the following pricing for t-shirts:
- $15 each for the first 5 t-shirts
- $12 each for the next 10 t-shirts
- $8 each for any additional t-shirts.

Copy the following axes and on it represent the pricing structure as a step graph.

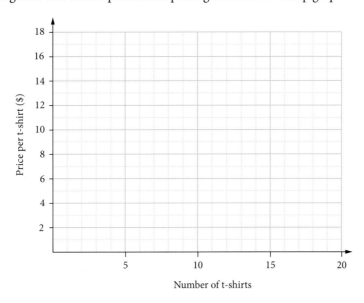

7 WORKED EXAMPLE 11 This step graph shows the cost of sending parcels of different weights by air freight.

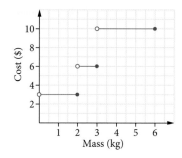

a Find the cost of sending
   i   one parcel weighing 3.4 kg
   ii one parcel weighing 1.7 kg
   iii two parcels, weighing 2 kg and 3 kg.
b What range of weights for a single parcel can be sent for $6?
c Why do you think there is no arrow on the $10 step?

Calculator-free

8 (1 mark) The *delivery fee* for a parcel, in dollars, charged by a courier company is based on the *weight* of the parcel, in kilograms.

This relationship is shown in the step graph for parcels that weigh up to 20 kg.

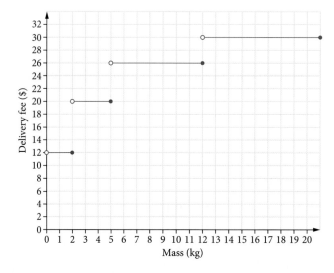

State which one (or more) of the following statements is true.

A The *delivery fee* for a 4 kg parcel is $20.
B The *delivery fee* for a 12 kg parcel is $26.
C The *delivery fee* for a 13 kg parcel is the same as the *delivery fee* for a 20 kg parcel.
D The *delivery fee* for a 10 kg parcel is $14 more than the *delivery fee* for a 2 kg parcel.
E The *delivery fee* for a 12 kg parcel is $18 more than the *delivery fee* for a 2 kg parcel.

9 (6 marks) The graph shows the volume of water in a water tank between 7 am and 5 pm on one day.

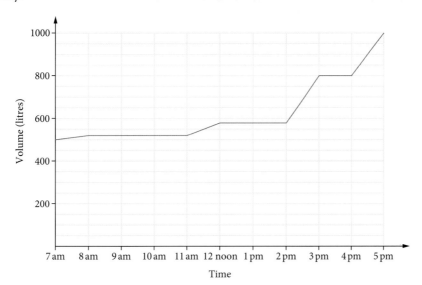

a What is the initial volume of water in the tank? (1 mark)
b Describe the behaviour of the volume of water in the tank between 8 am and 11 am. (1 mark)
c State the time the volume of water in the tank increases at the greatest rate. (1 mark)
d What is the change in the volume of water between 12 noon and 2 pm? (1 mark)
e Determine the rate of the water filling the tank between 4 pm and 5 pm. (2 marks)

10 (1 mark) The graph shows the number of people who attended a market over a seven-hour time period. For how many hours were there at least 600 people at the market?

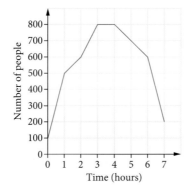

11 (2 marks) The annual *fee* for membership of a car club, in dollars, based on *years of membership* of the club is shown in the step graph.

In the Martin family:
- Hayley has been a member of the club for 4 years
- John has been a member of the club for 20 years
- Sharon has been a member of the club for 25 years.

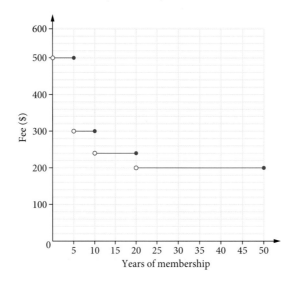

Determine the total fee for membership of the car club for the Martin family.

12 (2 marks) The graph shows the cost (dollars) of mobile telephone calls up to 240 seconds long.

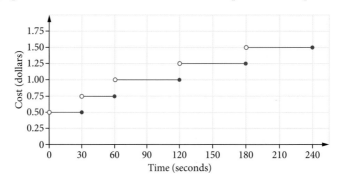

Determine the cost of making a 90-second call, followed by a 30-second call.

13 (1 mark) The graph shows a distance-time graph for a car travelling from home along a long straight road over a 16-hour period.

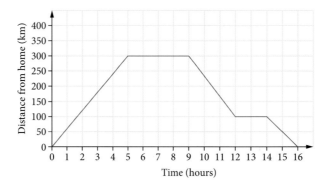

In which one of the time intervals is the speed of the car greatest?

14 (1 mark) A supermarket sells roasted chickens. For the first four hours after cooking, the roasted chickens are sold at full price. After this time, the selling price of each roasted chicken is reduced. The price of a roasted chicken, $P, at any time up to six hours after cooking is shown in the step graph.

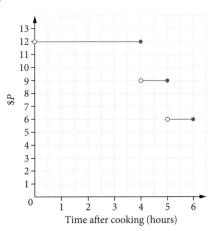

A roasted chicken is sold five hours after cooking. By how much has the full price of the roasted chicken been reduced?

Calculator-assumed

15 (4 marks) The piece-wise graph shows the income tax payable in a particular country.

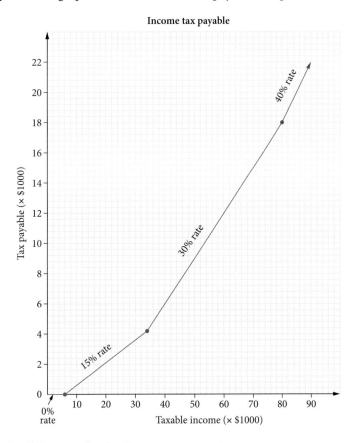

   a What is the income range for the first two tax rates? (2 marks)

   b Find the tax payable on each of the following incomes from the graph.

      i $30 000 (1 mark)

      ii $84 000 (1 mark)

16 (3 marks) Kyla owns and manages a truck and car care business. After a major repair on a truck, one of the mechanics took it on a long test drive. The test drive started at 12 noon. After four hours, the mechanic stopped to rest for one hour and then returned to the workshop. The graph shows the truck's distance from the workshop, d, in kilometres, and the number of hours of test driving, n, after 12 noon.

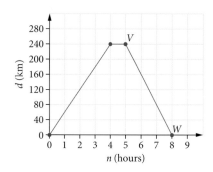

a At what time of the day did the mechanic arrive back at the workshop? (1 mark)

b Find the average speed, in kilometres per hour, of the truck during the first four hours of the test drive. (1 mark)

c On the return trip, the truck travelled at an average speed of 80 km/h. The equation of the line segment VW that represents this part of the test drive is of the form $d = k - 80n$. Find the value of k. (1 mark)

17 (3 marks) Customers who visit Rumi's shop park their cars in an underground car park. The graph shows the total charge for parking, in dollars, according to the number of hours parked in one visit.

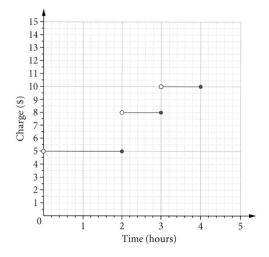

a On Monday, Leon parked in the underground car park for two hours and Lucy parked for one-and-a-half hours. What was the total combined charge for Leon and Lucy? (1 mark)

b Customers who park for more than four hours are charged $11.50 if they do not stay longer than five hours. Copy the graph and add this information. (1 mark)

c Another customer, Pam, parked in the underground car park on both Tuesday and Wednesday. Each day she parked for more than two hours but less than four hours. Pam was charged less than $20 in total for these two days. Write down the two possible amounts that Pam could have been charged. (1 mark)

Chapter summary

Simultaneous equations
- If we graph two linear equations on the same axes, the intersection of the two lines is the point that lies on both lines.
- The point of intersection of two lines is the solution to the two **simultaneous equations**.
- Simultaneous linear equations can be in the form $y = ax + b$ or $Ax + By = C$.
- Both forms of linear equations can be solved using CAS.
- Simultaneous equations can also be solved algebraically, either by **substitution** or **elimination**.
- To solve worded problems using simultaneous equations:
    1. Identify the unknowns and assign a **pronumeral** for each of them.
    2. Set up two equations by converting the given information into mathematical language.
    3. Solve the simultaneous equations using CAS.
    4. State the solution to the problem in sentence form.

Other linear graphs
- A **piece-wise linear graph** joins two or more straight line pieces. It is also known as a line segment graph.
- The rate for each line segment in a line segment graph is the slope of the line.
- **Distance-time graphs** are line segment graphs where the horizontal axis measures time and the vertical axis measures distance.
- The slope of the line segments in a distance-time graph measures speed, which is the rate at which distance is travelled.
- A **step graph** is a graph that has only horizontal straight-line pieces.

Cumulative examination: Calculator-free

Total number of marks: 18 Reading time: 2 minutes Working time: 18 minutes

1 a (4 marks) Maria earns 5% commission on all her sales. Find her commission on the following sales:

   i $100 (1 mark)

   ii $1000 (1 mark)

b Find Maria's rate of commission if she earns $12 commission on sales of $200. (2 marks)

2 (4 marks) The two matrices A and B are known to be equal matrices.

$$A = \begin{bmatrix} 1 & 3x + y \\ x - y & 5 \end{bmatrix} \qquad B = \begin{bmatrix} 1 & 9 \\ 7 & 5 \end{bmatrix}$$

Determine the value of x and the value of y.

3 (2 marks) The heights of players in a basketball team are measured. Describe the type of data using the terms continuous, discrete, numerical and/or categorical.

4 (4 marks) Copy the table and study the graph shown. Write down the solutions for the following simultaneous equations.

$y + x = 7$ $x + 4y = 4$	
$y = 3 + 3x$ $y = \dfrac{x}{2} - 2$	
$y = 3 + 3x$ $y + x = 7$	
$y = \dfrac{x}{2} - 2$ $x + 4y = 4$	

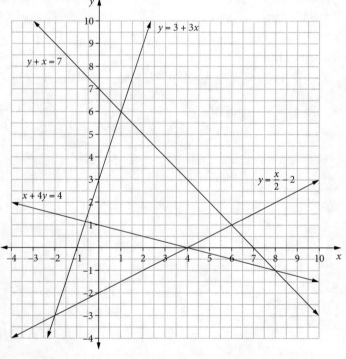

5 (4 marks) The sum of the current ages of Tim and Chris is 56. In 5 years' time, Chris will be twice as old as Tim. By setting up a pair of simultaneous equations, determine the current ages of Tim and Chris.

Cumulative examination: Calculator-assumed

Total number of marks: 32   Reading time: 4 minutes   Working time: 32 minutes

1 (3 marks) The matrix shows how five people, Alan (A), Bevan (B), Charlie (C), Drew (D) and Esther (E), can communicate with each other.

$$P = \text{Sender} \begin{array}{c} \\ A \\ B \\ C \\ D \\ E \end{array} \begin{array}{c} \text{Receiver} \\ \begin{bmatrix} A & B & C & D & E \\ 0 & 1 & 0 & 1 & 0 \\ 1 & 0 & 0 & 0 & 0 \\ 0 & 0 & 0 & 1 & 1 \\ 1 & 0 & 1 & 0 & 0 \\ 0 & 0 & 1 & 0 & 0 \end{bmatrix} \end{array}$$

A '1' in the matrix shows that the person named in that row can send a message directly to the person named in that column. For example, the '1' in row 3 and column 4 shows that Charlie can send a message directly to Drew.

a List the people that can communicate directly to Charlie. (1 mark)

b Esther wants to send a message to Bevan. Explain whether or not it is possible for Esther to do this by communicating with:

  i exactly one other person. (1 mark)

  ii exactly three other people. (1 mark)

2 (2 marks) An ice-cream dessert is in the shape of a hemisphere. The dessert has a radius of 5 cm. The top and the base of the dessert are covered in chocolate. Determine the surface area, correct to the nearest square centimetres, that is covered in chocolate.

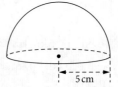

3 (4 marks) The plan of a cabin is drawn to scale below.

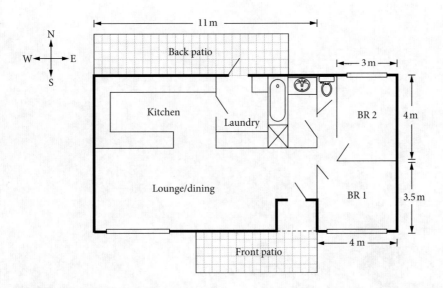

a Determine the scale in the form $1:a$, where a is a whole number. (1 mark)

b Determine the total area of the bedrooms. (2 marks)

c Determine the total cost of installing new flooring in the bedrooms if it costs $53/m². (1 mark)

4 (2 marks) The time spent by shoppers at a hardware store on a Saturday is approximately normally distributed with a mean of 31 minutes and a standard deviation of 6 minutes.

    a If 2850 shoppers are expected to visit the store on a Saturday, determine the number of shoppers who are expected to spend between 25 and 37 minutes in the store. (1 mark)

    b If 3000 shoppers are expected to visit the store on a Sunday, determine the number of shoppers who are expected to spend between 25 and 43 minutes in the store. (1 mark)

5 (2 marks) Samples of jellyfish were selected from two different locations, A and B. The diameter (in mm) of each jellyfish was recorded and the resulting data is summarised in the box plots shown.

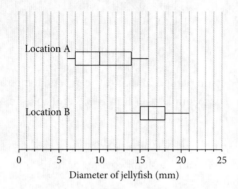

    a From the box plots, comment on the diameters of the jellyfish taken from location A compared to location B. (1 mark)

    b What is the percentage of jellyfish taken from location A with a diameter greater than 14 mm? (1 mark)

6 (6 marks) Town A is 6 km due south of town B. Town C is 9 km due east of town A.

    a Draw a labelled diagram to represent the situation. (2 marks)

    b Calculate the bearing, to the nearest degree, of town C from town B as a:

        i true bearing. (3 marks)

        ii compass bearing. (1 mark)

7 (13 marks) The cost of couriering a parcel weighing up to 5 kg with Speedy Mail is shown in the table below.

Weight, W (kg)	Cost, C ($)
$0 < W \leq 0.5$	12
$0.5 < W \leq 1$	15
$1 < W < 1.5$	25
$1.5 \leq W \leq 2.5$	30
$2.5 < W < 4.0$	35
$4.0 \leq W \leq 5.0$	45

a Copy a Cartesian plane similar to the one below and plot the step graph which represents the cost of couriering a parcel with Speedy Mail. (4 marks)

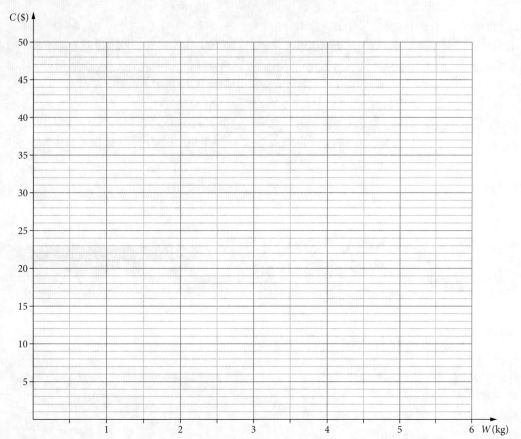

Another courier company, Lightning Courier, charges $6.25 per 500 g for the first 3 kg and then an additional $5 per 1 kg, or part thereof, for parcels weighing up to 5 kg.

b Copy and complete the piece-wise linear graph below, which describes the cost, C dollars, of couriering a parcel weighing W kilograms with Lightning Courier. (3 marks)

$$C = \begin{cases} \underline{\hspace{5cm}}, 0 < W \leq 3 \\ \underline{\hspace{5cm}}, \underline{\hspace{2cm}} \\ \underline{\hspace{5cm}}, \underline{\hspace{2cm}} \end{cases}$$

c Copy the axes above and on them plot the charges for Lightning Courier. (3 marks)

d Maria wants to send two parcels, weighing 2.5 kg and 4.2 kg respectively, to her friends. She wants to minimise the cost. Determine the courier company she should use. Justify your answer mathematically. (3 marks)

Answers

CHAPTER 1

EXERCISE 1.1

1 a $1330 **b** $2660 **c** $69 160

2 a $6875 **b** $3173.08 **c** $1586.54

3

$26.50	32	$848.00
$29.25	35	$1023.75
$23.23	35	$813.05
$23.23	40	$929.20
$25.30	32	$809.60
$26.50	20	$530.00
	Total wage bill	$4953.60

4 a $40.80 **b** $54.40 **c** $36.54
 d $48.72 **e** $36.90 **f** $49.20
 g $46.88 **h** $62.50

5 a 2 **b** $1092.75

6 a $67 600
 b A month is usually longer than 4 weeks
 c $5633.33 **d** $37.14

7 a $260.16 **b** 5 hours

8 a $879.23 **b** $1255.23

9 $15.68 **10** $4160

11 a $178 **b** 11
 c casual wage = $25 632, permanent wage = $31 616
 Permanent position pays more. Ashok should take the permanent job.

12 Job 1

EXERCISE 1.2

1 a $62 400 **b** $5200 **c** $31.58/h
 d $60 633.60

2 a $1152 **b** $2304 **c** $59 904

3 a $962.15 **b** $2290.88

4 $1218.63 **5** $554.40

6 a $18 **b** $61 **c** $159.50

7 a $17.50 **b** $65.00 **c** 400 letters

8 a $20/hr **b** 1000 bricks **c** $160

9 $78

10 a i $2 **ii** $20
 b 3%

11 a $98.10 **b** $174.40 **c** $272.50

12 a i $19 **ii** $28.50 **iii** $38
 iv $399 **v** $114 **vi** $228
 b $741

13 a $18 **b** $81

14 a $18 **b** $48.91 **c** $214

15 $2235

16 $88 **17** 412 700 **18** $210

19 a $980 **b** $133.30 **c** 1113.30

20 a $766.59 **b** $174.23 **c** $139.38
 d $1168.33

21 a $207.20 **b** 8:00 am **c** 1 hour
 d 12:00 pm **e** $103.60 **f** 11 am

EXERCISE 1.3

1 a $20 **b** $53 **c** $80

2 a $1134.62 **b** $2634.62

3 a $928.50
 b Nanna is underpaid $18.50/fortnight or $74 for 8 weeks.

4 $187.40 **5** $478.40

6 $338.40 **7** $2376.27

8 a $20 745.40 **b** $204.20

9 a $38 079.60
 b $25 625.60

10 The pension received is $11 975.60. The amount entitled is $14 635.92 so Rose has been underpaid $2660.32.

EXERCISE 1.4

1 $104.40

2 $931.50

3 a

Income		Expenses	
Office	$620	Rent	$280
Club	$215	Food	$60
		Mobile phone	$20
		Travel	$60
		Total fixed expenses	$420
		Entertainment	$60
		Savings	$315
		Clothes	$40
Total income	$835	**Total expenses**	$835

Others answers possible for entertainment, savings, clothes.

b

Income		Expenses	
Office	$620	Rent	$280
Club	$215	Food	$60
		Mobile phone	$20
		Car	$175
		Total fixed expenses	$535
		Entertainment	$50
		Savings	$210
		Clothes	$40
Total income	$835	**Total expenses**	$835

4 a $5800 **b** $111.54

5 a $312 **b** $41 080 **c** $16 224

6 **a** income = $420, expenses = $300
  b $120   **c** $1440
7 **a** $158   **b** $223
  c approx. 113 weeks   **d** Teacher to check

EXERCISE 1.5

1 **a** $200
  b

Income		Expenses	
Main job	$750	Rent	$250
Second job	$100	Food	$150
		Phone	$80
		Travel	$70
		Entertainment	$100
		Savings	$200
Total	$850	Total	$850

2 **a** increase, $0.70, 20%
  b decrease, 5 tomatoes, 25%
  c decrease, $10, 25%
  d increase, 1.5 kg, 60%
  e increase, 10 cm, 7.4%
3 **a** $12   **b** −$7.20   **c** −$5.40   **d** $21
4 **a** $75   **b** $51   **c** $63.60   **d** $55.20
5 **a** **i** 40%   **ii** 70%   **iii** 20%
  b **i** 20%   **ii** 10%   **iii** 50%
6 **a** $125.00   **b** $112.50   **c** $132.00   **d** $148.15
7 **a** **i** profit = $125
     ii profit percentage = 35.7%
  b **i** loss = $50
     ii loss percentage = 25%
  c **i** profit = $120
     ii profit percentage = 100%
  d **i** loss = $0.50
     ii loss percentage = 20%
8 **a** $600   **b** $800
9 **a** **i** $180   **ii** $1975
  b **i** $9   **ii** $98
  c **i** $144 642   **ii** $14 464
  d **i** $2591   **ii** $259
10 **a** **i** $0.0157 and $0.0148   **ii** 1 L
   b **i** $0.5556 and $0.5333   **ii** 15 cans
   c **i** $0.0168 and $0.0175   **ii** 500 g
11 **a** 15%   **b** $1050
12 **a** $44   **b** $22   **c** $77
13 **a** 20%   **b** $330
14 **a** **1** $8.90/kg   **2** $9.60/kg
       3 $9.04/kg   **4** $9.00/kg
       5 $9.13/kg
   b 2 kg for $17.80 is the best buy.

EXERCISE 1.6

1 **a** $6   **b** $46
2 **a** $40   **b** $440
3 **a** $3300   **b** $19 800   **c** $18 562.50
4 ClassPad
  a
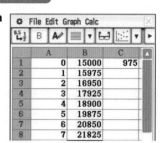
  b
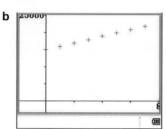
  TI-Nspire
  a
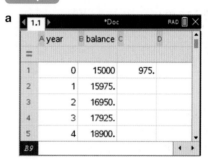
  b
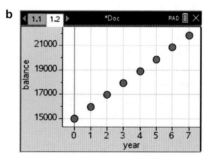
5 **a** 9 years   **b** 2.08%
6 **a** **i** 52   **ii** 468
     iii $\frac{3}{52}$% = 0.000 58
  b **i** 365   **ii** 3285
     iii $\frac{7}{365}$% = 0.000 19
  c **i** 12   **ii** 108
     iii $\frac{5}{12}$% = 0.004 17
  d **i** 4   **ii** 36
     iii $\frac{6}{4}$% = $\frac{3}{2}$% = 0.015
  e **i** 26   **ii** 234
     iii $\frac{8}{26}$% = $\frac{4}{13}$% = 0.0031

7 a i $P = 75\,000, r = 0.06, n = 12, t = 5$
  ii $101\,163.76   **iii** $26\,163.76
b i $P = 45\,000, r = 0.04, n = 4, t = 3$
  ii $50\,707.13   **iii** $5707.13
c i $P = 6000, r = 0.08, n = 2, t = 4$
  ii $8211.41   **iii** $2211.41
8 a 0.05   **b** $100

c

Year	Balance
0	2000
1	2100
2	2200
3	2300

9 a

n	Compound interest ($)	Value of investment ($)	Simple interest ($)	Value of investment ($)
0	–	3000	–	3000
1	$\frac{5}{100} \times 3000 = 150$	$3000 + 150 = 3150$	$\frac{5}{100} \times 3000 = 150$	$3000 + 150 = 3150$
2	$\frac{5}{100} \times 3150 = 157.50$	$3150 + 157.50 = 3307.50$	$\frac{5}{100} \times 3000 = 150$	$3150 + 150 = 3300$
3	$\frac{5}{100} \times 3307.50 = 165.38$	$3307.50 + 165.38 = 3472.88$	$\frac{5}{100} \times 3000 = 150$	$3300 + 150 = 3450$
4	$\frac{5}{100} \times 3472.88 = 173.64$	$3472.88 + 173.64 = 3646.52$	$\frac{5}{100} \times 3000 = 150$	$3450 + 150 = 3600$

b $3646.52   **c** $46.52

EXERCISE 1.7

1 a $1500   **b** $15\,000
  c $35\,000
2 a $P = 8000, r = 0.09, n = 4, t = 6$.
  b $13\,646.13   **c** $5646.13
3 a $666   **b** $980.50
  c $3175.68   **d** $878.38
4 a £108   **b** £2916
  c $866.67   **d** $10\,296.30
5 a $126.98   **b** 79.86

c	The cost of 1 barrel of oil in Australian dollars divided by 159. (There's 159 L of oil in a barrel).	79.86 c
	Excise (import tax)	44.2 c/L
	Production, transport and marketing costs	26.5 c
	Oil company profit	2.2 c
	Total	152.76 c
	GST on the total	15.276 c
d	**Final cost of 1 litre of petrol**	**$1.68**

6 a $24 \div 0.75 = 32$ **b** $32.96

7

Exchange rate	Calculation	Cost of supplies (to nearest $)
A$1 = US$0.55	420 ÷ 0.55 = 763.64	$764
A$1 = US$0.65	420 ÷ 0.65 = 646.15	$646
A$1 = US$0.75	420 ÷ 0.75 = 560	$560
A$1 = US$0.85	420 ÷ 0.85 = 494.12	$494
A$1 = US$1.05	420 ÷ 1.05 = 400	$400

8 a 80 c   **b** 60.8 c   **c** 16 c   **d** 3.2 c
9 a $0.65   **b** $0.48   **c** $0.47   **d** $0.72

EXERCISE 1.8

1 a £110   **b** £3300   **c** $1000   **d** $20\,000
2 A$1 = US$0.70
3 a $2730   **b** $22\,260   **c** $6930   **d** 31.1%
4 a 8.33%   **b** 3.43%   **c** 2.34%   **d** 1.91%
5 $3139.50   **6** 4.64
7 a $7.00   **b** $0.315   **c** $1.58   **d** 4.4
8 a $1.00   **b** $500   **c** $50   **d** 10%
9 a 7500   **b** $4125   **c** 17.2%
  d $180   **e** shares by $3945
10 a $20.00   **b** $2.40   **c** $6.00   **d** 3.3

CUMULATIVE EXAMINATION: CALCULATOR-FREE

1 a $2   **b** $22
2 $2000
3 a $260   **b** $180   **c** $80

CUMULATIVE EXAMINATION: CALCULATOR-ASSUMED

1 a $P = 35\,000, r = 0.08, n = 4, t = 5$
  b $52\,008.16
  c $17\,008.16
2 a December: $390, January: $529, February: $230
  b $1149
3 a 5000 shares   **b** $3750   **c** 12.9%
  d $217.50   **e** $3967.50

CHAPTER 2

EXERCISE 2.1

1 a −2   **b** 21   **c** 7   **d** −6
2 3
3 a $z = -8$ **b** $b = 2$
4 a D **b** $D = 6.16$
5 $A = 1005.31$
6 a 562   **b** 16
7 The first expression has a value of $\frac{3}{25}$. The second expression has a value of $\frac{3}{5}$. The first expression is one-fifth of the value of the second expression.
8 a 60   **b** 8

EXERCISE 2.2

1 B **2** D

3

F	32	60	104	120
C	0	15.56	40	48.89

4

C\B	2	5	8	14
4	1	2	3	5
A 7	2	3	4	6
13	4	5	6	8

5 a $A = 2, B = -4$

b When x and y increase by 1, the value of z decreases by 2.

6 a B1 = 1.5A1 − 20

b

	A	B
1	16	4
2	30	25
3	48	52
4	50	55

7 a The 100 represents the fact that the prices are per 100 g and the quantities C and N are measured in grams.

b $X = 12.5, Y = 22.5$

c $17.50, when the bag contains 400 g of chocolate and 150 g of nuts.

8 a $S = 0.25A + 0.6B$

b i $X = 1220$. When Stephen earns $2000 from Source A and $1200 from Source B, he will transfer $1220 to his savings.

ii $Y = 2800$

9 a 20.6

b i

BMI		H (m)	
	1.51	1.82	1.95
W (kg) 54	23.7	16.3	14.2
75	32.9	22.6	19.7
81	35.5	24.5	21.3

ii $\frac{5}{9}$ or 55.6%   **iii** 81 kg

EXERCISE 2.3

1 D **2** C

3 a 3×3 **b** 12 **c** 2×1 **d** −3

4 a 2×1; column matrix

b 3×3; square matrix/identity matrix (I_3)

c 1×4; row matrix/zero matrix

d 3×3; square matrix

e 3×1; column matrix/unit column matrix

f 4×4; square matrix

5 a $T = \begin{bmatrix} 172 & 67 & 30 \\ 3 & 139 & 10 \\ 0 & 65 & 9 \\ 11 & 15 & 17 \end{bmatrix}$; dim$(T) = 4 \times 3$

b $A = \begin{bmatrix} 3 & 139 & 10 \end{bmatrix}$; dim$(A) = 1 \times 3$

c $O = \begin{bmatrix} 67 \\ 139 \\ 65 \\ 15 \end{bmatrix}$; dim$(O) = 4 \times 1$

d $C = [9]$; dim$(C) = 1 \times 1$

6 a 60 **b** 5 **c** 1

d 3 **e** $1 \times 1, 2 \times 2, 3 \times 3$

7 a dim$(C) = 3 \times 2$ **b** 6 **c** 9

8 a $I_2 = \begin{bmatrix} 1 & 0 \\ 0 & 1 \end{bmatrix}$ **b** $\begin{bmatrix} 0 \\ 0 \end{bmatrix}$

9 $\begin{bmatrix} 1002 \\ 1081 \\ 1095 \end{bmatrix}, 3 \times 1$

10 $\begin{bmatrix} 15 & 19 & 22 \end{bmatrix}$

11 $C = \begin{array}{c} \\ S10 \\ S12 \\ S14 \\ S16 \end{array} \begin{array}{ccc} D & J & S \\ \begin{bmatrix} 2 & 0 & 3 \\ 2 & 4 & 0 \\ 2 & 3 & 1 \\ 1 & 5 & 0 \end{bmatrix} \end{array}$

12 a $\begin{bmatrix} 531 & 324 \end{bmatrix}$ **b** $\begin{bmatrix} 0 & 324 \\ 448 & 0 \end{bmatrix}$

EXERCISE 2.4

1 E **2** B

3 a not defined; the dimensions of A are not equal to the dimensions of B.

b defined; $\begin{bmatrix} 4 & -16 \end{bmatrix}$ **c** defined; $\begin{bmatrix} -1 & 12 \\ 0 & 1 \end{bmatrix}$

d not defined; the dimensions of D are not equal to the dimensions of B.

4 a $\begin{bmatrix} 2 & 4 & 0 \\ -3 & 5 & 10 \\ -1 & 25 & 1 \end{bmatrix}$ **b** $\begin{bmatrix} 10 & 10 & 10 \end{bmatrix}$

c $\begin{bmatrix} -2 \\ -4 \end{bmatrix}$ **d** $\begin{bmatrix} -6 \\ 4 \\ 2 \end{bmatrix}$

5 a $\begin{bmatrix} -2 & 1 \end{bmatrix}$ **b** not defined **c** $\begin{bmatrix} -6 & 3 \end{bmatrix}$

6 a $\begin{bmatrix} -3 & 9 \\ 15 & 21 \\ 3 & 21 \end{bmatrix}$ **b** $\begin{bmatrix} -6 & 15 \\ 24 & 33 \\ 5 & 34 \end{bmatrix}$

c $\begin{bmatrix} -16 & 18 \\ 20 & 22 \\ 6 & 32 \end{bmatrix}$ **d** $\begin{bmatrix} 18 & -9 \\ 0 & 9 \\ -3 & -6 \end{bmatrix}$

7 a $x = 4, y = -4, z = 7$

b $x = 9, y = -1, z = 1$

c $x = 15, y = 17, z = 14$

d $x = 7, y = -1.25, z = 12$

8 a defined; the dimensions of the matrix are the same

$\begin{bmatrix} 13 & -12 \end{bmatrix}$

b not defined; the dimensions of the matrix are not the same.

c not defined; the dimensions of the matrix are not the same.

d not defined; the dimensions of the matrix are not the same.

9 a $\begin{bmatrix} 12 & 36 \\ 0 & 24 \end{bmatrix}$ **b** $\begin{bmatrix} 16 & 8 \\ 4 & 20 \end{bmatrix}$

c $\begin{bmatrix} 3 & 4 \\ 3 & 26 \end{bmatrix}$ **d** $\begin{bmatrix} 4 & 3 \\ 4 & -5 \end{bmatrix}$

10 a $d = 10$ **b** $d = 7$

11 a 2×2 as in order for the addition to be defined, it must be the same dimensions as the other two matrices.

b $W = \begin{bmatrix} 0 & 2 \\ -1 & 2 \end{bmatrix}$ **c** $A = \begin{bmatrix} -4 & 2 \\ -1 & -2 \end{bmatrix}$

EXERCISE 2.5

1 C

2 A

3 a defined; $\begin{bmatrix} -3 & -10 & 2 \\ 6 & 20 & -4 \\ 9 & 30 & -6 \end{bmatrix}$; 3×3

b defined; $[11]$; 1×1

c not defined; B has 3 columns and C has 2 rows and $3 \ne 2$

d defined; $\begin{bmatrix} 15 & 50 & -10 \\ 6 & 20 & -4 \end{bmatrix}$; 2×3

4 a defined; 1×1; $[54]$

b defined; 3×3; $\begin{bmatrix} 0 & 0 & 0 \\ 3 & 4 & 10 \\ 15 & 20 & 50 \end{bmatrix}$

c not defined; B has 1 column and C has 2 rows and $1 \ne 2$

d not defined; B has 1 column and D has 3 rows and $1 \ne 3$

e defined; 3×2; $\begin{bmatrix} 4 & 3 \\ 6 & 2 \\ -1 & 1 \end{bmatrix}$

f defined; 2×2; $\begin{bmatrix} 3 & 0 \\ -6 & -1 \end{bmatrix}$

g not defined; D is not a square matrix

5 a $\begin{bmatrix} 25.8 & 32 \\ 44.4 & 52.4 \end{bmatrix}$

b $\begin{bmatrix} 3904 & 3270 \\ 4360 & 4449 \end{bmatrix}$

c $\begin{bmatrix} -436 & -413.4 \\ -265.6 & -280.2 \end{bmatrix}$

6 a $\begin{bmatrix} 0 \\ 3 \end{bmatrix}$ **b** $\begin{bmatrix} -4 \\ 0 \end{bmatrix}$

7 a $\begin{bmatrix} 0 & 0 & 0 & 0 \\ 0 & 0 & 15 & 0 \\ 0 & 0 & 0 & 0 \\ 0 & 0 & 0 & 0 \end{bmatrix}$ **b** $\begin{bmatrix} 2 \\ 2 \\ 2 \\ 2 \end{bmatrix}$

8 a 4×2 **b** 3×2 **c** 4×4

9 a $I = \begin{bmatrix} 1 & 0 & 0 \\ 0 & 1 & 0 \\ 0 & 0 & 1 \end{bmatrix}$ **b** $I = \begin{bmatrix} 1 & 0 \\ 0 & 1 \end{bmatrix}$

10 a possible; 1×1 **b** possible; 2×2

c not defined; the first matrix has 1 column and the second matrix has 3 rows.

d possible; 4×2 **e** possible; 3×4

11 a A and D

b AC exists as the number of columns in A is 2 and the number of rows in C is 2. $AC = \begin{bmatrix} 19 & 40 \\ 41 & 88 \\ 27 & 56 \end{bmatrix}$

c DB and AB

d C, as it is a square matrix.

EXERCISE 2.6

1 D

2 B

3 a $C = \begin{bmatrix} 79 \\ 199 \\ 399 \end{bmatrix}$

b $S = 1.3 \begin{bmatrix} 79 \\ 199 \\ 399 \end{bmatrix} = \begin{bmatrix} 102.70 \\ 258.70 \\ 518.70 \end{bmatrix}$

c $R = S + \begin{bmatrix} 15 \\ 15 \\ 15 \end{bmatrix} = \begin{bmatrix} 117.70 \\ 273.70 \\ 533.70 \end{bmatrix}$

d $D = R - \begin{bmatrix} 20 \\ 45 \\ 45 \end{bmatrix} = \begin{bmatrix} 97.70 \\ 228.70 \\ 488.70 \end{bmatrix}$

4 a $N = \begin{bmatrix} 133 & 98 \\ 75 & 62 \end{bmatrix}$ **b** $C = \begin{bmatrix} 3 \\ 5 \end{bmatrix}$

c $NC = \begin{bmatrix} 133 & 98 \\ 75 & 62 \end{bmatrix}\begin{bmatrix} 3 \\ 5 \end{bmatrix} = \begin{bmatrix} 889 \\ 535 \end{bmatrix}$;

$889 in Week 1 and $535 in Week 2.

d i $T = \begin{bmatrix} 1060 & 1555 \\ 3029 & 1124 \\ 889 & 535 \\ 896 & 2130 \end{bmatrix}$

ii $S = 1.55 \begin{bmatrix} 1060 & 1555 \\ 3029 & 1124 \\ 889 & 535 \\ 896 & 2130 \end{bmatrix}$

$= \begin{bmatrix} 1643 & 2410.25 \\ 4694.95 & 1742.20 \\ 1377.95 & 829.25 \\ 1388.8 & 3301.5 \end{bmatrix}$

e $P = S - T = \begin{bmatrix} 1643 & 2410.25 \\ 4694.95 & 1742.20 \\ 1377.95 & 829.25 \\ 1388.8 & 3301.5 \end{bmatrix} - \begin{bmatrix} 1060 & 1555 \\ 3029 & 1124 \\ 889 & 535 \\ 896 & 2130 \end{bmatrix}$

$= \begin{bmatrix} 583 & 855.25 \\ 1665.95 & 618.20 \\ 488.95 & 294.25 \\ 492.80 & 1171.50 \end{bmatrix}$

f $\begin{bmatrix} 1 & 1 & 1 & 1 \end{bmatrix} \begin{bmatrix} 583 & 855.25 \\ 1665.95 & 618.20 \\ 488.95 & 294.25 \\ 492.80 & 1171.50 \end{bmatrix}$

$= \begin{bmatrix} 3230.70 & 2939.20 \end{bmatrix}$

$\begin{bmatrix} 3230.70 & 2939.20 \end{bmatrix} \begin{bmatrix} 1 \\ 1 \end{bmatrix} = \begin{bmatrix} 6169.90 \end{bmatrix}$

Therefore, $6169.90 profit made over the two week period.

5 $C = \begin{array}{c} \\ P \\ Q \\ R \\ S \\ T \\ U \end{array} \begin{bmatrix} P & Q & R & S & T & U \\ 0 & 1 & 0 & 0 & 0 & 0 \\ 1 & 0 & 1 & 0 & 1 & 0 \\ 0 & 1 & 0 & 0 & 0 & 0 \\ 0 & 0 & 0 & 0 & 1 & 0 \\ 0 & 1 & 0 & 1 & 0 & 1 \\ 0 & 0 & 0 & 0 & 1 & 0 \end{bmatrix}$

6 a Ahmed can send to Beth. Beth can send to Ahmed, Crystal and Daniella. Crystal can send to Daniella. Daniella can send to Ahmed and Crystal.

b Because a person cannot send a direct message to themselves; it is a redundant communication link.

c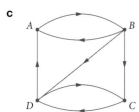

d Crystal can send a message to Daniella, and then Daniella sends the message to Ahmed.

7 $M^4 = \begin{bmatrix} 0 & 0 & 1 & 0 \\ 0 & 0 & 1 & 1 \\ 1 & 1 & 0 & 1 \\ 1 & 0 & 0 & 0 \end{bmatrix}^4 = \begin{bmatrix} 3 & 2 & 1 & 2 \\ 4 & 3 & 1 & 3 \\ 3 & 1 & 5 & 3 \\ 1 & 0 & 2 & 1 \end{bmatrix}$

The largest number of four-step communications occur from system B to system A (ignoring from C to C). There are 4 possible four-step communications.

8 a Because there is a '1' in the leading diagonal in the third row, third column.

b It is not a square matrix.

9 a $m_{23}^2 = 0$. There are no possible two-step communications from Beth to Casper. That is, there is no possible way Beth can send a message to Casper through exactly one other person.

b $m_{14}^2 = 1$. There is one possible two-step communication from Ariana to Dodge. That is, there is one possible way Ariana can send a message to Dodge through exactly one other person.

c $m_{33}^2 = 2$. There are two possible two-step communications from Casper back to Casper. That is, there are two possible ways Casper can receive a message back that he originally sent to exactly one other person.

10 a 4 × 4

b $C = \begin{array}{c} \\ S \\ T \\ U \\ V \end{array} \begin{bmatrix} S & T & U & V \\ 0 & 1 & 0 & 1 \\ 0 & 0 & 0 & 1 \\ 0 & 1 & 0 & 1 \\ 1 & 1 & 1 & 0 \end{bmatrix}$

c $C^2 + C = \begin{bmatrix} 1 & 2 & 1 & 2 \\ 1 & 1 & 1 & 1 \\ 1 & 2 & 1 & 2 \\ 1 & 3 & 1 & 3 \end{bmatrix}$

This matrix shows the total number of communications between two people that go through at most one other person. That is, the total number of one- and two-step communications.

11 a The total cost (in dollars) of the apples and bananas that Peter bought.

b $NC = \begin{bmatrix} 2.83 \end{bmatrix}$; that is, $2.83.

c $\begin{bmatrix} 3 & 4 \end{bmatrix} \begin{bmatrix} 1 \\ 1 \end{bmatrix} = \begin{bmatrix} 7 \end{bmatrix}$.

The total number of pieces of fruit that Peter bought.

12 a MP, as the columns of M and the rows of P both represent the type of bed (Classic or Deluxe).

b $MP = \begin{bmatrix} 174 & 252 & 420 \\ 249.75 & 364.50 & 553.50 \end{bmatrix}$

c $553.50

13 a 2 × 3

b The rows represent the different stores; that is, row 1 is Idla and row 2 is AGI. The columns represent the three different cheeses; that is, column 1 is Cheddar, column 2 is Gouda and column 3 is Blue.

c $W = \begin{bmatrix} 131.30 \\ 130.75 \end{bmatrix}$

It would cost $131.30 to purchase the cheeses from Idla or $130.75 to purchases the cheeses from AGI.

d The cheeses should be bought from AGI, as it costs 55 c less; however, the quality and brand of the cheeses at both stores should be considered.

14 a Ben and Elka

b Elka can receive Cheng's message from Amara or Dana.

15 a 0 ways **b** 7 ways

16 a $R = \begin{array}{c} \\ A \\ B \\ C \\ D \\ E \end{array} \begin{bmatrix} A & B & C & D & E \\ 2 & 1 & 0 & 0 & 1 \\ 1 & 0 & 1 & 0 & 0 \\ 0 & 1 & 0 & 1 & 1 \\ 0 & 0 & 1 & 0 & 2 \\ 1 & 0 & 1 & 2 & 0 \end{bmatrix}$

b The routes are two-way, meaning they can be travelled in two directions.

c $R^2 = \begin{array}{c} \\ A \\ B \\ C \\ D \\ E \end{array} \begin{bmatrix} A & B & C & D & E \\ 6 & 2 & 2 & 2 & 2 \\ 2 & 2 & 0 & 1 & 2 \\ 2 & 0 & 3 & 2 & 2 \\ 2 & 1 & 2 & 5 & 1 \\ 2 & 2 & 2 & 1 & 6 \end{bmatrix}$

$r_{11}^2 = 6$, meaning there are six different routes from A to A, passing through exactly one town. For example, A–A–A (4 different ways), A–B–A, A–E–A.

CUMULATIVE EXAMINATION: CALCULATOR-FREE

1 a 0   **b** 25
2 $V = 54\pi$
3 a $y = 1$ **b** $\begin{bmatrix} 5 \\ x \\ 5 \end{bmatrix}$

  c $x = 30$ **d** $x = 11, y = -7$

4 a $C = \begin{bmatrix} 3.50 & 4.20 & 1.90 \end{bmatrix}$

  b $N = \begin{bmatrix} 3 \\ 2 \\ 1 \end{bmatrix}$ **c** $CN = \begin{bmatrix} 20.80 \end{bmatrix}$; $20.80

CUMULATIVE EXAMINATION: CALCULATOR-ASSUMED

1 $576.25

2 a When $C = 0$, $4AC = 0$, and so $X = \dfrac{-4 + 4}{2A}$ which is always 0.

  b

X	C			
		-1	0	1	2
A	-1	0.27	0	-0.24	-0.45
	1	0.24	0	-0.27	-0.59
	2	0.22	0	-0.29	-1

3 a A person cannot follow themselves on the social media platform.
  b Dana follows Cassie.
  c $\begin{bmatrix} 1 & 2 & 1 & 1 \end{bmatrix}$.
  The matrix represents the number of followers within the group of four people, that each person has.

4 a 2×3
  b P has one column and Q has two rows and $1 \neq 2$.
  c $M = \begin{bmatrix} 145\,978 \\ 171\,848.50 \end{bmatrix}$.
  The total amount of money made at each shopping centre per month for selling products A, B and C.

CHAPTER 3

EXERCISE 3.1

1 a 0.436 81 km   **b** 7 000 000 cm³   **c** 50 000 cm²
  d 0.6 L   **e** 50 000 000 000 mm³
  f 4200 L   **g** 2500 cm²   **h** 0.68 m²
  i 1 500 000 mm

2 a 1 520 000 cm²   **b** 3 000 000 cm²   **c** 10 cm²
3 a 1 000 000   **b** 1000   **c** 0.001
4 a 4530 mm²   **b** 560 000 mm²
  c 7 100 000 mm²
5 a 0.56 L   **b** 12 670 L   **c** 8600 L

EXERCISE 3.2

1 D   **2** D
3 a 22.67 mm   **b** 27.20 cm   **c** 30.41 cm
  d 18.73 cm   **e** 5.66 cm   **f** 7.62 m

4 a 6.3 km   **b** 9.0 cm   **c** 98.6 m
5 a 125 m   **b** 566 cm   **c** 1825 m
6 19 m
7 a 21.9 cm   **b** 15.5 cm
8 17 m   **9** 20 m   **10** 8.06 m
11 2 of the pencils would not fit as the maximum length is 21.36 cm.
12 a 18.8 m   **b** 19 m
13 $\sqrt{25^2 - 15^2} = 20$ or $25^2 - 15^2 = 400$, $\sqrt{400} = 20$.
  Incorrect: $25^2 - 15^2 = \sqrt{400} = 20$.

14 a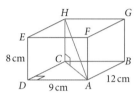

  b   **c** 15 cm

  d   **e** 17 cm

EXERCISE 3.3

1 C   **2** D
3 a i 3200 cm   **ii** 640 000 cm²
  b i 2500 cm   **ii** 348 600 cm²
  c i 63 cm   **ii** 314 cm²
  d i 1699 cm   **ii** 110 111 cm²
  e i 9400 cm   **ii** 3 750 000 cm²
  f i 754 cm   **ii** 45 239 cm²
  g i 8 cm   **ii** 3 cm²
  h i 80 cm   **ii** 360 cm²
  i i 2600 cm   **ii** 200 000 cm²
  j i 2712 cm   **ii** 424 000 cm²
  k i 780 cm   **ii** 31 500 cm²
  l i 160 cm   **ii** 1200 cm²
4 a i 3.67 m   **ii** 10.67 m
  iii 6.41 m²
  b i 202.32 mm   **ii** 454.32 mm
  iii 12 746.07 mm²
  c i 148.59 cm   **ii** 202.99 cm
  iii 2020.83 cm²
5 a i 94.00 cm   **ii** 500.00 cm²
  b i 36.57 m   **ii** 65.13 m²
  c i 11.60 m   **ii** 7.92 m²
  d i 68.00 m   **ii** 111.00 m²
  e i 99.78 m   **ii** 523.36 m²
  f i 66.00 cm   **ii** 208.00 mm²

6 a 97.0 m²   **b** 85.8 cm²
   c 2375.0 cm²   **d** 298.5 m²
   e 28.9 km²   **f** 4 426 820.3 mm²

7 a

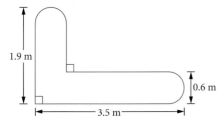

   b $2803   **c** $566

8 109.72 m   **9** 4200 cm²   **10** 35 cm²

11 a 1.5 m²   **b** 2.25 m²

12 a 8 m²   **b** 12.8 m

13 a $\sqrt{2.2^2 + 2.3^2} = 3.182... \approx 3.2$
   b 17.3 km
   c 17 km²

14 a 59.7 m
   b $2577
   c $597 (including gate)

EXERCISE 3.4

1 D   **2** C

3 a $V = 250$ cm³, $C = 250$ mL
   b $V = 2.0$ m³, $C = 2000$ L
   c $V = 30\,846$ cm³, $C = 30\,846$ mL
   d $V = 3036$ cm³, $C = 3036$ mL
   e $V = 17$ m³, $C = 17\,000$ L
   f $V = 127$ m³, $C = 127\,000$ L

4 a $V = 31\,000$ m³, $C = 31\,000\,000$ L
   b $V = 6568$ cm³, $C = 6568$ mL
   c $V = 2$ m³, $C = 2000$ L
   d $V = 248$ cm³, $C = 248$ mL
   e $V = 27$ m³, $C = 27\,000$ L
   f $V = 427$ cm³, $C = 427$ mL

5 a 1 243 547 mm³   **b** 1437 cm³
   c 8379 mm³   **d** 18 m³
   e 18 m³   **f** 1030 cm³

6 $\frac{1}{3} \times 20 \times 20 \times \sqrt{200}$

7 0.576 m³   **8** 1304 m³   **9** 643 cm³

10 24 cm   **11** 960 m³

12 a 4250 m²   **b** 1000 m³   **c** 90.6 m

13 a 0.048 m²   **b** 6.6 m³

EXERCISE 3.5

1 A   **2** A

3 a 2225.00 cm²   **b** 982.86 cm²   **c** 58 106.90 mm²
   d 1392.00 cm²   **e** 301.59 cm²   **f** 5400.00 cm²

4 a 30.22 cm²   **b** 24.13 cm²   **c** 510.00 cm²
   d 204.00 m²   **e** 384.8 cm²   **f** 1445.13 mm²

5 20 m²   **6** $\pi r^2 + 2\pi rh + \pi rs$

7 224 m²   **8** 67 cm²

9 39 042 cm²

10 a 5755 mm²   **b** 214 mm

11 a 2800 cm³   **b** 1055.6 cm²

12 a $OM = \sqrt{3.4^2 - 3^2} = 1.6$ m

   b Area of front face = area of rectangle + area of triangle

$$\text{Area} = l \times w + \frac{1}{2}bh$$
$$= 2.2 \times 6 + \frac{1}{2} \times 6 \times 1.6$$
$$= 18 \text{ m}^2$$

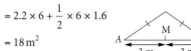

   c 180 m³
   d i 208 m²   **ii** 13 litres

CUMULATIVE EXAMINATION: CALCULATOR-FREE

1 a $C = \begin{bmatrix} 2.50 \\ 3.20 \\ 1.90 \end{bmatrix}$

   b $W = \begin{bmatrix} 2 & 1 & 3 \end{bmatrix}$

   c $WC = \begin{bmatrix} 2 & 1 & 3 \end{bmatrix} \begin{bmatrix} 2.50 \\ 3.20 \\ 1.90 \end{bmatrix} = \begin{bmatrix} 13.90 \end{bmatrix}$

   Sean's total fruit purchase costs $13.90.

2 2 × 1 000 000   **3** 10 m²   **4** 10 cm

5 a 6.5 m²   **b** $52   **c** 1500 L

6 256 mL

7 a 5 m × 24 m   **b** 520 m²   **c** 65 L

CUMULATIVE EXAMINATION: CALCULATOR-ASSUMED

1 a i 8   **ii** 7   **iii** 4
   iv 5   **v** $18.32   **vi** $27.48
   vii $27.48   **viii** $36.27   **ix** $146.56
   x $192.36   **xi** $109.92   **xii** $181.35

   b $630.19

2 a 62 m   **b** 854 m

3 762.78 mm

4 a 69 m
   b length $AB = \sqrt{23^2 - 11.5^2} = 19.9$ m
   c 229 m²   **d** 24 429 cm³
   e $h = \frac{1244}{36\pi} = 11$ cm   **f** 933 cm³

5 a 374.73 mm   **b** 180 744.59 mm²

6 a 2788 cm³   **b** 2.8 L

7 5730 litres

CHAPTER 4

EXERCISE 4.1

1. a i 0.6   ii 7.2 m
   b i 1.4   ii 40.0 m
   c i 0.625   ii 6.0 cm
   d i 0.8   ii 20.0 cm

2. a Similar because the shapes are squares so the scale factors of the matching sides will always be the same.
   b Similar because the shapes are rectangles and $\frac{14}{4} = \frac{7}{2} = 3.5$.
   c Not similar because not all the scale factors are the same; $\frac{12}{4} = 3$ and $\frac{12}{3} = 4$

3. a Similar because $\frac{30}{20} = \frac{15}{10} = 1.5$ and the angle between the two sides is 90° for both triangles (SAS). $\triangle ABC \sim \triangle DEF$
   b Similar because $\frac{12}{16} = \frac{9}{12} = \frac{6}{8} = 0.75$ (SSS). $\triangle MNO \sim \triangle QPR$
   c Similar because two matching angles are equal (AA); $\triangle XYZ \sim \triangle DFE$

4. a 0.75   b 2.7 cm

5. Triangle P and Triangle Q

6. a $\angle B = \angle B$, $\angle A$ and $\angle D$ are corresponding angles and therefore are equal. Similar due to AA, $\triangle ABC \sim \triangle DBE$.
   b 0.75   c 18 cm

7. a 0.8   b 7.2 cm

8. a scale factor = 1.25; Dimensions of second triangle are 65 cm, 60 cm, and 25 cm.
   b 150 cm

9. a $2\frac{2}{3}$ b 21.3 cm

EXERCISE 4.2

1. 130.5 cm   2 33 m
3. a i 1.5   ii 21.3 cm²   iii 3164.1 cm³
   b i 0.833   ii 33.3 cm²   iii 1620 cm³
4. 16   5 400 cm²
6. a 576 cm²   b 279 cm³
7. a 6 L   b 7.5 mL
8. a 2   b 271.25 cm²

EXERCISE 4.3

1. 27   2 90 cm²
3. a 3.85 m by 3.3 m   b $660.66
4. a 4.4 m by 4.3 m
   b i 18.92 m²   ii $1135.20
5. a Master bedroom is 6 m × 4.8 m, Second bedroom is 4.8 m × 3.6 m.
   b i 46.08 m²   ii $2304

6. a 8.4 m²   b 140 tiles
7. a 1:150
   b

   780 m / 400 m / 300 m / 1020 m

   c 57.63 ha
8. a 3:200 000   b 5.7 km   c 5.3 km

CUMULATIVE EXAMINATION: CALCULATOR-FREE

1. a The matrices have different orders and so cannot be added together.
   b Both matrix products are defined, as dim$(P) = 3 \times 1$ and dim$(Q) = 1 \times 3$. PQ would form a 3×3 matrix and QP would form a 1×1 matrix.
   c $PQ = \begin{bmatrix} 6 & -9 & -12 \\ 10 & -15 & -20 \\ -2 & 3 & 4 \end{bmatrix}$

2. a 6 cm³   b 8 700 000 cm³   c 5 600 000 000 cm³
3. a 2   b 28 cm   c 4
4. a 2   b 65 680 cm³
5. a 25 cm   b 45 cm, 60 cm and 75 cm   c 4
6. a 3.42 m by 3.4 m   b $584.82
7. a 1:75   b 22.4 m²   c $1120

CUMULATIVE EXAMINATION: CALCULATOR-ASSUMED

1. a $2250   b $40   c $6000
   d $12 500   e $6570   f $570
2. 45.4 cm
3. a 11.4 m²
   b i 2.25   ii 25.7 m²
4. a 13.4 m   b 111 m
5. 78.5 km²
6. a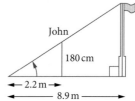

   b yes, triangles are similar due to AA; both have a right angle, and the angle the sun makes with John and the flagpole is a corresponding angle and therefore equal.
   c $\frac{8.9}{2.2} = 4.05$
   d 7.3 m
7. a 1:20   b 120 cm   c 144.2 cm

CHAPTER 5

EXERCISE 5.1

1

```
                    Data
              /            \
        Numerical        Categorical
         /    \           /      \
   Continuous  Discrete  Ordinal  Nominal
```

2
- **a** i numerical ii discrete
- **b** i numerical ii continuous
- **c** i categorical ii ordinal
- **d** i numerical ii continuous
- **e** i numerical ii continuous
- **f** i categorical ii ordinal
- **g** i categorical ii nominal
- **h** i categorical ii ordinal
- **i** i numerical ii discrete
- **j** i categorical ii ordinal
- **k** i numerical ii discrete
- **l** i categorical ii nominal
- **m** i numerical ii continuous
- **n** i categorical ii ordinal
- **o** i numerical ii continuous
- **p** i categorical ii nominal

3
- **a** i 6.5 screens
  - ii 4 and 5 screens
  - iii 13 screens
- **b** i 15 marks
  - ii 15 and 18 marks
  - iii 11 marks
- **c** i 8 min
  - ii 3 min
  - iii 17 min
- **d** i 6 cm
  - ii 5 cm and 6 cm
  - iii 2 cm

4 continuous numerical data

5 a categorial, nominal **b** there is no median

6 a categorical, nominal **b** categorical, ordinal
c numerical, discrete **d** numerical, continuous
e numerical, continuous

7 both categorical variables

8 a 7 **b** 8

9 a −4.4% **b** 8.4%

10 a i 23.6°C
  ii No mode because every data value occurs once.
  iii 9.9°C
- **b** i 24.0°C ii 24.0°C iii 10.0°C
- **c** Rounding early in a calculation makes a difference to your answers. 23.6°C in part **a** is the more accurate of the two medians because you lose accuracy if you round before the last step of a calculation.
- **d** The 24°C in part **b** gives the most helpful information about the mode. Data rounded to a large number of decimal places will often give no information about the mode because the level of accuracy is so great, data values rarely repeat.

EXERCISE 5.2

1 D **2** D

3 a true **b** false **c** true **d** false
e false **f** true **g** true

4 a i The modal interval is 40–50.
  ii range = 50
- **b** i The modal interval is 0–<5.
  ii range = 25
- **c** i The modal interval is 20–<30.
  ii range = 60

5 a 8 **b** 75 **c** 15.4%
d positively skewed with a possible outlier
e 60–<120 minutes

6 ClassPad

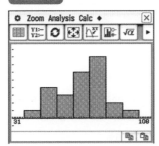

TI-Nspire

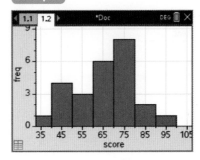

7 a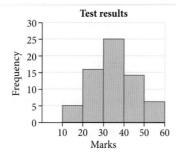

b The histogram is approximately symmetric.

8 a positively skewed **b** 54

9 a negatively skewed with a possible outlier
   b students who had more than 9 but less than 10 hours of sleep

10 a 14 countries **b** 16.4%

11 a 50
   b
   c 4 **d** 8%
   e Burger Heaven employs a greater proportion of younger people, particularly 15- to 24-year-olds.

EXERCISE 5.3

1 E **2** E

3 a i 2 **ii** 5 **iii** 2 **iv** 2.44
   b approximately symmetric

4 a i 31 **ii** 54 **iii** 34 **iv** 37.24
   b Positively skewed with 63 and 77 possible outliers

5 a 50 **b** Labor **c** 50
   d 480 **e** 31.25%

6 a positively skewed **b** negatively skewed
   c positively skewed **d** negatively skewed

7 a
Stem	Leaf
3	4 5 7 8
4	0 1 4 9
5	1 3 3 3 4 6 7
6	0 2 2 8
7	3 5 9

   Key: 6 | 8 = 68

   b 22 matches **c** 79 **d** 59%
   e approximately symmetric

8 a 19 **b** 16% **c** 38% **d** 29.5%

9 a north-west **b** 16%

10 a 1 **b** 2

11 a 26 **b** 36

12 a mode = 78 **b** range = 9

13 a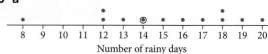

b i 15.5 **ii** 92%

14 a i 25.0 years **ii** 28.2 years
   b i 26.0 years **ii** 30.5 years

15 a 3 stars **b** 150 **c** 13%

EXERCISE 5.4

1 E **2** C

3 a i 13 **ii** 3.5
   b i 50 **ii** 2.9

4 $\bar{x} = 38.07$, $\sigma = 4.74$

5 $\bar{x} = 7.31$, $\sigma = 1.83$

6 $\bar{x} = 26.1$, median = 26, mode = 24, range = 11 and $\sigma = 3.0$

7 a i $\bar{x} = 19.27$, $\sigma = 1.98$ **ii** 7 data values
   b i $\bar{x} = 27.31$, $\sigma = 9.47$ **ii** 9 data values

8 a 170 g **b** 173 g

9 a 70% **b** 1.15

10 a $\bar{x} = 2.07$ **b** $\sigma = 1.53$

EXERCISE 5.5

1 D **2** B

3 a yes **b** no **c** no **d** yes

4 a 34% **b** 0.15% **c** 49.85% **d** 50%

5 a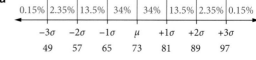

b 50% **c** 99.7% **d** 2.5% **e** 99.85%
f i 68 **ii** 290

6 a
```
 0.15% 2.35% 13.5%  34%   34%  13.5% 2.35% 0.15%
   -3σ   -2σ   -1σ    μ    +1σ   +2σ   +3σ
    12    15    18   21    24    27    30
```

b yes **c** no **d** yes **e** yes
f yes **g** yes **h** no **i** yes
j yes **k** yes

7 a 0.227 **b** 0.296

8 a k = 73.4 kg **b** a = 7.2 kg

9 a 68% **b** 81.5%

10 a 3400 **b** 250

11 a 84%
   b 68% since it is 1 standard deviation either side of the mean

12 a 50% **b** 68%

13 a 670 hours **b** 15 hours

14 a 8.55 hours **b** 248

15 a 3 cups **b** 25 cups
16 a 84% **b** 97.5%
17 a i 20°C **ii** 23.3%
 b 97.5%
18 a 2.5% **b** 2375 eggs

EXERCISE 5.6

1 A **2** D
3 a −2
 b $z = -2$ means the teacher's height is two standard deviations below the mean.

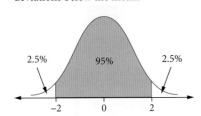

From the diagram, 97.5% of Year 12 teachers are taller than this particular teacher.

 c 160.65 cm
4 a The z-scores are 1.2, −0.5, 3, −1, −1.2, so the student's best subject was Hospitality.
 b Using the following diagram, the student was in the bottom 16% for Psychology and Systems Engineering.

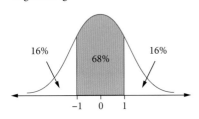

5 a −0.3 **b** 34%
6 a −1.1 **b** 82.3
7 a 59 beats per minute. **b** 84%
 c 37
8 a 1.8 **b** 87
9 a −0.8
 b i 16% **ii** 163

CUMULATIVE EXAMINATION: CALCULATOR-FREE

1 a $20/hr **b** 20 trees **c** $160
2 13 cm
3 a number of children
 b sex, type of car and postcode
 c 1
4 a 3.1 km² **b** 6.9 km²
5 a 92% **b** positively skewed
6 a 84%
 b 81.5%
 c $0.975 \times 456 = 444.6 \therefore 444$ people

d Brett and Sanjeeva are both below the mean weight for players in the basketball competition. Sanjeeva weighs more than Brett. Brett weighs more than 68 kg. More than 50% of the players in the basketball competition weigh more than Sanjeeva.
7 a actual birth weight = $3300 - 0.75 \times 550$
 b 0.15% **c** 489

CUMULATIVE EXAMINATION: CALCULATOR-ASSUMED

1 a 3×1
 b $1.05 \begin{bmatrix} 2800 \\ 1700 \\ 2400 \end{bmatrix} = \begin{bmatrix} 2940 \\ 1785 \\ 2520 \end{bmatrix}$
 c $\begin{bmatrix} 2800 \\ 1700 \\ 2400 \end{bmatrix} + \begin{bmatrix} 2940 \\ 1785 \\ 2520 \end{bmatrix} = \begin{bmatrix} 5740 \\ 3485 \\ 4920 \end{bmatrix}$
 $\begin{bmatrix} 1 & 1 & 1 \end{bmatrix} \begin{bmatrix} 5740 \\ 3485 \\ 4920 \end{bmatrix} = \begin{bmatrix} 14145 \end{bmatrix}$
 d $\begin{bmatrix} 5 & 6 & 8 \end{bmatrix} \begin{bmatrix} 5740 \\ 3485 \\ 4920 \end{bmatrix} = \begin{bmatrix} 88970 \end{bmatrix}$. Therefore, the cost of all imports is $88 970
2 a 38.8 m **b** $466.19
3 a 0.5 **b** 75 g
4 a type of mammal
 b mean = 9.2, standard deviation = 4.1
 c 31.6% **d** 5.4 hours
5 a i 17.8 mm **ii** 0 mm
 b i 16 days **ii** 10%
 c

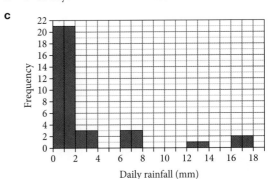

6 a 38 cm
 b i 1 **ii** 1
7 a positively skewed
 b i 26 **ii** $\frac{20}{103} \times 100 = 19.4\%$
8 a i mean = 39.43, standard deviation = 9.47
 ii 9 data values
 b 68% **c** 64.3%

CHAPTER 6

EXERCISE 6.1

1 a min = 35, Q_1 = 45.5, median = 60, Q_3 = 70.5, max = 80

b

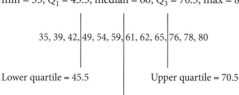

  i $\dfrac{3}{12} = 25\%$ **ii** $\dfrac{6}{12} = 50\%$ **iii** $\dfrac{9}{12} = 75\%$

2 a lower fence: $Q_1 - 1.5 \times$ IQR = 78.5 − 1.5 × 17.5
  = 52.25

  upper fence: $Q_3 + 1.5 \times$ IQR = 96 + 1.5 × 17.5
  = 122.25

  52 is less than 52.25, so it's a possible outlier.
  110 isn't greater than 122.25, so it's *not* an outlier.

b lower fence: $Q_1 - 1.5 \times$ IQR = 21.5 − 1.5 × 5.5
  = 13.25

  upper fence: $Q_3 + 1.5 \times$ IQR = 27 + 1.5 × 5.5
  = 35.25

  9 is less than 13.25, so it's a possible outlier.
  33 isn't greater than 35.25, so it's *not* an outlier.
  35 isn't greater than 35.25, so it's *not* an outlier.

3 a 2   **b** 1   **c** 4   **d** 3

4 a **i** 34   **ii** 29   **iii** 42.5   **iv** 13.5

  b yes; upper fence = 62.75 so 63 and 77 are possible outliers

5 lower fence: $Q_1 - 1.5 \times$ IQR = 37 − 1.5 × 12.5 = 18.25
  upper fence: $Q_3 + 1.5 \times$ IQR = 49.5 + 1.5 × 12.5 = 68.25
  12 is less than 18.25, so it's a possible outlier.
  23 isn't less than 18.25, so it's not an outlier.
  71 is greater than 68.25, so it's a possible outlier.

6 a **i** 30   **ii** 15   **iii** 40   **iv** 25

  b upper fence = $Q_3 + 1.5 \times$ IQR = 40 + 1.5 × 25 = 77.5
  60 < 77.5, so 60 is **not** an outlier.

7 a 10

  b lower fence = $Q_1 - 1.5 \times$ IQR = 90 − 1.5 × 10 = 75
  60 < 75, so 60 is an outlier.

8 a 12

  b lower fence = $Q_1 - 1.5 \times$ IQR = 84.5 − 1.5 × 12 = 66.5
  60 < 66.5, so 60 is an outlier.

EXERCISE 6.2

1 median = 121.5   **2** IQR = 27.5

3 a ClassPad

b ClassPad

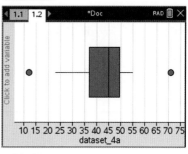

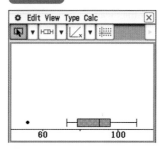

c ClassPad

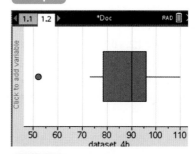

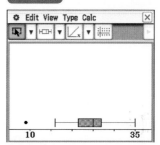

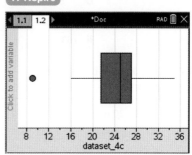

4 a **i** min = 4, Q_1 = 5, median = 6, Q_3 = 7, max = 9
  ii 25% **iii** 100% **iv** 25% **v** 15
  vi scores less than 2   **vii** scores greater than 10

b **i** min = 5, Q_1 = 7, median = 9, Q_3 = 10, max = 14
  ii 75% **iii** 75% **iv** 25% **v** 0
  vi scores less than 2.5   **vii** scores greater than 14.5

c **i** min = 3, Q_1 = 5, median = 7, Q_3 = 9, max = 10
  ii 50% **iii** 100% **iv** 25% **v** 15

vi No scores at the lower end would be considered outliers.

vii No scores at the upper end would be considered outliers.

5 a positively skewed; the box and whisker in the positive direction are longer than the box and whisker in the negative direction; no outliers

b approximately symmetric; median approximately in the middle of the box and whiskers are about the same length; one outlier shown by dot

c negatively skewed; the box and whisker in the negative direction are longer than the box and whisker in the positive direction; two outliers shown by dots

6 a positively skewed with outliers

b approx. 455 seconds   **c** 20

7 a median age = 38   **b** IQR = 10 years

8 a 12   **b** 15   **c** positively skewed

d 50%   **e** 90

9 a $Q_3 = 52$ **b** IQR = 52 − 28 = 24   **c** 76

d The upper fence is $Q_3 + 1.5 \times \text{IQR} = 52 + 1.5 \times 24 = 88$. An outlier is a value greater than the upper fence. The dot shown is 76 which is less than 88, meaning it should not be drawn as an outlier.

EXERCISE 6.3

1 D   **2** C

3 a

Camera 1		Camera 2
9 8 8 4	6	
6 5 5 4 3 2	7	
9 8 4 4 3 2 2	8	4 8
5 1	9	2 6 6 7
5	10	0 0 3 9 9 9
	11	0 0 1 4 5 6 8 9
	12	
	13	
Key: 4 | 8 = 84 km/h	14	Key: 8 | 4 = 84 km/h

Other numbers can be used as keys.

b negatively skewed

c Camera 1: median = 79 km/h, range = 41 km/h, IQR = 13.5 km/h

Camera 2: median = 109 km/h, range = 35 km/h, IQR = 16 km/h

d The second road has the greater speeding problem. The median speed for the first road is 79 km/h, which is 1 km/h below the 80 km/h speed limit. The median speed for the second road is 109 km/h, which is 9 km/h above the 100 km/h speed limit.

4 a

Irina		Steven
	1	
	1	6
4 2	2	2
9 9 8 7	2	5 6
3 3 2 2	3	0 2 2 3
7 6	3	8 8 9 9
Key: 2 | 3 = 32 pamphlets		Key: 3 | 2 = 32 pamphlets

Other numbers can be used as keys.

b negatively skewed

c Irina: median = 30.5, range = 15, IQR = 5.5

Steven: median = 32, range = 23, IQR = 12.5

d Although the medians (30.5 and 32) are similar, Irina's range (15) and IQR (5.5) are considerably lower than Steven's range (23) and IQR (12.5). This means Steven's deliveries have more variability and are less consistent than Irina's deliveries, which indicates that Irina is the better delivery person.

5 ClassPad

TI-Nspire

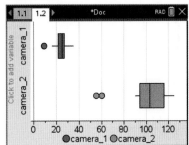

6 a Colebrook   **b** Ashville

c Ballinga   **d** Ballinga

e Colebrook has noticeably higher average July temperatures than Ashville and Ballinga. Colebrook's median (26°C) is much higher than Ashville's (10°C) and Ballinga's (12°C).

7

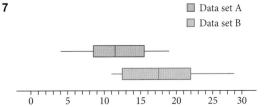

Data set A has a median of 11.5 compared to set B having a median of 17.5. Set A has a much lower minimum of 4 compared to 11 for set B. The maximum of set B is also much higher at 28 compared to 19 for set A. Both data sets have a similar range of 15 and 17 respectively. While the IQR is slightly higher for set B at 9.5 compared to 7.5 for set A, both sets are approximately symmetrical and neither set has an outlier. The Q_1 value for set B is 12.5 compared to the median of 11.5 for set A. This indicated that more than 75% of the data for set B is greater than 50% of the data for set A.

8 a 15%   **b** 6

c The female smoking rate is lower and less variable than the smoking rates for males.

9   The inner suburbs has an outlier of 10 000, while the outer suburbs has no outlier. The median for the outer suburbs is lower at 1800 compared to a median of 4200 for the inner suburbs. Comparing the median of the outer suburbs with the minimum value for the inner suburbs, at 1500, we can see that almost half of the outer suburbs have a lower population than any of the inner suburbs. The shape of the distribution for the outer suburbs is positively skewed compared to symmetrical for the inner suburbs, indicating that there is more variability for the outer suburbs.

10 As conditions change from hot to mild to cool, the rate of growth for these trees.

    increases on average and becomes less variable.

11 a The wind direction with the lowest recorded wind speed was south-east.

       The wind direction with the largest range of recorded wind speeds was north-east.

    b north-east

    c north-west, south-west and west

    d north-west and south-west

EXERCISE 6.4

1 B   2 E

3
	A	B
Mean	3	3.5*
σ	2.2	1
Minimum	1	1
Q_1	1	2.5
Median	2.5	3.5
Q_3	4	4.5
Maximum	10	6
IQR	3	2

*estimated

A is positively skewed, while B is approximately symmetrical. The mean is similar for both but A has a larger range, IQR and standard deviation, suggesting that the data is more spread out. The minimum is the same for both sets but the Q_1 for B is the same as the median for A. This shows that the first quartile is clustered and the second quartile is more spread out for A compared to B as can be seen from the graphs. Set A has a possible outlier. (Set A: upper fence = 4 + 1.5 × 3 = 8.5, so 10 is a possible outlier because 10 > 8.5.)

4 Possible answer below:

The mean for set B is higher at 27.3, while set A has a mean of 18.6. The median is also significantly higher in set B at 26.5, compared to 18 for set A. The range for set B is larger with 31, while set A is only 5. Data set B also has a higher IQR of 16.5, compared to 2.5 for set A, suggesting that set B is more spread out with higher values. Neither set has outliers or gaps in the data.

5 Possible answer below:

The mean and standard deviation are both lower for group 2 at 2.88 and 1.81 respectively compared to 3.46 and 2.44. The minimum value of 0 and the median of 3 are the same for both groups but group 1 has a higher max at 8 compared to group 2. The IQR is also higher for group 1 at 4, compared to 3 for group 2. Group 2 is approximately symmetrical while group 1 is positively skewed. There were no outliers for either group but group 1 had a gap for 1 movie, this could be due to the lower number of students in this group compared to group 2. Group 1 also had a higher proportion of students that streamed more movies than group 2.

6
	Class 1	Class 2
Mean	64.5	75.25
σ	6.02	6.63
Minimum	52	62
Q_1	62	72.5
Median	64	75
Q_3	70.5	81
Maximum	72	85
IQR	8.5	8.5

Class 2 performed significantly better than class 1. The mean was higher at 75.25 for class 2, compared to 64.5 for class 1. The standard deviation, IQR and range are similar for both, suggesting that the spread of scores are similar for both classes. The Q_1 for class 2 is 72.5 compared to the maximum scores of 72 for class 1, showing that 75% of the students in class 2 outperformed all the students in class 1.

7
	Group 1	Group 2
Mean	188.1	207.9
σ	26.74	29.7
Minimum	152	158
Q_1	161.5	185
Median	185	205
Q_3	215	236.5
Maximum	230	256
IQR	53.5	51.5

Group 2 has a faster reaction time. The mean is higher for group 2 at 207.9 compared to 188.1 for group 1. The standard deviation for group 2 is slightly higher at 29.7 compared to 26.7 for group 1, while the IQR is slightly lower at 51.5 for group 2 compared to 53.5 for group 1. The range is also larger for group 2 at 98 compared to 78 for group 1. This suggests that the upper and lower quartiles are more spread out for group 2. The Q_1 for group 2 is 185, compared to the same value as the median for group 1, showing that 75% of the reaction times in group 2 are faster than 50% of the reaction times in group 1.

EXERCISE 6.5

Check with teacher

CUMULATIVE EXAMINATION: CALCULATOR-FREE

1 $120

2 a 1×3

b i $\begin{bmatrix} 2 & -2 & 2 \end{bmatrix}$ **ii** $\begin{bmatrix} 2 & 1 \\ 3 & 1 \end{bmatrix}$

c $\begin{bmatrix} 7 \\ 14 \end{bmatrix}$, $m = 7$. $\begin{bmatrix} -3 \\ -6 \end{bmatrix}$, $n = -3$.

d $AB = \begin{bmatrix} -2 & 2 & -2 \\ -4 & 4 & -4 \end{bmatrix}$

3 Yes, it is a right-angled triangle because Pythagoras' theorem holds true, $9^2 + 12^2 = 15^2$.

4 1600 cm^2

5 a 2%

b approximately symmetrical

c 55-60 and 60-65

d 60

6 a numerical and categorical variables respectively.

b 23°C

c approximately symmetric with possible outliers.

7 a 1980 **b** 1990 and 2000 **c** $15/hr

8 a 26 min **b** The mean would decrease.

9 a Outliers at 31 and 44.

b
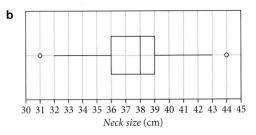
Neck size (cm)

10
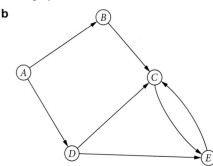

CUMULATIVE EXAMINATION: CALCULATOR-ASSUMED

1 $10 600

2 a Cynthia only needs to communicate with Elina for the project.

b

c Alison, as nothing can be delegated to her.

d $M^2 = \begin{bmatrix} 0 & 0 & 2 & 0 & 1 \\ 0 & 0 & 0 & 0 & 1 \\ 0 & 0 & 1 & 0 & 0 \\ 0 & 0 & 1 & 0 & 1 \\ 0 & 0 & 0 & 0 & 1 \end{bmatrix}$

e Cynthia can be delegated a role from Benji, Daithi or Elina, which all can be delegated from Alison.

3 192 cm^2

4 Scale factor = 1.375, dimensions of second triangle = 55 cm, 48.125 cm and 23.375 cm.

5 a 31

b 61%, since $\frac{49 + 45}{153} \times 100 = 61.4379...$

6 a 30.8 mm

b upper fence = $31.6 + 1.5 \times 1.3 = 33.55$

Since $33.5 < 33.55$, 33.5 mm is not an outlier.

7 a estimated: 10 cm; actual: 26 cm

b estimated: 34 cm; actual: 11 cm

c estimated: 21 cm; actual: 4 cm

d estimated: positively skewed

actual: approximately symmetrical

e The estimates median of 10 cm is considerably less than the actual median of 26 cm. This supports Edgar's theory that many people tend to underestimate the length of a piece of string.

8 a i 25.0 years **ii** 28.2 years

b 1.1 years

c $Q_1 - 1.5 \times IQR = 29.9 - 1.5 \times 1.1 = 28.25$

$26.0 < 28.25$, so the age of 26.0 would be shown on a box plot as an outlier.

9 a 3 **b** 84%

c The maximum value is equal to Q_3.

d 45.88 metres

10 a $Q_1 = 148$, $Q_2 = 151.5$, $Q_3 = 159$

IQR = 159 − 148 = 11

b Yes, 179 cm is an outlier.

CHAPTER 7

EXERCISE 7.1

1 a $\frac{7}{25} = 0.28$; the length of the opposite side is 28% of the hypotenuse.

b $\frac{24}{25} = 0.96$; the length of the adjacent side is 96% of the hypotenuse.

c $\frac{7}{24} = 0.2917$; the length of the opposite side is 29.17% of the adjacent side.

2 a $\beta = 62°$

b $\alpha = 17°$

Answers 421

3 a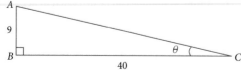

b 41

c $\sin\theta = \dfrac{9}{41}, \cos\theta = \dfrac{40}{41}$

4 a $\tan 49° = \dfrac{d}{18}, d = 18\tan 49° = 20.71$

b $\cos 50° = \dfrac{u}{11}, u = 11\cos 50° = 7.07$

c $\tan 50.6° = \dfrac{p}{11}, p = 11\tan 50.6° = 13.39$

d $\tan 38° = \dfrac{12.4}{d}, d = \dfrac{12.4}{\tan 38°} = 15.87$

e $\cos 65° = \dfrac{22.1}{q}, q = \dfrac{22.1}{\cos 65°} = 52.29$

f $\sin 30° = \dfrac{24.6}{y}, y = \dfrac{24.6}{\sin 30°} = 49.20$

5 a 7.63 b 17.27 c 11.01

6 a 64° b 42° c 39° d 38°
  e 84° f 19° g 34° h 41°

7 a $\dfrac{4}{5}$ b $\dfrac{3}{5}$

c $\dfrac{4}{3}$, $\tan\theta = \dfrac{8}{6} = \dfrac{4}{3}$. $\tan\theta = \dfrac{\sin\theta}{\cos\theta}$

8 a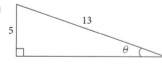

b $\cos\theta = \dfrac{12}{13}$

9 a $\tan\theta = 0.5882$; the length of side *LM* is 58.82% of the side *KM*.

b $\theta = 30.5°$

10 a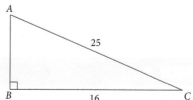

b $AB = 19.21$ cm

c $\angle ACB = 50.2°, \angle BAC = 39.8°$

EXERCISE 7.2

1 D 2 A 3 2.5 m

4 25.9° 5 $\sin^{-1}\left(\dfrac{0.6}{1.4}\right)$ 6 43 m

7 7 km 8 132.8 m 9 11.3°

10 12°

11 a 2.06 m b 3.04 m c 5.43 m

12 18.1°

EXERCISE 7.3

1 B 2 E

3 49°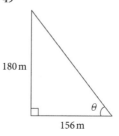

4 109 m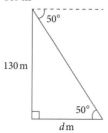

5 a 7.1° b 403.1 m

6 a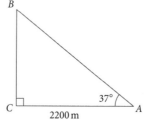

b 1658 m

7 a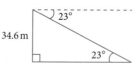

b 81.5 m

8 a 1469 m b 1232 m

9 34.8°

EXERCISE 7.4

1 70 m 2 49°

3 a 043° b 138° c 113° d 345°
  e 083° f 308° g 026° h 206°

4 a 143° b 25 km

5 a N 43° E b S 42° E c S 67° E
  d N 15° W e N 83° E f N 52° W
  g N 26° E h S 26° W

6 a N 54° W b S 54° E

7 a 132° b N 48° W

c No, as no lengths are known and it cannot be assumed that *Q* is due east of *T*.

8 a

b i N 45° W ii 315°

9 a 82 m   b 57 m   c N 55° W
10 a 25 m   b 4.8°
11 a 80 m   b 066°
12 a

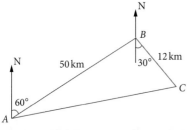

  b The bearing of B from A is N 60° E meaning that the bearing of A from B is S 60° W, by alternate angles. The bearing of C from B is S 30° E and so 60 + 30 = 90°.
  c 51 km
  d 253°

EXERCISE 7.5

1 A   2 E
3 a 82.6 cm²   b 23.8 cm²   c 73.2 cm²
   d 127.9 m²   e 31.3 m²
4 a 21 m²   b 9 cm²   c 756 mm²
5 a $A = \frac{1}{2}ab\sin C$.
      Two sides and the included angle are given.
   b 25°
6 If $C = 90°$, then $A = \frac{1}{2}ab\sin 90° = \frac{1}{2}ab \times 1 = \frac{1}{2}ab$.
   This is the same as the area formula, $A = \frac{1}{2}bh$, where b is the base and a is the height, h.

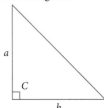

7 a Because the angle of 38° is not the included angle between the two sides labelled 8 cm and 6 cm.
   b $C = 180 - 55 - 38 = 87°$.
      $A = \frac{1}{2}(8)(6)\sin(87°) = \frac{48}{2}\sin(87°) = 24\sin(87°)$.
8 a 16.3 m²   b 185 m³
9 a 60 m   b 7.2 km
10 a $P = 42 + x$
   b $s = 21 + \frac{x}{2}$
   c $A = \sqrt{\left(21+\frac{x}{2}\right)\left(3+\frac{x}{2}\right)\left(\frac{x}{2}-3\right)\left(21-\frac{x}{2}\right)}$
   d $x = 12$ cm

EXERCISE 7.6

1 D   2 B
3 a 7.47   b 4.12   c 9.58
4 47°
5 a 19°   b 55°
6 a 9.6   b 15.8   c 10.7
7 a 123°   b 36°
8 a 11   b 91°
9 177.68 m
10 a 10 km   b 229°   c 62 km²
11 6.4 km
12 a

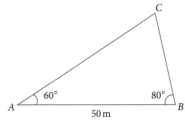

   b 77 m
13 237 m
14 a

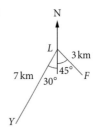

   b 75°
   c 6.87 km
15 118.2°
16 3 m
17 a 142 m   b 087°
18 a 33 km   b 57 km
   c i S 16° E   ii 164°
19 a 42.7°
   b $NT^2 = 10^2 + 13^2 - 2(10)(13)\cos(65°)$
      $NT^2 = 159.11 …$
      $NT = 12.61$
      $NT \approx 12.6$ m
   c 69°
20 a 3.23 km   b S 38° E   c 322°
21 a 475 m
   b i 908 m
      ii $\cos\theta = \frac{950^2 + 908^2 - 1400^2}{2(950)(908)}$
          $\theta = \cos^{-1}\left(\frac{950^2 + 908^2 - 1400^2}{2(950)(908)}\right) = 97.75°$
          $97.75 - 60 = 37.75°$
          S 38° E or 142°

CUMULATIVE EXAMINATION: CALCULATOR-FREE

1 a 600   **b** B2 = 1.5(C1 − A1)

2 Both variables are categorical.

3 a 29   **b** 26   **c** 37   **d** 11

4 a S 80° W   **b** 146°

CUMULATIVE EXAMINATION: CALCULATOR-ASSUMED

1 a $2044   **b** $6440

2 a Rhianna scored 8 points in match J.

   b 3 players scored no points in some of their matches.

   c $PU = \begin{bmatrix} 14 \\ 21 \\ 15 \\ 17 \\ 10 \end{bmatrix}$.

Each row of the matrix PU shows the total amount of points scored by each individual across the five matches.

3 a $24.48 \, m^2$ **b** $612

4 lower fence = $Q_1 − 1.5 \times IQR = 26 − 1.5 \times 11 = 9.5$.
There are no outliers on the lower end of the data.
upper fence = $Q_3 + 1.5 \times IQR = 37 + 1.5 \times 11 = 53.5$
60 > 53.5, so 60 is an outlier.

5 a 123.3   **b** 56.5   **c** 32.0

6 a 65°   **b** 158 m

7 a 246 m   **b** $16\,278 \, m^2$

CHAPTER 8

EXERCISE 8.1

1 a −2   **b** 15   **c** 9

   d 4   **e** 4   **f** $\frac{7}{2}$

2 a −5   **b** −3   **c** 4

   d $-\frac{3}{2}$ **e** $\frac{7}{2}$ **f** −5

3 a $\frac{1}{2}$ **b** 12   **c** −4

   d 7   **e** −1   **f** $-\frac{25}{18}$

4 a $\frac{7}{5}$ or 1.4   **b** $-\frac{3}{2}$ or −1.5   **c** 7

   d 1.62   **e** 3.07   **f** −0.92

   g 1.74   **h** −1.92   **i** 1.56

5 a −7   **b** $-\frac{2}{3}$ **c** −2

   d 10   **e** 4   **f** 11

6 a 2   **b** $-\frac{10}{3}$ **c** 0   **d** −1

   e $\frac{22}{7}$ **f** $\frac{4}{3}$ **g** 4   **h** 10

7 a 2   **b** $-\frac{10}{3}$ **c** 0   **d** −1

   e $\frac{22}{7}$ **f** $\frac{4}{3}$ **g** 4   **h** 10

8 a 2   **b** 5   **c** 12

9 a 16   **b** −22.92   **c** 5.42

EXERCISE 8.2

1 a 5   **b** 6

2 a $\frac{9}{5}$ or 1.8   **b** 4   **c** $\frac{30}{11}$ or 2.73

3 The three numbers are 12, 13 and 14.

4 The four numbers are 15, 17, 19 and 21.

5 The number is −4.

6 The number is −1.

7 The area is $675 \, m^2$.

8 The ice cream costs $3.50 each and the medium popcorn costs $6.50 each.

9 The number of male students is 15.

10 Martha is currently 46 years old and Alex is currently 22 years old.

11 The father is currently 40 years old and the daughter is currently 20 years old.

12 a Let x = the first of the three consecutive even numbers

$x + (x + 2) + (x + 4) = 42$

The three numbers are 12, 14 and 16.

   b Let x = the number

$\frac{1}{3}(x − 3) = 39$

The number is 120.

13 a Let p = Mark's weekly wage

$\frac{p}{4} + \frac{p}{3} + 210 = p$

Mark's weekly pay is $504.

   b Let a = Heather's current age

$\frac{1}{2}(a + 2) + \frac{1}{3}(a − 3) = 20$

Heather is currently 24 years old.

EXERCISE 8.3

1 $18.24\,\text{m}^2$

2 a i $y = -5, y = -2, y = 1, y = 4, y = 7$

ii
x	-2	-1	0	1	2
y	-5	-2	1	4	7

iii
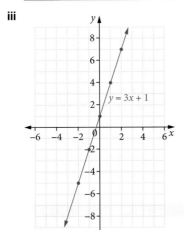

b i $y = 8, y = 6, y = 4, y = 2, y = 0$

ii
x	-2	-1	0	1	2
y	8	6	4	2	0

iii
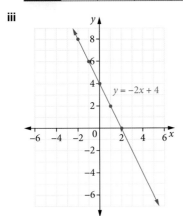

c i $y = 11, y = 7, y = 3, y = -1, y = -5$

ii
x	-2	-1	0	1	2
y	11	7	3	-1	-5

iii
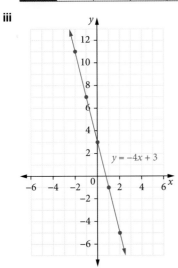

d i $y = -10, y = -8, y = -6, y = -4, y = -2$

ii
x	-2	-1	0	1	2
y	-10	-8	-6	-4	-2

iii
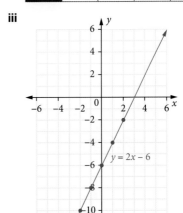

3 a ClassPad
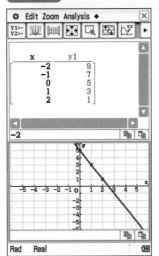

TI-Nspire
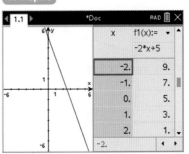

b ClassPad

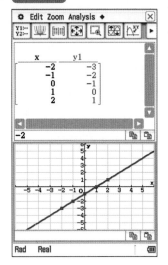

TI-Nspire

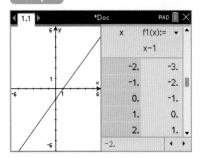

c ClassPad

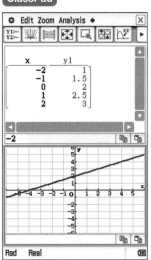

TI-Nspire

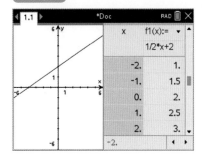

d ClassPad

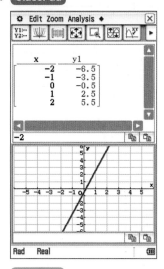

TI-Nspire

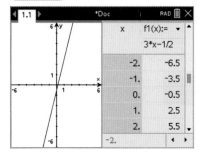

4 a $(-1, 5)$ doesn't lie on the line $y = 15x - 5$ as $5 \neq -15 - 5$.

b $(-1, 5)$ doesn't lie on the line $y = -5x - 1$ as $5 \neq 5 - 1$.

c $(-1, 5)$ lies on the line $y = -10x - 5$ as $5 = 10 - 5$.

5 a i negative   **ii** $-\dfrac{1}{2}$

  b i negative   **ii** $-\dfrac{5}{2}$

  c i negative   **ii** $-\dfrac{1}{4}$

  d i positive   **ii** 2

  e i zero

  f i not defined

6 a -2 **b** -1 **c** 3

7 a $y = 2x + 3$ **b** $y = 3x - 1$ **c** $y = 4x$

  d $y = -2x + 2$ **e** $y = 1$

8 It is linear because the power of the variable x is 1.

9 $\dfrac{1}{2}$ **10** $y = 2$

11 a gradient = $\frac{1}{2}$; y-intercept = 2

b

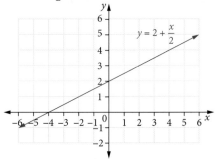

12 a −2 **b** (2, 0) and (0, 4) **c** $y = -2x + 4$

13 $x = 8$ **14** Point C

EXERCISE 8.4

1 $y = x + 2$

2 a slope = 5, y-intercept = −2

b The point (7, 32) does not lie on the line because $32 \neq 5 \times 7 - 2$.

3 a 1000 **b** 3 **c** $C = 1000 + 3k$
d $1000 **e** $3 **f** $13 600
g 3667 kilolitres

4 a 184 chirps/min **b** 8 chirps/min
c 4°C is an unrealistic temperature for an Australian summer.

5 a revenue = 5.5n **b** profit = 1.5n − 110
c $190 **d** loss of $35
e 474; The profit needs to be *at least* $600 and selling 473 key rings won't quite make that profit. So, the number of key rings that need to be sold is 474.

6 $C = 250t + 500$ **7** $V = 5000 - 2t$

8 a 12.5 L/min **b** $V = 2000 - 12.5t$

9 $1980 **10** 52

11 $C = 0.02E + 30$

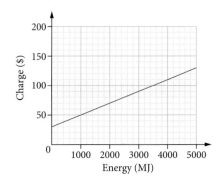

12 a 18 000 yen **b** 90 yen

13 a Substituting (20, 567) into the equation gives $567 = M \times 20$. Solving this equation gives $M = 28.35$.
b 5.67 g **c** 8 ounces

14 a profit = $24n - (6n + 260) = 18n - 260$
b 43 students

15 a $C = 80t + 22$ **b** C **c** $14.62
d 22; starting cost of call (at 0 min)
e $2.40 **f** 7 min

CUMULATIVE EXAMINATION: CALCULATOR-FREE

1 19.5 cm²

2 a 25%
b The box plot is approximately symmetrical with two outliers at $72 000 and $78 000.

3 a Line 1: $y = 4$
Line 2: $y = -3x + 7$
Line 3: $y = 7x + 12$
Line 4: $x = -3$
b $y = -6x - 2$

CUMULATIVE EXAMINATION: CALCULATOR-ASSUMED

1 $2875

2 $1.3 \begin{bmatrix} 230 \\ 290 \\ 310 \end{bmatrix} + \begin{bmatrix} 20 \\ 20 \\ 20 \end{bmatrix} = \begin{bmatrix} 319 \\ 397 \\ 423 \end{bmatrix}$

3 2869 cm³ **4** 150 cm³

5 a 3.48 m by 4.56 m **b** $872.78

6 a 6.8 **b** 1.86 **c** 15

7 a 75% **b** 21, 27.4, 28.7, 30, 35.9

8 a $A = \frac{1}{2}ab\sin C$
$= \frac{1}{2}(12)(13)\sin(30°)$
$= \frac{1}{2}(12)(13)\left(\frac{1}{2}\right)$
$= 39 \text{ unit}^2$

b $A = \sqrt{s(s-a)(s-b)(s-c)}$
$a = 4, b = 6, c = 8$
$s = \frac{4+6+8}{2} = \frac{18}{2} = 9$
$A = \sqrt{9(9-4)(9-6)(9-8)} = \sqrt{9531} = \sqrt{135}$
$\sqrt{121} = 11$ and $\sqrt{144} = 12$
$\sqrt{121} < \sqrt{135} < \sqrt{144}$
∴ $11 < A < 12 \text{ units}^2$

9 $\cos(38°) = \frac{x}{34}$
$x = 34\cos(38°)$
$x = 26.8 \text{ m}$

10 a Grace is currently 29 years old and Marcia is currently 5 years old.
b i $W = -5t + 80$ **ii** 44 g
iii 11 hours **iv** 16 hours

CHAPTER 9

EXERCISE 9.1

1 a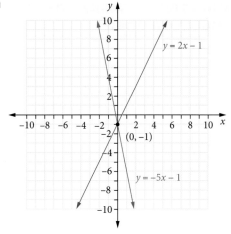

b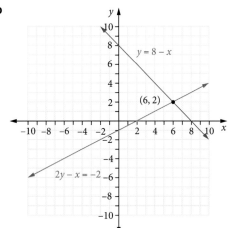

c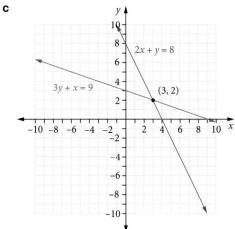

d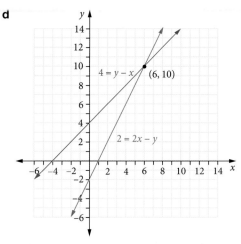

2 a $x = -1, y = 2$

ClassPad

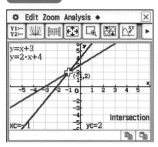

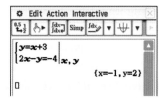

TI-Nspire

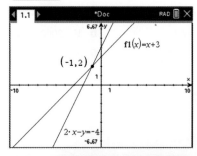

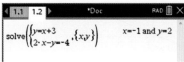

b $x = -2, y = 4$

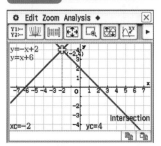
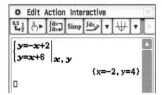
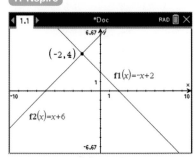
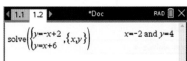

c $x = 1, y = 1$

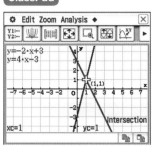
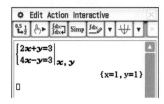
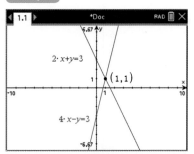
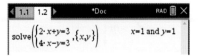

d $x = 2, y = 2$

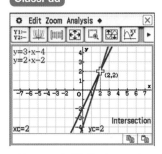
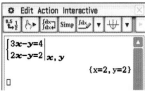
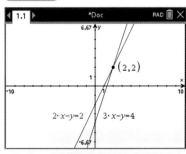
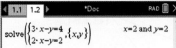

3 a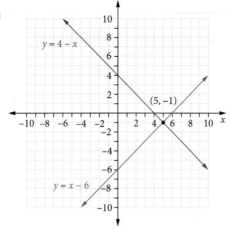

b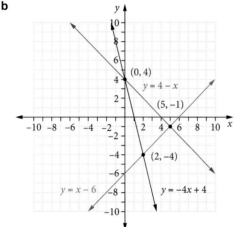

c i $x = 0, y = 4$ or $(0, 4)$ **ii** $x = 2, y = -4$ or $(2, -4)$

4 a ClassPad

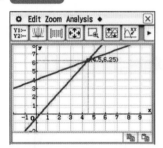

TI-Nspire

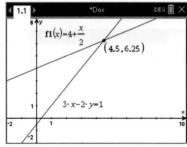

The solution is $x = \dfrac{9}{2}, y = \dfrac{25}{4}$ or $\left(\dfrac{9}{2}, \dfrac{25}{4}\right)$.

b ClassPad

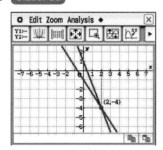

TI-Nspire

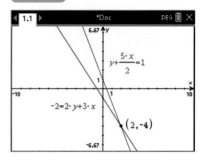

The solution is $x = 2, y = -4$ or $(2, -4)$.

EXERCISE 9.2

1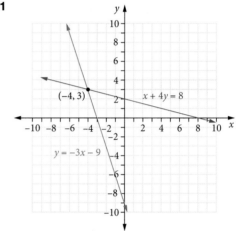

2 ClassPad

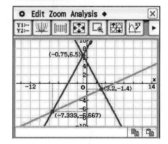

TI-Nspire

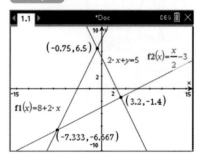

Points of intersection: $\left(-\dfrac{3}{4}, \dfrac{13}{2}\right), \left(\dfrac{16}{5}, -\dfrac{7}{5}\right), \left(-\dfrac{22}{3}, -\dfrac{20}{3}\right)$

3 a $x = 3, y = 2$ **b** $x = 0, y = 4$
  c $x = 2, y = 3$ **d** $x = 3, y = 4$
  e $x = 2, y = 0$ **f** $x = 6, y = 6$
  g $x = -27, y = -39$ **h** $x = 6, y = 13$

4 a $x = 5, y = 0$ **b** $x = 2, y = 3$
  c $x = -1, y = 5$ **d** $x = -5, y = 14$
  e $x = -\dfrac{21}{4}, y = \dfrac{27}{14}$ **f** $x = 2, y = 3$

5 **a** $x=-2, y=4$ **b** $x=2, y=4$
   c $x=1, y=-3$ **d** $x=-2, y=-1$
   e $x=4, y=7$ **f** $x=1, y=-2$

6 **a** $x=5, y=2$ **b** $x=-1, y=4$
   c $x=6, y=0$ **d** $x=2, y=0$
   e $x=0, y=-3$ **f** $x=1, y=1$

7 **a** $x=2, y=1$ **b** $x=\frac{31}{11}, y=\frac{16}{11}$
   c $x=\frac{55}{23}, y=\frac{12}{23}$ **d** $x=2, y=4$
   e $x=\frac{21}{5}, y=\frac{6}{5}$ **f** $x=\frac{20}{7}, y=\frac{24}{7}$
   g $x=-12, y=8$ **h** $x=-50, y=-60$
   i $x=\frac{19}{2}, y=22$ **j** $x=-\frac{57}{2}, y=-41$
   k $x=\frac{36}{23}, y=\frac{60}{23}$ **l** $x=3, y=-2$

8 **a** $x=\frac{10}{3}, y=-2$ **b** $x=-1, y=4$

9 **a** $x=0.57, y=1.32$ **b** $x=1.69, y=2.62$

EXERCISE 9.3

1 $x=2, y=-1$

2 **a** $y=3-2x$
   b
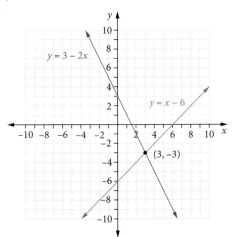

3 The price of a magnet is $2.55 and the price of a tea towel is $4.

4 There are 75 chickens on the farm.

5 **a** $3x + 4y = 67$ and $2x + 5y = 57.5$
   b The cost of one adult ticket is $15 and the cost of one child ticket is $5.50.

6 The cost of one doughnut and two buns is $4.55.

7 **a**
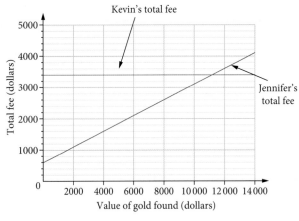

b $11\,200

8 **a** wattle tree; 1 metre **b** 30 months
   c 4 metres
   d **i** $h=\frac{2}{15}t$ **ii** $h=1+\frac{1}{10}t$
   e No, trees don't always grow at the same rate and the graph may only apply for the time period given.

9 **a** Let x = the number of cards Jack has and y = the number of cards Jill has.
   $x + y = 180$
   $x + 20 = y$
   b Jack has 80 cards and Jill has 100 cards.

10 The two numbers are 14 and 30 respectively.

11 Ben bought 3 pairs of shorts and 6 t-shirts.

12 **a** **i** $c = 80 + 0.15d$ **ii** $c = 0.55d$
   b
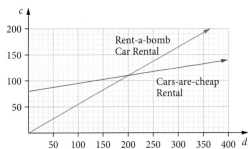

   c 200 km, $110
   d Cars-are-cheap Rental will charge $132.50 for 350 km while Rent-a-bomb Car Rental will charge $192.50 for 350 km. Martin should rent from Cars-are-cheap Rental because it will cost $60 less.

EXERCISE 9.4

1 $x = -2, y = -6$ **2** $3.40

3
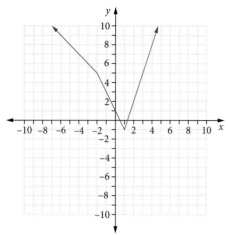

4 a The graph is made up of four different line segments and each line segment has a different slope. This means the rainwater tank fills at four different rates.
  b 18 000 litres **c** 200 hours
  d approximately 80 hours to 120 hours
  e 25 litres per hour

5 a They started their journey at the park 1 km from Volker's house.
  b The slope of the line segment after 45 minutes is greater than the slope of the line segment before, so Volker and Herb started walking faster.
  c The horizontal line indicates that Volker and Herb were the same distance (3.5 km) from Volker's house during this time, so we can conclude they are at Herb's house from the information provided in the question.
  d 15 minutes **e** 3.5 km **f** 7 km/h

6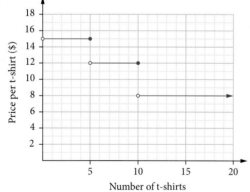

7 a i $10 **ii** $3 **iii** $9
  b more than 2 kg up to 3 kg
  c 6 kg is the heaviest item that can be sent in this type of parcel delivery.

8 True: A, B, C, D
9 a 500 L **b** The volume of water is constant.
  c between 2 pm and 3 pm
  d 0 **e** 200 l/h
10 4 hours **11** $940 **12** $1.50
13 9 to 12 hours **14** $3
15 a 0% for up to $6000 and 15% for over $6000 up to $34 000.
  b i $3600 **ii** $19 600
16 a 8 pm **b** 60 km/h **c** $k = 640$ km
17 a $10
  b
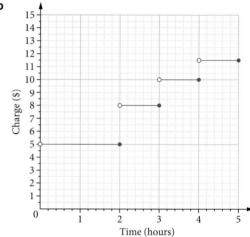
  c $16 and $18

CUMULATIVE EXAMINATION: CALCULATOR-FREE

1 a i $5 **ii** $50
  b 6%
2 $x = 4, y = -3$
3 The data is numerical continuous.
4

$y + x = 7$ $x + 4y = 4$	$(8, -1)$
$y = 3 + 3x$ $y = \dfrac{x}{2} - 2$	$(-2, -3)$
$y = 3 + 3x$ $y + x = 7$	$(1, 6)$
$y = \dfrac{x}{2} - 2$ $x + 4y = 4$	$(4, 0)$

5 Currently, Tim is 17 years old and Chris is 39 years old.

CUMULATIVE EXAMINATION: CALCULATOR-ASSUMED

1 a Drew and Esther

  b i Not possible, as $p^2_{52} = 0$.

  ii Possible, as $p^4_{52} = 1$, meaning that there is one way that it can occur.

2 236 cm^2

3 a $1:160$ **b** 26 m^2 **c** $1378

4 a 1938 **b** 2445

5 a The jellyfish taken from location A have a diameter that is less than the diameters of the jellyfish taken from location B and they are more variable.

  b 25%

6 a

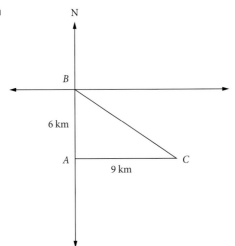

  b i Let $\angle ABC = \theta$.

  $\tan(\theta) = \dfrac{9}{6}$

  $\theta = \tan^{-1}\left(\dfrac{9}{6}\right)$

  $\theta = 56.31$

  $\theta \approx 56°$

  $180° - 56° = 124°$

  C is on a true bearing of 124° from B.

  ii S 56° E

7 a

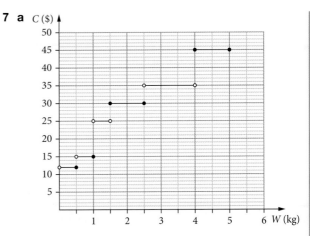

  b $C = \begin{cases} 12.5W, 0 < W \leq 3 \\ 37.5, 3 < W \leq 4 \\ 42.5, 4 < W \leq 5 \end{cases}$

  c

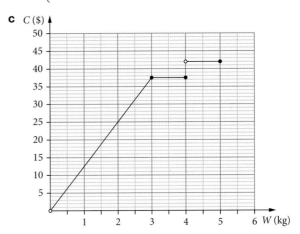

  d Lightning Courier will charge $73.75 while Speedy Mail will charge $75. Therefore, Maria should use Lightning Courier as the cost is $1.25 cheaper.

Glossary and index

additive identity *See* **zero matrix**.

adjacent side The side of a right-angled triangle that is next to the reference angle, that is not the hypotenuse. (p. 279)

algebraic expression A collection of algebraic terms representing the basic operations of arithmetic: addition, subtraction, multiplication and division. Algebraic expressions can be called linear when the power of each variable is 1 or non-linear when they are not. (p. 51)

algebraic techniques Solving simultaneous equations by manipulating the equations algebraically to find the values of the unknown. (p. 371)

allowances Extra pay for doing unpleasant work, for working under difficult conditions, or to cover expenses such as uniform and travel. (p. 9)

angle of depression The angle made by a line of sight downwards from the horizontal. (p. 293)

angle of elevation The angle made by a line of sight upwards from the horizontal. (p. 292)

apex The point at one end of a cone and the point where all the triangular faces of a pyramid meet. (p. 128)

arc A part of the circumference of a circle formed by two radiuses. (p. 117)

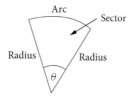

arc length The length of part of a circle's circumference. (p. 117)

area The amount of space inside a two-dimensional shape. (p. 113)

asymmetric *See* **skewed distribution**.

back-to-back stem plot A statistical graph used when dealing with two sets of data values for the same variable where the original data values are visible. (p. 250)

balance The value of an investment or loan at any time. (p. 17)

bar chart A graphical display used for categorical data, where the frequency of each different category is shown using a vertical column or a horizontal bar. (p. 197)

base The face of a three-dimensional shape from which the perpendicular height is measured. Typically this is considered the bottom of the shape. (p. 126)

bearing The direction of a fixed point or path of an object from another point of observation. (p. 295)

bell-shaped distribution A graphical representation of normally distributed data shaped as a bell curve. (p. 211)

bonus Extra pay for doing good work, reaching targets or meeting deadlines. (p. 9)

box plot (box-and-whisker plot) A graphical display of numerical data based on the five-number summary, IQR and outliers. (p. 245)

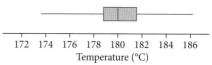

break-even point The point at which revenue begins to exceed the cost of production. (p. 358)

budget An organised list of expected income and expenses used to help manage money. (p. 17)

capacity The amount of liquid a three-dimensional object can hold. (p. 126)

cardinal points The reference directions of north, east, south and west on a compass. (p. 295)

categorical data Data involving categories. (p. 178)

centre of a distribution The single value that best represents the distribution. (p. 180)

circumference The perimeter of a circle. (p. 114)

column matrix A matrix with m rows and 1 column. (p. 63)

e.g. $\begin{bmatrix} 8 \\ 1 \\ 5 \end{bmatrix}$

commission A percentage of the value of the items an individual sells. (p. 10)

communication matrix A square matrix that shows the direct communication links between people or systems with 1s and 0s, whereby a '1' represents that the communication is possible and '0' represents that the communication is not. (p. 88)

communication network A network diagram representing the direct communication links between people or systems, where labelled vertices represent the people or systems and lines (edges) between vertices represent the possible communication. (p. 88)

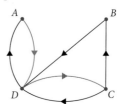

commutative The property of an operation that means order does not matter. For example, $A + B = B + A$. (p. 69)

compass bearing A bearing taken from either north or south, on some acute angle in either an easterly or a westerly direction. (p. 298)

complementary angles Two angles that sum to 90°. (p. 279)

composite shape A shape formed by combining two or more shapes. (p. 118)

compound interest Interest that is added to the principal, where the interest for the next time period is calculated using this new balance. (p. 31)

compounding period The length of the time period before interest compounds. (p. 31)

cone A three-dimensional object with a circular base and a curved surface that joins the base to an apex. (p. 128)

constancy of the trigonometric ratios The relationship that says that similar triangles have equivalent trigonometric ratios. (p. 281)

constant rate of change (rate) A measure of the change in y as x changes for a linear function. (p. 356)

continuous data Numerical data that can be measured to an increasing level of accuracy. (p. 178)

coordinates A set of values showing an exact position. (p. 345)

cosine rule A rule in trigonometry that is an extension of Pythagoras' theorem for non-right-angled triangles. (p. 313)

cost The amount of money that has been used to produce something or deliver a service. (p. 358)

cost price (CP) The price at which a product is purchased. (p. 24)

cross-section The two-dimensional shape that is visible when a solid object is cut parallel to its base. (p. 126)

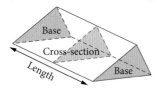

cylinder A three-dimensional figure that has congruent circles at each end, joined by a curved surface. (p. 126)

data Information collected through observation that can be used to make informed decisions. (p. 177)

diameter The distance from one side of a circle to the other through the centre. (p. 114)

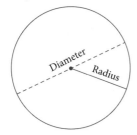

dilation An enlargement or reduction of a shape to produce a similar figure. (p. 154)

dimensions (of a matrix). *See* **order**.

discount A percentage price reduction. (p. 22)

discrete data Numerical data that can't be measured to an increasing level of accuracy. (p. 178)

discretionary expenses Items on which income is spent but which aren't essential, such as entertainment or magazines. (p. 17)

distributive law Multiply a single value with two or more than two values within a set of parentheses. (p. 339)

dividend A share of a company's profits. (p. 38)

dividend yield The dividend as a percentage of a share's market price.

dividend yield = $\dfrac{\text{dividend}}{\text{share price}} \times 100\%$ (p. 39)

dot plot Graphical display for categorical or numerical discrete data. (p. 194)

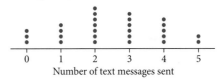

double time Twice the normal pay rate. (p. 6)

elements (of a matrix) The numerical entries within a matrix. The value of an element in the ith row and jth column in a matrix A is represented by a_{ij}. (p. 62)

elimination An algebraic technique in solving simultaneous equations. It involves manipulating the equations by adding or subtracting them in a way that eliminates one of the variables, allowing for the determination of the remaining variable. (p. 371)

equation A statement that shows that values of two mathematical expressions are equal (indicated by the sign =). Equations can be linear or non-linear and have either one variable or multiple variables. (p. 52)

exchange rate The amount of overseas currency we can get for 1 Australian dollar (AUD). (p. 35)

expenses All the ways an individual, business or company spends their money. (p. 17)

five-number summary A five-number summary is a method of summarising a set of data using the minimum value, the lower or first quartile (Q_1), the median, the upper or third quartile (Q_3) and the maximum value. It forms the basis for a box plot. (p. 239)

fixed expenses Costs that are essential and must be paid. (p.17)

formula An equation that has a specific purpose. (p. 53)

frequency table A table used to organise large amounts of data, with data values in one column and the corresponding frequencies in another. (p. 184)

Goods and Services Tax *See* **GST**.

gradient *See* **slope**.

graphical method Solving simultaneous equations by plotting the equations on a Cartesian plane and finding the points of intersection. (p. 371)

grouped frequency table A frequency table where numerical data has been grouped into regular intervals. (p. 184)

GST Currently a 10% tax on most sales and services in Australia. (p. 25)

hemisphere Half a sphere. (p. 128)

Heron's rule The measurement formula used to calculate the area of a triangle knowing three sides. (p. 304)

histogram Graphical display for numerical data (discrete or continuous) with vertical, joined columns. (p. 185)

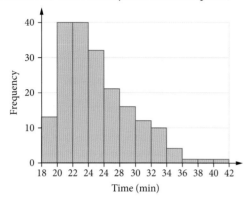

hypotenuse The longest side of a right-angled triangle. (p. 105)

identity matrix The multiplicative identity matrix, I_n, is a square matrix of dimensions $n \times n$ in which all the elements in the leading diagonal are 1s and the remaining elements are 0s. (p. 63)

included angle The angle formed between two sides of a triangle. (p. 303)

income All the money someone earns. (p. 17)

initial value The value of the asset or item at the start, i.e. when $x = 0$. (p. 356)

interest The fee for using someone else's money, usually a financial institution such as a bank. (p. 29)

interquartile range (IQR) The measure of the spread of the middle 50% of the data values. (p. 241)

intersection The point where two graphs intersect or cross each other. (p. 371)

inverse trigonometric ratio The inverse operation that is used to find the value of a reference angle using the ratio of two sides. (p. 284)

leading diagonal The diagonal of a square matrix that runs from the top left corner to the bottom right corner. (p. 63)

linear combination An expression that combines the operations of addition, subtraction and scalar multiplication. (p. 71)

linear equation (linear function) A linear equation in one variable, such as x, is an equation of the form $ax + b = 0$, for example, $3x + 1 = 0$. A linear equation in two variables, such as x and y, is an equation of the form $ax + by + c = 0$, for example, $2x - 3y + 5 = 0$. (pp. 52, 337)

linear expression An algebraic expression where the power of each of the variables is 1. (p. 51)

location The notion of central or the 'typical value' in a sample distribution. (p. 252) *See also* **mean**, **median** and **mode**.

loss (L) The amount of money lost when a product is sold at a price less than the cost price. (p. 24)

lower fence The value below which a data point may be considered an outlier. (p. 241)

lower quartile The data point that has 25% of the data below it. (p. 239)

lowest common denominator The smallest number that can be divided exactly by all the numbers below the lines in a group of two or more fractions. (p. 337)

mark-up A percentage price increase. (p. 22)

matrix (plural **matrices**) A rectangular array of elements displayed in m rows and n columns. (p. 62)

matrix multiplication A row-by-column method of multiplication for multiplying two matrices. (p. 76)

mean The arithmetic mean of a list of numbers is the sum of the data values divided by the number of values in the list. In everyday language, the arithmetic mean is commonly called the average. For example, for the following list of five numbers: 2, 3, 3, 6, 8, the mean equals
$$\frac{2+3+3+6+8}{5} = \frac{22}{5} = 4.4$$
In more general language, the mean of n observations $x_1, x_2 \ldots x_n$ is $\bar{x} = \frac{\Sigma x_1}{n}$. (p. 204)

median The median is the value in a set of ordered data values that divides the data into two parts of equal size. When there are an odd number of data values, the median is the middle value. When there are an even number of data values, the median is the average of the two central values. (p. 180)

modal category The category with the highest frequency. (p. 180)

modal interval The interval with the highest frequency. (p. 184)

mode The mode is the most frequently occurring value in a data set. (p. 180)

multimodal Data with more than one mode. (p. 180)

multiplicative identity The multiplicative identity for numbers is 1, the number which when multiplied by another number does not change the result. For matrices, the identity matrix I is the multiplicative identity. (p. 63)

negatively skewed or **negatively-skewed distribution** A distribution that has a tail at the lower end. (p. 186)

net A two-dimensional figure that shows all the faces of a solid and can be folded up to form that solid. (p. 136)

nominal data Categorical data that doesn't have a natural order, even when numbers are involved. (p. 178)

non-linear expression An algebraic expression where the power of at least one of the variables is not 1. (p. 51)

normal distribution or **bell-shaped distribution** A distribution with a bell shape that is symmetric about the mean, peaks in the centre and tails off towards zero on both sides. (p. 211)

numerical data Data involving numbers that have a mathematical meaning such as counting or measuring. (p. 178)

opposite side The side of a right-angled triangle that is directly opposite the reference angle. (p. 279)

ordinal data Categorical data that has a natural order, but doesn't involve counting anything and can't be measured to an increasing level of accuracy. (p. 178)

order (of a matrix) The number of rows and columns that a matrix has. For example, if A has m rows and n columns, then $\dim(A) = m \times n$. (p. 62)

outlier An outlier in a set of data is an observation that appears to be inconsistent with the remainder of that set of data. An outlier is a surprising observation. (p. 186)

overtime Working beyond usual working hours or days, and paid at a higher rate than the normal rate per hour. (p. 6)

parallel box plot A graph where two or more box plots are shown on the same axis. (p. 252)

per annum (p.a.) Per year. (p. 29)

percentage change The amount of increase or decrease of a quantity written as a percentage of the quantity. (p. 21)

perimeter The total distance around the outside of a shape. (p. 113)

piece-wise linear graph A graph made up of more than one straight line piece. (p. 383)

piecework A type of work where a person is paid per item produced or processed. (p. 10)

positively skewed or **positively skewed distributions** A distribution that has a tail at the upper end. (p. 186)

post-multiplication If matrix A is post-multiplied by matrix B, then the matrix product is AB. (p. 79)

pre-multiplication If matrix A is pre-multiplied by matrix B, then the matrix product is BA. (p. 79)

price earnings (P/E) ratio A ratio that gives investors an indication of how expensive the company's shares are in relation to their profits.

$$\text{P/E ratio} = \frac{\text{market price per share price}}{\text{annual earnings per share}}$$ (p. 40)

principal The amount of money invested or borrowed. (p. 29)

prism A three-dimensional object with straight edges that has the same cross-section along its full length. (p. 126)

profit (P) The amount of money made when a product is sold at a price greater than the cost price, the difference between revenue and cost. (pp. 24, 358)

pronumeral A letter or symbol used to represent a quantity that can have many different values in a particular situation. (p. 341)

Pythagoras' theorem A rule for calculating the third side of a right-angled triangle given the length of the other two sides. (p. 105)

quadrilateral A shape with four straight sides. (p. 114)

quartile A statistical value that divides a data set into 4 equal parts. (p. 215)

radius The distance from the centre of a circle to the circumference. (p. 114) See **diameter** for diagram.

range A measure of the spread of the data; the difference between the largest and smallest observations. (p. 180)

rate See **constant rate of change**.

redundant link A communication link with the same sender and receiver, which are found in the leading diagonal of the communication matrix and will always be 0s. (p. 88)

reference angle The angle defined by a trigonometric ratio. (p. 279)

retainer A set payment that does not depend on sales. (p. 10)

revenue The amount of money generated by the sale of goods and services. (p. 358)

right-angle triangle A triangle with exactly one angle of size 90°. (p. 105)

row matrix A matrix with 1 row and n columns. (p. 63)
e.g. $\begin{bmatrix} 4 & 12 & 5 \end{bmatrix}$

salary A fixed amount per year, paid by an employer, that does not depend on the number of hours worked. (p. 3)

scalar A number that is not in a matrix. (p. 70)

scalar multiplication The process of multiplying a matrix by a scalar (numerical value), which multiplies every entry in the matrix by that scalar value. (p. 70)

scale factor A measurement of how much a shape needs to be enlarged or reduced to produce a similar shape. (p. 154)

sector The part of a circle formed by two radiuses and the arc between them. (p. 117) See **arc** for diagram.

selling price (SP) The price at which a product is sold. (p. 24)

semi-perimeter (of a triangle) Half of the value of the perimeter of a triangle. (p. 304)

shape A two-dimensional figure. (p. 113)

shape of a distribution A description of data as symmetric, positively skewed or negatively skewed. (p. 186)

share An investment that purchases part of a company. (p. 38)

share market The place where shares in companies are bought and sold. Also called the stock market. (p. 38)

similar figure Two-dimensional figures that have the same shape but are different sizes. The symbol for similar shapes is ~. (p. 154)

simple interest The fixed amount of interest paid at regular time periods calculated as a percentage of the amount of money invested or borrowed. (p. 29)

simultaneous equations Two or more linear equations, each with two or more variables, which are being solved to find values that are common solutions to all the equations. (p. 371)

sine rule A rule in trigonometry that says for any triangle, the proportion of each side length to the sine of its opposing angle is the same for all three sides. (p. 309)

size (of a matrix) See **order**.

skewed distribution A distribution that is asymmetric and has a tail. (p. 186)

slant length The distance from a point on the perimeter of the base to the apex of a pyramid or cone. For a pyramid, the distance is measured along the centre of a triangular face. (p. 129)

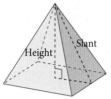

slope (gradient) The slope or gradient of a line describes its steepness, incline or grade. Slope is normally described by the ratio of the 'rise' divided by the 'run' between two points on a line. (p. 347)

solid A three-dimensional object. (p. 126)

sphere A round, three-dimensional figure where all the points on the surface are the same distance from its centre. (p. 128)

spread of a distribution How much data varies around the centre of a distribution. (p. 180)

square matrix A matrix with the same number of rows as columns. (p. 63)

standard deviation The standard deviation is a measure of the variability, or spread, of a data set. It gives an indication of the degree to which the individual data values are spread around their mean. (p. 205)

standardised values or **z-scores** Values calculated using the formula

$$\text{standardised value} = \frac{\text{actual value} - \text{mean}}{\text{standard deviation}}$$

that allow us to compare values from different normal distributions. (p. 219)

statistical investigation process The statistical investigation process is a cyclical process that begins with the need to solve a real-world problem and aims to reflect the way statisticians work. One description of the statistical investigation process in terms of four steps is as follows.

Step 1. Clarify the problem and formulate one or more questions that can be answered with data.

Step 2. Design and implement a plan to collect or obtain appropriate data.

Step 3. Select and apply appropriate graphical or numerical techniques to analyse the data.

Step 4. Interpret the results of this analysis and relate the interpretation to the original question; communicate findings in a systematic and concise manner. (p. 265)

Mathematics Applications ATAR Course Year 11 syllabus p.23 © SCSA

stem plot (stem-and-leaf plots) A graphical display for numerical data that is either discrete or continuous. (p. 195)

step graph A piece-wise linear graph that has only horizontal straight line pieces. (p. 383)

subject The variable used to represent the purpose of a formula. (p. 53)

substitution An algebraic technique in solving simultaneous equations by rearranging one equation for one variable and substituting that expression into the other equation. (p. 371)

surface area The sum of all the areas of the faces of a three-dimensional object. (p. 136)

symmetric distribution A distribution that has the same shape on both sides of the median. (p. 186)

three-figure bearing See **true bearing**.

time-and-a-half 1.5 times the normal pay rate. (p. 6)

trigonometric ratio A comparison of the lengths of two sides in a right-angled triangle. (p. 279)

trigonometry The study of the measurement of triangles. (p. 279)

true bearing A bearing that is measured from the North line, in a clockwise direction, written as a three-digit number between 000° and 360°. (p. 295)

unit column matrix A column matrix where all entries are 1s. (p. 63)

unit cost method A way of solving problems by calculating the value of one unit. (p. 26)

unit row matrix A row matrix where all entries are 1s. (p. 63)

upper fence The value above which a data point may be considered an outlier. (p. 241)

upper quartile The data point that has 75% of the data below it. (p. 239)

variable A quantity that can have many different values. (pp. 51, 177)

volume The amount of space a three-dimensional object takes up. (p. 126)

wage An amount paid by an employer to an employee for each hour worked. (p. 3)

whiskers Parts of a box plot that show the minimum and maximum values if there are no outliers. (p. 245) See **box plot** for diagram.

y-intercept The point where the graph crosses the y-axis. (p. 351)

z-score A statistical measure that allows comparison of values from different normal distributions. (p. 219) See also **standardised values**.

zero matrix A matrix of any size with all entries as 0s. (p. 63)